Introduction to Semiconductor Integrated Optics

For a complete listing of the *Artech House Optoelectronics Library*,
turn to the back of this book.

Introduction to Semiconductor Integrated Optics

Hans P. Zappe

Artech House
Boston • London

Library of Congress Cataloging-in-Publication Data
Zappe, Hans.
Introduction to semiconductor integrated optics / Hans P. Zappe.
p. cm.
Includes bibliographical references and index.
ISBN 0-89006-789-9 (alk. paper)
1. Integrated optics. 2. Semiconductors. I. Title.
TA1660.Z37 1995 95-21287
621.36'93–dc20 CIP

British Library Cataloguing in Publication Data
Zappe, Hans P.
Introduction to Semiconductor Integrated Optics
I. Title
621.38152

ISBN 0-89006-789-9

685 Canton Street
Norwood, MA 02062

International Standard Book Number: 0-89006-789-9
Library of Congress Catalog Card Number: 95-21287

10 9 8 7 6 5 4 3 2 1

For Fran

Contents

Acknowledgments

Many people have kindly helped to review this book's material; provided suggestions, stimulation, and intellectual sparring; and occasionally preempted the appearance of careless foolishness in print. It is my particular pleasure to thank Donat As of the Universität Gesamthochschule Paderborn, Germany; Clifton Fonstad of the Massachusetts Institute of Technology, Cambridge, Massachusetts; Carl Galewski of Genus, Santa Clara, California; Dominic Gallagher of Photon Design, Oxford, England; Wolfgang Jantz of the Fraunhofer Institute for Applied Solid State Physics, Freiburg, Germany; Jack Lee of the University of Texas, Austin, Texas; John Ralston of SDL, San Jose, California; Josef Rosenzweig of the Fraunhofer Institute for Applied Solid State Physics, Freiburg, Germany; Jürgen Söchtig of the Paul Scherrer Institute, Zurich, Switzerland; and Katalin Voros of the University of California, Berkeley, California; for their investment of time and effort in the reading of parts of the manuscript and assisting with their comments, criticisms, and corrections in the production of a better book. I am indebted to my graduate students, Daniel Hofstetter (who provided some of the nice device data) and Bernd Maisenhölder, both of whom provided a valuable student's-eye view of the text. To my colleagues at the Paul Scherrer Institute, who supported me with discussions, tips, and encouragement, I am likewise grateful. Thanks to Fritz Senn of the Zurich James Joyce Foundation for helping prove that Joyce did not invent the word "laser." Any errors, unclarities, ambiguities, or imbecilities that happen to remain are solely my own doing.

And to Fran I promise less ordinary menus and a less ornery mate.

Acknowledgments

Preface

Semiconductor technology has reached an advanced stage of maturity, with its products appearing in everything from spacecraft to cuddly toys, and it has captured the popular imagination through exquisitely colorful photographs on the covers of magazines. Integrated optics, in which optical waveguides guide photons in a manner analogous to the guiding of electrons by electric wires, most recently reached the public eye when CD players began to lurk in every household. The book before you intends to cover the overlap between these two disciplines.

This text is designed, on one hand, for the semiconductor engineer who might know little about optics but would like to begin applying his or her expertise to integrated optical structures. It may also, on the other hand, be employed by the engineer well versed in bulk or integrated optics who would like to probe into the world of semiconductors for new possibilities. And on the third hand, for those of us with three hands, the text should be useful for the beginning graduate student interested in both fields. It is hoped that all will benefit from reading it.

For the reader who dives in at the start, the approach includes a discussion of basic semiconductors and specific semiconductor optical material systems, primarily the III-Vs and silicon, plus a close look at their optical properties. The waveguide, fundamental to a wide variety of integrated optical devices, is studied in depth. Once these two areas have been solidly presented, the text turns its attention to lasers, modulators, and detectors—the active devices that are essential for an integrated optical system. Finally, after reading the survey of integration and hybridization with which the book concludes, the assiduous student should have a firm grasp on the techniques required for the fabrication of a semiconductor integrated optical circuit.

The book is intended to be an engineering text, so we have striven to provide an adequate balance between theoretical development and concrete application. In most subject areas, derivations of the basic concepts are undertaken to provide a basic understanding of the physics involved. These concepts are then developed in a way that will allow the reader to solve real-world problems and learn to appreciate the engineering aspects of the issues. For this reason, the derived equations and relationships are written in such a way that the reader can plug in a few numbers and come up with an answer, and also understand what it means. It is the hope of the author that someone could pick up the book, read a few chapters and proceed directly to the design of a semiconductor integrated optical structure that would actually work.

To facilitate understanding, a numerical example is occasionally stuck into the development, when suitable and not utterly trivial. This helps to see what kind of numbers might emerge from the barrage of mathematical gymnastics just presented. All variables will have their units defined [in square brackets] at their first introductions. The mks system will semirigorously be adhered to; exceptions are made for those terms for which common usage dictates otherwise (i.e., current density in A/cm^2, carrier density in cm^{-3}, etc.). This should keep clarity from succumbing to pedantry. Standard engineering notation (e.g., $\sqrt{-1} = j$) is employed.

The prospective reader should have a thorough, undergraduate-level knowledge of physics, particularly electromagnetics, fields, and waves, plus a familiarity with the mathematical tools that are commonly taught concomitantly. A rudimentary knowledge of basic semiconductor physics and the associated electrical devices would help, as this text does not even approach the completeness of standard works in the field.

Zurich, July 1995

Introduction

The following study is broadly divided into three sections: semiconductors, optics, and devices.

Semiconductors

We start with a review of basic electromagnetics and then begin a presentation of basic semiconductor theory, looking at the band structure, doping, and transport in bulk silicon and III-V materials. The pn junction and the quantum well, two important semiconductor structures, are also presented and analyzed. In the following chapter, we turn to the materials' optical properties, focusing on those that are important for the eventual fabrication of devices, such as refractive index and propagation constant. From there, a discussion of commonly used optical materials, including silicon and the numerous members of the III-V family, and their characteristics will bring about an understanding of those aspects of material design necessary for the development of an optical device. The remainder of this section then deals with the actual fabrication, in which we present the basics of III-V growth and semiconductor processing techniques, from cleaning to packaging.

Optics

An important initial discussion in the area of optics concerns the reflection and transmission of optical waves at boundaries, breaking the ground for presentation of the basis to almost any integrated optical structure, the waveguide. We present the one-dimensional slab waveguide and the widely employed two-dimensional channel waveguide, learning how to calculate their modes, design monomode structures, and characterize them. A discussion of popular waveguide configurations, in both the III-V and silicon-based material systems, provides an overview of the real-life devices that have been fabricated. By this point, we will have seen a synthesis of material and optical concepts.

Devices

We turn our attention in last third of the text to the active and passive optoelectronic devices, based on the previously presented materials and waveguide structures, which form the backbone of semiconductor integrated optics. Light emitters, namely lasers and light-emitting diodes (LEDs), are introduced, with considerable attention paid to the physics of coherent light emission. Waveguide detectors, both PIN and MSM photodiodes, form the opposite end of an integrated optical system, converting an optical signal back into an electrical one. The penultimate chapter delves into modulator structures, with an emphasis on electro-optics, quantum wells, and interferometers, which permit control of optical fields. To conclude, we take a look at hybridization and monolithic integration, as well as the development of photonic integrated circuits, which are following in the footsteps of their purely electrical cousins.

Chapter 1

Basic Electromagnetics

We have heard these words before; and we shall hear them again.
W. H. Auden and Christopher Isherwood, 1936 [1]

A modest set of equations and relations is useful for the study and analysis of most optical systems. Students are likely to have had these items pounded into their heads incessantly at some point in their educations, but a brief discussion to stimulate the memory of the reader is useful. This chapter provides a brief review and summary of the most important electromagnetic equations and relationships. Standard texts on electromagnetics will do a more thorough job,[1] but, if nothing else, this section will serve as an introduction to the notation and units of the variables that we will encounter again and again.

1.1 GENERAL RELATIONSHIPS

Almost everything we are interested in, at least between the covers of this book, has to do with electric and magnetic fields, currents, and charges in one form or another. Let us briefly outline the most important relations.

1.1.1 Maxwell's Equations

Gauss' (1.1), Faraday's (1.2), and Ampère's (1.3) laws relate these quantities to each other, in space and time, as

$$\nabla \cdot D = \rho \tag{1.1}$$

$$\nabla \cdot B = 0 \tag{1.2}$$

$$\nabla \times \mathcal{E} = -\frac{dB}{dt} \tag{1.3}$$

[1] See, for example, the classic Jackson, J., *Classical Electrodynamics*, 2nd Edition, New York: John Wiley & Sons, 1975.

$$\nabla \times H = J + \frac{dD}{dt} \tag{1.4}$$

These are known collectively as Maxwell's equations, here in mks units. These equations relate the electric field $\mathscr{E}$ [V/m], magnetic field H [A/m], charge density ρ [C/m^3], and current density J [A/cm^2], to one another. We recall that

$$\nabla = \left(\frac{\partial}{\partial x}, \frac{\partial}{\partial y}, \frac{\partial}{\partial z}\right) \tag{1.5}$$

and further that the electric flux density D [C/m^2] and the induced electrical polarization P [C/m^2] are related to $\mathscr{E}$ by

$$D = \epsilon_m \mathscr{E} + P \tag{1.6}$$

Likewise, the magnetic field vector H is related to the magnetic flux density B [T = Wb/m^2] by

$$B = \mu_m H \tag{1.7}$$

where we have ignored the magnetization, which does not usually play a role in the materials that we shall consider. Finally, in the previous two equations, ϵ_m and μ_m represent the permitivity [F/m] and permeability [H/m], respectively. We will see what roles these various quantities play in more detail presently.

1.1.2 Current Continuity

We may also relate the current density J [A/cm^2] to the electric field $\mathscr{E}$ by Ohm's law, namely

$$J = \sigma \mathscr{E} \tag{1.8}$$

where σ represents the conductivity [1/Ωm]. Furthermore, the current continuity equation tells us that what goes in must come out, or

$$\nabla \cdot J = -\frac{\partial \rho}{\partial t} \tag{1.9}$$

This equation states that any change in the current flowing into or out of an infinitesimal volume must be accounted for by generation or recombination of carriers.

With these few relations, we are already armed for the study of the propagation of waves in space and in materials. In the appendix, we have a summary of the all of the variables in use in this text, including the field quantities we have just introduced, along with their mks units.

1.1.3 The Wave Equation

Maxwell's equations describe the variation of electromagnetic fields with time implicitly, since the spatial variation of $\mathcal{E}$ is related to a time variation of H and vice versa. We may describe the propagation of electromagnetic waves explicitly by means of a wave equation, which we shall now derive from Maxwell's equations. Let's take the curl of Equation 1.3

$$\nabla \times \nabla \times \mathcal{E} = -\mu_m \frac{\partial}{\partial t}(\nabla \times H) \tag{1.10}$$

which we may rewrite, by plugging in Ampère's law (Equation 1.4), as

$$\nabla \times \nabla \times \mathcal{E} = -\mu_m \left[\frac{\partial J}{\partial t} + \frac{\partial^2 D}{\partial t^2} \right] \tag{1.11}$$

$$= -\mu_m \left[\sigma \frac{\partial \mathcal{E}}{\partial t} + \epsilon_m \frac{\partial^2 \mathcal{E}}{\partial t^2} + \frac{\partial^2 P}{\partial t^2} \right] \tag{1.12}$$

By applying the vector identity

$$\nabla \times \nabla \times \mathcal{E} = \nabla(\nabla \cdot \mathcal{E}) - \nabla^2 \mathcal{E} \tag{1.13}$$

where $\nabla \cdot \mathcal{E} = 0$ for the case where $\rho = 0$, we can replace the left hand side of Equation 1.12 by $-\nabla^2\mathcal{E}$. We also recall that

$$\nabla^2 = \left(\frac{\partial^2}{\partial x^2}, \frac{\partial^2}{\partial y^2}, \frac{\partial^2}{\partial z^2} \right) \tag{1.14}$$

By assuming for the moment that $\sigma = 0$, and that P may be ignored, as would be the case for free space, Equation 1.12 then yields the wave equation

$$\nabla^2 \mathcal{E} = \epsilon_m \mu_m \frac{\partial^2 \mathcal{E}}{\partial t^2} \tag{1.15}$$

If we define the velocity v [m/s] of the wave as $v^2 = 1/(\epsilon_m \mu_m)$, we can rewrite this as

$$\nabla^2 \mathcal{E} = \frac{1}{v^2} \frac{\partial^2 \mathcal{E}}{\partial t^2} \tag{1.16}$$

We thus have an explicit relation between the time and spatial variation of the electric field; an equivalent relationship for H can be easily derived in the same manner. In free space, where $\epsilon_m = \epsilon_0 = 8.87$ pF/m and $\mu_m = \mu_0 = 400\pi$ nH/m, $v = c = 3 \times 10^8$ m/s, as we expect; in a material with different ϵ_m and μ_m, the velocity is correspondingly slower. We will consider the case of nonzero σ and nonignorable P, important for propagation in most materials, in Chapter 3.

1.1.4 TEM Waves and Polarization

The electric and magnetic fields of an electromagnetic wave are always orthogonal to each other, and both are orthogonal to the direction of propagation. For that reason, we refer to the propagating wave as a transverse electromagnetic, or TEM, wave, as sketched schematically in Figure 1.1(a). By proper choice of axes, we can describe such a wave as having components along the **x** and **y** axes. ($\mathcal{E}_x$, H_y), for example, would imply that the wave is linearly polarized in the **x** direction; we illustrate this in Figure 1.1(b). With such a definition, we can derive a subset of wave equations in which the field and propagation directions are explicit and one-dimensional. For the case in which $\sigma = \rho = 0$, we have the pair of wave equations

$$\frac{\partial^2 \mathcal{E}_x}{\partial z^2} = \epsilon_m \mu_m \frac{\partial^2 \mathcal{E}_x}{\partial t^2} \tag{1.17}$$

$$\frac{\partial^2 H_y}{\partial z^2} = \epsilon_m \mu_m \frac{\partial^2 H_y}{\partial t^2} \tag{1.18}$$

The directions associated with **x**, **y**, or **z** are arbitrary, at least in free space, and we can usually choose a convenient orientation. The case where the electric field direction changes during propagation, by the electric field vector rotating about the origin as shown in Fig. 1.1(c), is known as circular polarization. Finally, if the magnitude of the field during this rotation changes, the most general case, elliptical polarization, is obtained. A more thorough discussion can be found in [2].

1.2 SOLVING THE WAVE EQUATION

We can convince ourselves by substitution that solutions to the wave equation (Equation 1.15 or the subset 1.17–1.18) are sinusoids and can take the form, for propagation in the **z** direction,

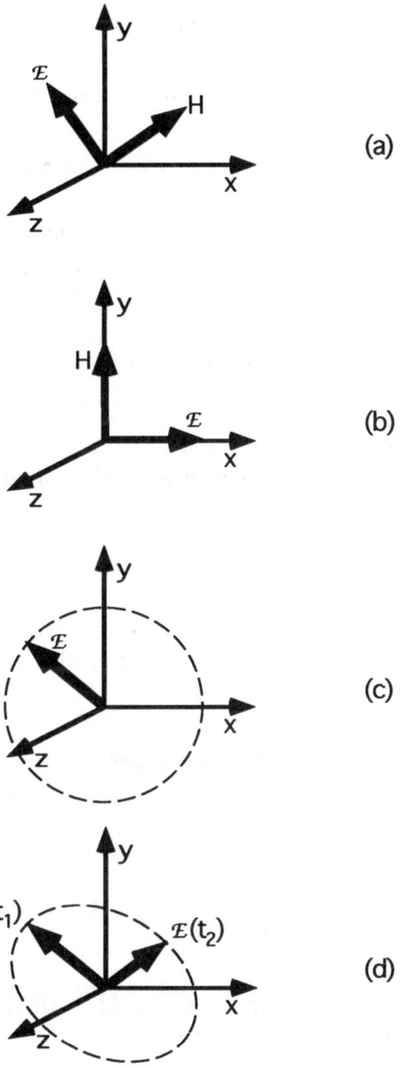

Figure 1.1 (a) The electric and magnetic field vectors are orthogonal to each other and the propagation direction, the latter assumed to be **z**. By proper choice of axes, we can have $\mathcal{E}$ along the **x** axis and H along the **y** axis (b); this is an example of linear polarization. If the $\mathcal{E}$ vector rotates as the wave propagates (c), circular polarization is obtained, which is actually a special case of elliptical polarization (d), which occurs when the magnitude of the field varies as it rotates.

$$(\mathscr{E}, H) = (\mathscr{E}_0, H_0)[\cos(kz \pm \omega t) + \sin(kz \pm \omega t)] \tag{1.19}$$

which may alternatively be written in the exponential format as

$$(\mathscr{E}, H) = (\mathscr{E}_0, H_0) \cdot \exp[j(kz \pm \omega t)] \tag{1.20}$$

The arguments of the sinusoids or exponentials are defined below. We can, without any loss of generality, take only the real part of this exponential expression for the fields, so that

$$(\mathscr{E}, H) = (\mathscr{E}_0, H_0) \cdot \mathrm{Re}(\exp[j(kz \pm \omega t)]) = (\mathscr{E}_0, H_0)\cos(kz \pm \omega t) \tag{1.21}$$

where Re represents the real part. This will be understood from now on, and we will not explicitly refer to the real part when writing the exponential form for the propagating wave.

1.2.1 Propagation Constant

The argument of the exponential is the phase ϕ [rad] of the electromagnetic wave, namely

$$\phi = (kz \pm \omega t) \tag{1.22}$$

Upon taking the time derivative, this yields

$$\left|\frac{\partial \phi}{\partial t}\right| = \omega = 2\pi f \tag{1.23}$$

for angular frequency ω [rad/s] or frequency f [Hz]; this is simply the frequency of the wave. The spatial derivative

$$\frac{\partial \phi}{\partial z} = k \tag{1.24}$$

defines the wavevector, or propagation constant k [m^{-1}]. This latter term is related to wavelength λ [m] and frequency as

$$k = \frac{2\pi}{\lambda} = \frac{\omega}{v} \tag{1.25}$$

for wave velocity v, as presented above. In summary: ω represents the change in phase of a wave as a function of time, and k represents the change of the phase as a function of position.

The term k is an important parameter for describing a wave propagating in free space or in a material. The propagation constant is a vector and, in general, complex, defining the propagation direction and velocity of a wave. In free space (ϵ_0, μ_0), k is usually replaced by the parameter k_0 [m^{-1}] where

$$k_0 = \frac{\omega}{c} \tag{1.26}$$

and $c = 3 \times 10^8$ m/s. k and k_0 are related by the index of refraction, n [dimensionless] as

$$k = nk_0 \tag{1.27}$$

This index n contains the effect of a material in which a wave propagates; we will discuss this in greater detail in Chapter 3.

We may thus summarize: the wave equation for electromagnetic waves has sinusoids as solutions, and these take the form of Equation 1.20. The phase of the sinusoid contains ω, the frequency, which defines time variation, and k, the wavevector, which defines the spatial variation of the wave. When the wave is propagating in free space, $k = k_0$, and in a material they are related by the index of refraction, n.

1.2.2 Maxwell for Sinusoidal Fields

Maxwell's equations may be written in a slightly more practical and simple form for sinusoidally varying electric and magnetic fields. In addition, the directional components are explicit.[2] For propagation in the **z** direction, we have the fields

$$\mathcal{E} = \mathcal{E}_0 \exp[j(\omega t - k_z z)] \tag{1.28}$$

$$H = H_0 \exp[j(\omega t - k_z z)] \tag{1.29}$$

If we plug these directly back into Maxwell's equations, we obtain three spatial components of the electric field

[2]For a derivation see C. T. A. Johnk, *Engineering Electromagnetic Fields and Waves,* Chapter 8, New York: John Wiley & Sons, 1975.

$$\mathcal{E}_x = \frac{1}{j\omega\epsilon_m}\left(\frac{\partial H_z}{\partial y} + k_z H_y\right) \tag{1.30}$$

$$\mathcal{E}_y = \frac{-1}{j\omega\epsilon_m}\left(\frac{\partial H_z}{\partial x} + k_z H_x\right) \tag{1.31}$$

$$\mathcal{E}_z = \frac{1}{j\omega\epsilon_m}\left(\frac{\partial H_y}{\partial x} - \frac{\partial H_x}{\partial y}\right) \tag{1.32}$$

and the three spatial components of the magnetic field

$$H_x = \frac{-1}{j\omega\mu_m}\left(\frac{\partial \mathcal{E}_z}{\partial y} + k_z \mathcal{E}_y\right) \tag{1.33}$$

$$H_y = \frac{1}{j\omega\mu_m}\left(\frac{\partial \mathcal{E}_z}{\partial x} + k_z \mathcal{E}_x\right) \tag{1.34}$$

$$H_z = \frac{-1}{j\omega\mu_m}\left(\frac{\partial \mathcal{E}_y}{\partial x} - \frac{\partial \mathcal{E}_x}{\partial y}\right) \tag{1.35}$$

With these scalar equations, we can easily derive the electric or magnetic field components if we know one or the other.

1.3 ENERGY

The energy density $\mathcal{W}$ [J/m^3] of an electromagnetic wave consists of an electric component

$$\mathcal{W}_E = \frac{1}{2}\epsilon_m \mathcal{E}^2 \tag{1.36}$$

and a magnetic one

$$\mathcal{W}_H = \frac{1}{2}\mu_m H^2 \tag{1.37}$$

such that the total energy in an electromagnetic wave is $\mathcal{W}_T = \mathcal{W}_E + \mathcal{W}_H$. The electric and magnetic fields are related by the impedance of the medium (or free space)

$$\frac{\mathcal{E}}{H} = \sqrt{\frac{\mu_m}{\epsilon_m}} \equiv Z \tag{1.38}$$

and we see that the energy is equally split between the electric and magnetic components, or $\mathcal{W}_E = \mathcal{W}_H$.

The impedance Z [Ω] relates the magnitudes of the electric and magnetic fields in any medium, so that we can simply write $\mathcal{E} = ZH$. In free space, this becomes

$$Z_0 = \sqrt{\frac{\mu_0}{\epsilon_0}} = 377\Omega \tag{1.39}$$

where Z_0 is the so-called impedance of free space.

1.3.1 Poynting Vector

The propagation of power in a wave is given by the Poynting vector S [W/m^2], which is defined as the vector product of the electric and magnetic field vectors

$$S = \mathcal{E} \times H \tag{1.40}$$

and indicates both the magnitude and direction of power flow.

The total power S' [W] flowing through a surface, A, is given by the integral over that surface, or

$$S' = \oint_A S \, dA \tag{1.41}$$

In addition, we can define a time average of power flow <S> as the integral of S over one period T of a cycle (recalling that $\mathcal{E}$ and H are sinusoids)

$$\langle S \rangle = \frac{1}{T}\int_0^T S(t')dt' = \frac{1}{T}(\mathcal{E} \times H^*) \tag{1.42}$$

The above expressions may be simplified slightly by employing the impedance to write

$$S = ZH^2 = \frac{1}{Z}\mathcal{E}^2 \tag{1.43}$$

and thus the evaluated time averages as

$$\langle S\rangle = \frac{1}{2}ZH^2 = \frac{1}{2Z}\mathcal{E}^2 \tag{1.44}$$

Finally, we can include the effect of a dielectric medium by adding the refractive index so that

$$\langle S\rangle = \frac{1}{2\text{n}}Z_0H^2 = \frac{\text{n}}{2Z_0}\mathcal{E}^2 \tag{1.45}$$

It is important to note that <S> is the actual observable intensity of a wave, such that physical measurements are usually of <S> and not S, or W or $\mathcal{E}$ or H. It is difficult to measure a field value ($\mathcal{E}$ or H) directly, but most detectors measure power averaged over some time and a finite space, implying a measurement of <S′>.

1.4 INTERFERENCE

If two or more TEM waves meet in space, the resultant field intensities are the sum of the individual intensities. The waves are said to *interfere* with one another, and this phenomenon is quite important for a number of integrated optical structures, most notably interferometers [3]. For simplicity, we can illustrate interference by considering two waves of identical polarization, ($\mathcal{E}_1$, H_1) and ($\mathcal{E}_2$, H_2). The total intensity at the point where these waves meet is found from

$$S_T = S_1 + S_2 = [(\mathcal{E}_1 + \mathcal{E}_2) \times (H_1 + H_2)] = \frac{1}{Z}(\mathcal{E}_1 + \mathcal{E}_2)^2 \tag{1.46}$$

We may expand this to yield the time average, and thus observable, intensity

$$\langle S_T\rangle = \frac{1}{2Z}[|\mathcal{E}_1|^2 + |\mathcal{E}_2|^2 + 2\text{Re}(\mathcal{E}_1\mathcal{E}_2)] \tag{1.47}$$

If we now consider the electric fields to be of our standard sinusoidal form (Equation 1.20) and take one field to be shifted in phase by ϕ with respect to the other, and assume they have the same intensity $\mathcal{E}_0$, we have

$$\mathcal{E}_1 = \mathcal{E}_0\exp[j(\omega t - k_1z)] \tag{1.48}$$

$$\mathcal{E}_2 = \mathcal{E}_0\exp[j(\omega t - k_2z + \phi)] \tag{1.49}$$

These represent two waves, one with propagation constant k_1 and the other with k_2, which we have assumed to propagate in the **z** direction. Equation 1.47 then becomes

$$\langle S_T \rangle = \frac{1}{2Z}\{2\mathcal{E}_0^2 + 2\mathcal{E}_0^2 \exp[j((k_2 - k_1) - \phi)]\} \tag{1.50}$$

$$= \frac{\mathcal{E}_0^2}{Z}\{1 + \cos[(k_2 - k_1) - \phi]\} \tag{1.51}$$

$$= 2\frac{\mathcal{E}_0^2}{Z}\cos^2\left[\frac{1}{2}(k_2 - k_1) - \frac{\phi}{2}\right] \tag{1.52}$$

$$= S_0 \cos^2\left[\frac{\Delta\phi}{2}\right] \tag{1.53}$$

where S_0 corresponds to the maximum combined intensity of the two waves and $\Delta\phi$ the phase difference between them.

From Equation 1.53, we note that there are two interesting conditions for $\Delta\phi$: when $\Delta\phi/2 = i\pi$, where i = 0,1,2, . . . , $<S_T> = S_0 = 2\mathcal{E}_0^2/Z$, the maximum intensity; this is the point of maximum constructive interference. When $\Delta\phi/2 = (i + 1/2)\pi$, $<S_T> = 0$, and a maximum of destructive interference or a minimum of intensity occurs. Between these two extremes, the intensity of the interference varies with the well-known ($\cos^2$) dependence, as is shown in Figure 1.2.

For two waves with $k_1 = k_2$, meaning two identical waves traveling in the same medium, that have a relative, time-independent phase shift of $\Delta\phi$, the observable intensity of the interference, $<S_T>$ becomes a standing wave. This standing wave has an amplitude $S_0\cos^2(\Delta\phi)$, which is also time independent. We will study interferometers further in Chapter 11.

A final parameter useful for characterizing interference is the coherence length, L_C [m], defined as

$$L_C = \frac{c}{\Delta\nu} \tag{1.54}$$

where c is the speed of light, as usual, and $\Delta\nu$ [Hz = s^{-1}] the linewidth of the signal. Linewidth is the range of frequencies present in a practical source of TEM waves; even lasers, referred to as *single frequency*, are not, and they have a small but nonzero linewidth, as we shall see in Chapter 9. L_C then defines the distance over which interference can take place or, alternatively, the maximum path difference between two interfering waves; this can limit the physical size of an interferometer.

Example

For an optical linewidth of $\Delta\nu$ = 1 MHz, which would be a good value for a laser, we can see immediately from Equation 1.54 that interference could take place over

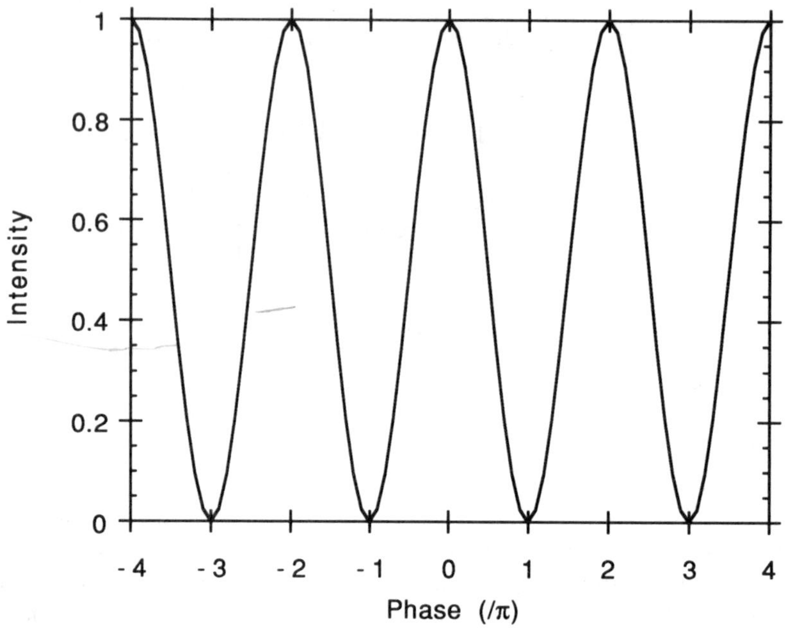

Figure 1.2 The merging of two electromagnetic waves gives rise to an interference pattern. The intensity varies as ($\cos^2$) of the phase difference, as shown in the plot for two waves of identical amplitude and polarization.

300 m, which is larger than most interferometers anyway. Thus lasers are good optical sources for interferometry. For semicoherent sources, such as an LED, with $\Delta\nu = 1$ THz (about 20 nm), $L_C \approx 300\ \mu$m, a much more stringent restriction. If one tries to generate interference with a path difference longer than the coherence length, the interference pattern will fade and eventually vanish.

REFERENCES

[1] Auden, W. H., and Isherwood, C., *The Ascent of F.6,* London: Faber and Faber, 1986.

[2] Kraus, J., and K. Carver, *Electromagnetics,* New York: McGraw Hill, 1973.

[3] Hariharan, P., *Principles of Interferometry,* San Diego: Academic Press, 1992.

Chapter 2

A Summary of Semiconductors

Uber Halbleiter sollte man nicht arbeiten, das ist eine Schweinerei; wer weiss ob es überhaupt Halbleiter gibt.[1]

W. Pauli, 1931 [1]

Despite Wolfgang Pauli's apparent dislike for semiconductors, the past several decades have seen remarkable progress in their development and application. Whereas semiconductors have been fruitful in unveiling the secrets of solid state physics to researchers in the pure sciences, they have also so successfully infiltrated all aspects of our lives that now only the fewest philistines, the author among them, would dream of brewing coffee without a silicon-based microprocessor imbedded in a coffee maker to accomplish the task for them.

Semiconductors are only a subdiscipline of solid state physics, but nonetheless, the field is vast. We will now attend to a brief review of only the most salient aspects of this area. In this chapter, we will review, very much in general, the structural, crystalline, and electrical properties of semiconductors, and we will discuss how one determines carrier densities and Fermi levels. The mechanics of doping and transport will be followed by a discussion of both PN- and hetero-junctions as basic device elements. There are many excellent standard texts[2] about semiconductors and the interested reader is referred to those for more details. Throughout this section, we will use silicon and gallium arsenide as generic representatives of the various optically interesting materials that we shall meet in Chapter 4.

2.1 SILICON AND GALLIUM ARSENIDE

In the substantial zoo of semiconductor materials, two beasts have become dominant in the world of integrated optics: the first is the III-V system and the second is silicon

[1] "One should not work on semiconductors, they are a mess; who knows if semiconductors even exist."

[2] The literature is extensive; the author likes: Ridley, B. K., *Quantum Processes in Semiconductors*, 2nd edition, Oxford, England: Clarendon Press, 1988; Madelung, O., *Introduction to Solid State Theory*, Berlin: Springer-Verlag, 1978; Kittel, C., *Quantum Theory of Solids*, New York; John Wiley & Sons, 1963; Ashcroft, N. W., and N. D. Mermin, *Solid State Physics*, Philadelphia: Saunders, 1976

(Si), although the latter plays a more marginal role in integrated optics than the former. Silicon, the second most abundant element on the planet after oxygen, has been in use since the dawn of time, or maybe longer, primarily as an ingredient in glass. The start of its more modern application, in the realm of electronics, could arguably be defined as concomitant with the invention of the transistor [2]. Si is now unmatched in its development for electrical applications, but its inability to efficiently emit light has hampered its use in integrated optics. Nevertheless, its relatively low price and well-advanced technology have resulted in its use for much passive optics and packaging, and its use in integrated optics may increase in the future.

The family of III-V semiconductors, with its patriarch gallium arsenide (GaAs), is actually an extended community of interrelated compounds. These do not have quite such an extensive history or large market as silicon: GaAs, for example, was first synthesized in the 1920s [3] and, in Switzerland, more gallium is used as a component in tooth fillings than in electronics. Since the invention of the semiconductor laser [4], however, the III-Vs have taken a firm hold on most semiconductor optical applications, as their efficient light emission and flexible optical properties have given rise to both the cheap lasers found in CD players and complex optical circuits for telecommunications. In this chapter, we will refer to GaAs as an able representative for the entire III-V material system, deferring the treatment of its near and distant relations to Chapter 4.

2.1.1 Crystalline Properties

Silicon is an elemental semiconductor, sitting in column IV of the periodic table. Si has a tetrahedral crystalline structure, implying that each Si atom has four nearest neighbors, at equal distances, at the corners of a regular tetrahedron, as shown in Figure 2.1. The entire structure thus has a cubic symmetry. The lattice structure of Si has a diamond unit cell, as shown in Figure 2.2; the unit cell represents the smallest unit of Si atoms that can be repeated in all directions to form a bulk crystal of silicon. The diamond lattice is equivalent to two face-centered-cubic (fcc) unit cells, one of which is shown on the left of the figure, where one fcc lattice is displaced along the cube diagonal by a quarter of the length of this diagonal from the other. The atoms of both fcc unit cells are Si and have a lattice spacing a [m] of $a = 0.543095$ nm [5].

The III-V semiconductors are compound semiconductors combining elements from columns III and V of the periodic table; hence their clever name. Their crystalline structure is, however, quite similar to that of silicon. Gallium arsenide has a zincblende lattice structure, which is identical to the diamond lattice, save that of the two fcc sublattices, one consists only of gallium atoms and the other only of arsenic. The gallium arsenide unit cell is shown in Figure 2.3, showing the slightly larger gallium

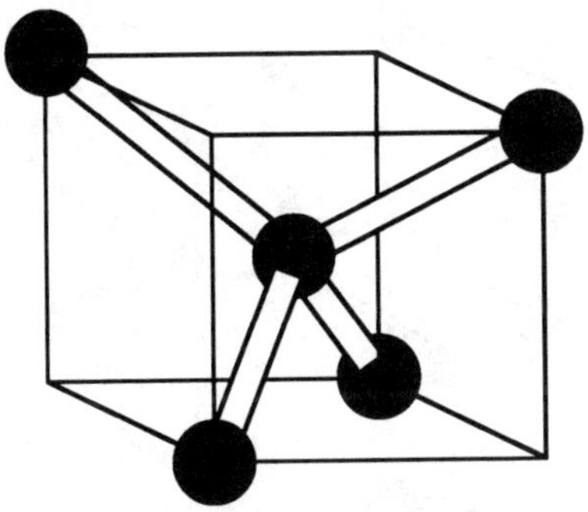

Figure 2.1 The four nearest neighbors of a Si atom in a crystal lattice showing the tetrahedral symmetry.

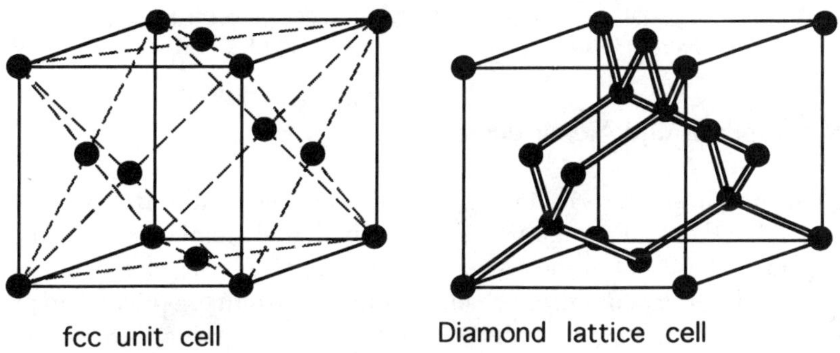

Figure 2.2 The Si unit cell has the diamond structure, which consists of two fcc sublattices displaced along the cube diagonal. A single fcc lattice is shown on the left.

atoms each with four arsenide nearest neighbors. The lattice spacing of this material is a = 0.56533 nm [5].

The atoms in both Si and GaAs are held together by covalent bonds, meaning that the electrons in the outer orbitals rearrange themselves in response to the neighboring atoms such that the electrons are effectively shared between two atoms. Like most covalent solids, these materials are quite brittle and hard, and they have a high melting temperature. The first property can be a concern for manufacturing: whereas a Si wafer can survive a drop onto the cleanroom floor intact, abusing a GaAs wafer in the same manner usually results in a search for the vacuum cleaner. Because GaAs is composed of two different elements, there is a small ionic component (about 4% based on electronegativity difference) to the atomic bonding, meaning that the material is polar, which has implications for some of its optical and electrical properties.

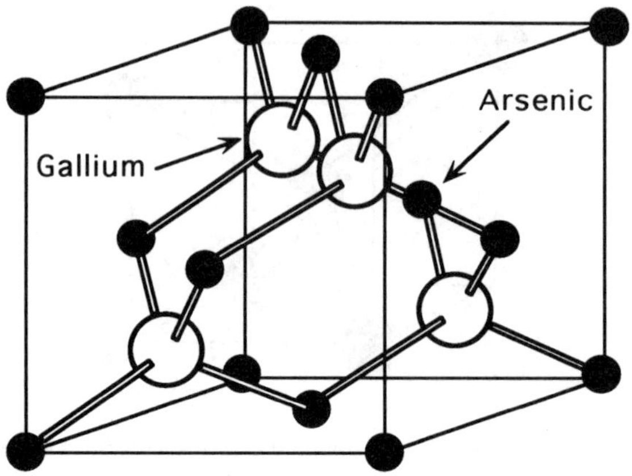

Figure 2.3 GaAs has a zincblende structure, equivalent to the diamond but with a Ga and an As fcc sublattice. The relative sizes of Ga (0.126 nm) and As (0.120 nm) are exaggerated for clarity.

2.1.2 Miller Indices and Atomic Planes

The regular atomic spacings and periodic repetitions evident in the lattice structures of Figures 2.2 and 2.3 show that we can usefully define atomic planes and directions in the crystal. These definitions are important because the electrical and optical properties can differ depending on which direction a current or optical field propagates through the crystal. The cubic symmetry lends itself to use of a Cartesian coordinate system as shown in Figure 2.4. We define the **x**, **y**, and **z** axes as the principal axes of the crystal, namely along the edges of the cubes drawn in Figures 2.2 and 2.3. The plane of atoms, for example, in the **yz** plane intersects the **x** axis at a point one lattice spacing, a, from the origin; this is thus denoted, using the notation (xyz), the

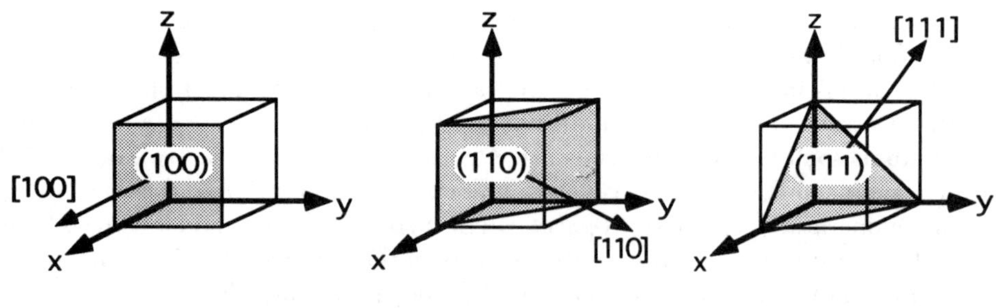

Figure 2.4 Definition of the Miller indices for the planes and directions in a crystal; the (100), (110), and (111) planes are shown, with the associated [100], [110], and [111] directions. For notation, see Table 2.1.

(100) plane. The associated crystalline *direction*, namely that normal to the plane, is labeled the [100] direction. These labels are known as the Miller indices of the atomic lattice.

We could as easily have chosen the **xz** or **xy** planes, with indices (010) and (001), respectively. Because these are both equivalent to the (100) plane, we often refer to the family of planes, {100}, as those with an intercept along one primary axis of the crystal. The associated family of directions is then labeled <100>. A summary of the notation for Miller indices is given in Table 2.1.

We can define two further families of planes and directions: the (110) direction has an intercept along two of the primary axes of the crystal and the [110] direction makes an angle of 45° with those two axes along the face of the unit cell. Finally, the (111) plane intersects all three axes implying that the [111] direction projects along the body diagonal of the unit cell. Although one can define more sets of planes and directions [6], often using more than one unit cell, the ones we have discussed above are typically the most important for our materials of interest. Which of these planes actually forms the crystal surface has important electrical and physical implications. The surface of both Si and GaAs wafers is typically (100), although (111) surfaces are sometimes used for specific applications. Wafers are frequently broken into smaller pieces, either before or after processing, to individually package chips or generate laser mirrors, for example. This is typically accomplished not by dropping them on the floor, but through a precise incision and then the application of pressure normal to the wafer plane. The substrate then cleaves precisely along one crystal plane: GaAs cleaves best along the {110} planes and Si along the {111}. Determining the crystal surface of a semiconductor wafer is discussed in Chapter 4.

2.2 ENERGY BANDS

A semiconductor, by virtue of its name, is such due the intermediate magnitude of its conduction; it lies between that of a conductor and that of an insulator. Metals typically have resistances of less than 10^{-6} Ω-cm. This can be better understood if

Table 2.1
Notation Employed for the Miller Indices to Define Planes and Directions, plus Families of Equivalent Planes and Directions

Notation	*Meaning*
(xyz)	Plane
{xyz}	Family of equivalent planes
[xyz]	Direction
<xyz>	Family of equivalent directions

we examine the schematic plots of Figure 2.5, where we see that the outer shell electrons from the metal atoms have many empty states to which they can easily transfer and be moved by an applied electric field. The insulator, on the other hand, requires its electrons to jump a large energy gap, typically several electronvolts (eV) [1 eV = q × energy in Joules = 1.6×10^{-19} J], before reaching empty states in which they can be driven about by an electric field; hence insulators have resistivities of 10^{14} Ω-cm or greater.

Semiconductors lie in between: as the figure shows, the electrons still need enough energy to surmount an energy gap, but the magnitude of this gap is smaller than that for insulators so that thermal excitation at room temperature is often sufficient to populate the upper, free states. Clearly the magnitude and shape of this energy gap define the characteristics of a semiconductor so that we should discus it in some detail.

2.2.1 Band Structure Calculations

By solving the Schrödinger equation, which relates the kinetic energy of a particle to the potential energy distribution of its surroundings, in a region of constant potential energy, the relationship between energy and momentum of an electron is found to be

$$E = \frac{\hbar^2 k^2}{2m} \tag{2.1}$$

for energy E [J or eV] and mass m [kg]. The term k [m^{-1}] is known as the wavevector, or crystal momentum; the wavevector is related to the momentum p [kg m/s] as

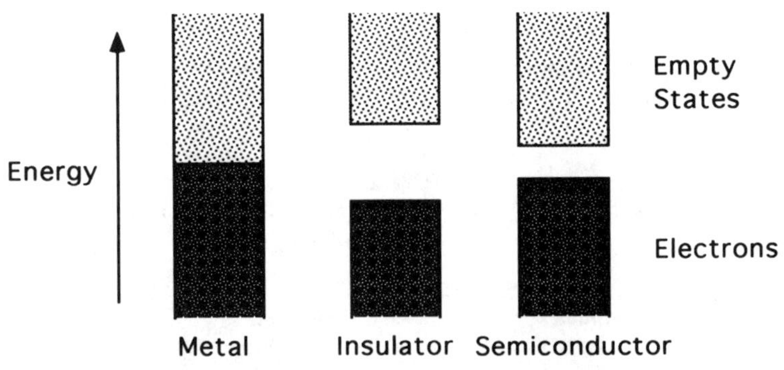

Figure 2.5 A schematic representation of the distinction between metals, insulators, and semiconductors. The lower region is filled with electrons, and the upper is empty. Metals have virtually no energy gap, and that for insulators is so large that excitation is unlikely under normal conditions.

$$k = \frac{p}{\hbar} = \frac{mv}{\hbar} \tag{2.2}$$

for electron velocity v [m/s] and Planck's constant $\hbar = h/2\pi$ [J s]. The energy momentum relationship of Equation 2.1 would hold, for example, in free space.

In the crystal of a semiconductor, however, the relationship between energy and velocity, or momentum, is no longer so simple. The periodically spatially varying potential energy due to the atomic cores of the crystal (i.e., the atoms minus the electrons that are floating around doing all the conducting) strongly affects the motion of the electrons. Solving for the energy of an electron in a crystal is in general quite a complicated procedure, even for a handful of atoms let alone the 10^{23} or so which make up a real crystal. Various approaches to the solution of the Schrödinger equation for crystal structures, usually based on good approximate mathematical models, have been undertaken. A comprehensive survey of these is well beyond the scope of the current text and we will merely note that the band structures we will sketch schematically below are based on such calculations.[3]

2.2.2 Energy Bands

By using one of these approximate techniques to solve for the relation between energy and momentum of an electron in a crystal, the simple free-space relation of Equation 2.1 is replaced by the rather complex dependence shown in Figure 2.6. We see that the electron has certain allowed energy states in a Si or GaAs crystal, the two examples shown. The relationship between E and k is known as the energy band diagram. This diagram is commonly drawn with electron energy increasing along the ordinate and the momentum k for one of several crystal directions plotted along the abscissa from 0 to the value $1/a$. We have plotted in this figure the momentum along the [111] and [100] directions.

The range of k values from 0 to $1/a$ is known as the Brillouin zone. Just as the crystal lattice is periodic in space, with period a, the energy versus wavevector band diagram is periodic in inverse space, or k-space. Then, just as we can draw only the unit cell to represent the smallest unit of the crystal lattice in real space, we can draw the Brillouin zone as representing the smallest unit of our band diagram. The diagram from 0 to $1/a$ is identical to the one from $1/a$ to $2/a$, and so forth. We can thus think of the Brillouin zone as the unit cell in k-space. The center of the Brillouin zone, where $k = 0$, is labeled the Γ point and the endpoints of the two crystalline directions, where $k = \pm 1/a$, are labeled L for the [111] and X for the [100] directions.

Although the E-k relationship for both Si and GaAs has been calculated to higher and lower energies than indicated in the plot, we have concentrated on the

[3]For example: Cohen, M. L. and J. R. Chelikovsky, *Electronic Structure and Optical Properties of Semiconductors*, Berlin: Springer-Verlag, 1988.

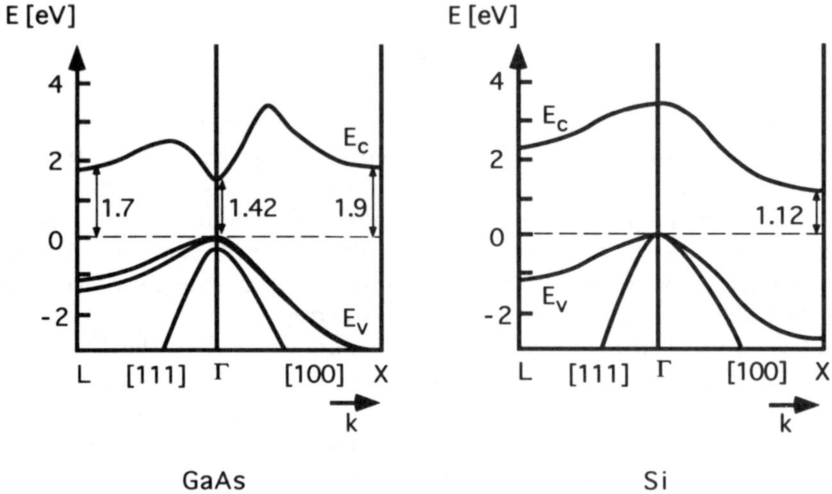

Figure 2.6 The Brillouin zones of GaAs and Si relating electron energy E to the wavevector k; the [100] and [111] directions have been drawn. The corresponding zone edges, where $k = \pm 1/a$, are labeled X and L, respectively, and the zone center is denoted Γ.

region shown since that is the energy regime where most of the interesting things happen. The upper energy level, labeled E_c and known as the conduction band, has no electrons at T = 0 K and thus plenty of empty electron states. The lower level, the valence band denoted E_v, is filled with electrons at zero temperature and has no empty electron states. Between the two lies the energy gap, E_g, which has a magnitude E_g = 1.12 eV for Si and E_g = 1.42 eV for GaAs; the energy gap is defined as the closest spacing between E_c and E_v and is a forbidden region for electrons. Most electronic and optical properties depend on the electron transitions between these two bands.

The band diagram for Si shows two valence bands, with varying curvature, which meet at the Γ point to make the valence band maximum. The conduction band minimum is at the Brillouin zone edge, at the X point. For GaAs, three valence bands appear, two meeting at the zone center forming the valence band maximum and one, the split-off band, peaking at an energy 0.34 eV below the maximum. The conduction band has its lowest minimum at the zone center and a neighboring minimum, at slightly higher energy, at the L point.

2.2.3 Electron Transitions between Bands

As we will see in more detail in the next section, when the temperature rises above zero some of the electrons in the valence band can surmount the energy gap and populate the conduction band; they leave behind a "missing electron" or hole. Likewise, electrons in the conduction band can fall back into the valence band, giving off the difference in energy, typically E_g, as a photon. One of the major distinctions between Si and GaAs becomes apparent when we compare the two energy band diagrams of Figure 2.6. Although both Si and GaAs have the valence band maximum at Γ, GaAs also has the conduction band minimum at the Brillouin zone center, whereas Si has its conduction band minimum at the X point. Based on the band structure, Si is known as an indirect bandgap semiconductor and GaAs a direct bandgap material.

The band structure implies that an electron leaping from the conduction band in GaAs need only emit a photon of the proper energy such that both energy and momentum are conserved, since the difference in momentum is zero. For Si, on the other hand, the conduction band minimum lies at the X point, such that a momentum contribution is required in order for both energy and momentum to be conserved. Photons have a very small momentum component, such that the remainder typically comes from interaction with the crystal lattice through lattice vibrations, or phonons. The transition of an electron from X to Γ in the Si lattice then requires both a photon and several phonons, and because the probability of such an interaction is small, these transitions are very unlikely. Thus, most electrons in silicon drop into the valence band nonradiatively, meaning they do not emit a photon and instead donate their energy to a succession of phonons: for that reason, Si is a very inefficient emitter of light.

A final characteristic of the GaAs band structure is of interest. In addition to the conduction band minimum at the Brillouin zone center, there is a further minimum at the L point in the [111] direction. Although this is at a higher energy above the valence band, and it requires a nonzero momentum contribution to reach from the Γ point, electrons excited to higher energies in the conduction band can often enter this neighboring minimum. This gives rise to numerous interesting electrical properties, such as the Gunn effect.

2.2.4 Effective Mass

We can show that the form of the energy-momentum relationship given by the band structure of a semiconductor strongly affects the transport properties of the electron. If we assume that the energy in a small region of the band structure varies roughly parabolically with momentum, such that Equation 2.1 holds, we can take the first and second derivatives

$$\frac{dE}{dk} = \frac{\hbar^2 k}{m^*} = \hbar v \qquad (2.3(a))$$

$$\frac{d^2E}{dk^2} = \frac{\hbar^2}{m^*} \tag{2.3(b)}$$

and show that electron velocity v is related to the band slope and the mass m^* to the curvature. The mass m^* is an effective mass of the electron that defines how the dynamics of the electron are affected by the presence of the crystal lattice; we use the notation m^* to distinguish it from a normal electron rest mass, since the two can be very different. The effective mass is defined by rewriting Equation 2.3(b)

$$m^* = \hbar^2\left(\frac{\partial^2 E}{\partial k^2}\right)^{-1} \tag{2.4}$$

from which we see clearly that a region of the energy band with high curvature (high $\partial^2E/\partial k^2$) has a smaller m^* than flatter regions of the band.

The effective mass m^* is strictly a tensor since the E-k relationship, and thus effective mass, is direction dependent. Electron mass is related to the curvature of the conduction band, and hole effective mass to the curvature of the valence band. It is clear from the energy band diagrams that the electron and hole effective masses vary with position in the Brillouin zone: the conduction band of GaAs has a low-mass minimum at the Γ point and a higher mass minimum, due to its lower curvature, at the L point. Both materials have a light hole and heavy hole valence band, the former having a higher curvature than the latter.

The effective mass m^* is usually related to the free electron mass ($m_0 = 9.1 \times 10^{-31}$ kg) as $m^* = m_x \times m_0$ where m_x is a dimensionless factor that merely indicates how the effective mass relates to that of an electron in free space. The subscript x refers to which mass we are talking about; Table 2.2 gives a summary of typical values for Si and GaAs [7]. Si has two electron masses, GaAs has only one; both materials have several different effective hole masses due to the energy band structure of the valence band near the Γ point.

Table 2.2
Effective Mass Values (m_x) for Different Parts of the Brillouin Zone in Si and GaAs

	Si			*GaAs*	
Mass	Name	m_x	Mass	Name	m_x
Electron:					
Lateral	m_{el}	0.92	Electron	m_e	0.07
Transverse	m_{et}	0.19			
Heavy hole	m_{hh}	0.52	Heavy hole	m_{hh}	0.68
Light hole	m_{lh}	0.16	Light hole	m_{lh}	0.12
			Split-off hole	m_{so}	0.20

Because most of the relevant electron transitions and activity take place near the valence band maximum and conduction band minimum, we can use the effective mass picture to simplify the band structure for our purposes. At the band extrema, we can approximate the energy-momentum relationship as a quadratic one, replacing the mass m in Equation 2.1 by m^* to yield

$$E = \frac{\hbar^2 k^2}{2m^*} \tag{2.5}$$

We can then draw the center of the GaAs Brillouin zone as shown in Figure 2.7, which allows us to display the electron transitions for small variations of k. For many applications, we can ignore the rest of the energy band.

In many cases, the variation of energy with momentum is less relevant than the behavior of the energy gap and the energy bands in real space, i.e., x rather than k. For that reason, we can make a further simplification and ignore k altogether and draw the energy bands as shown in Figure 2.8. The effective mass is assumed constant and we have to keep the restrictions imposed by momentum conservation in mind: thus, if we use a diagram like that of Figure 2.8 for Si, we must remember that the

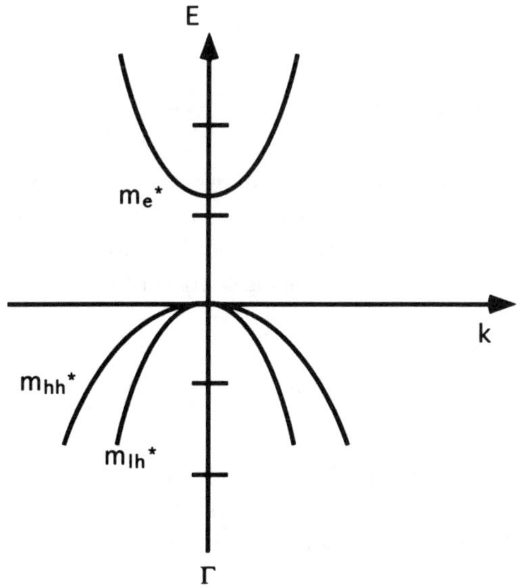

Figure 2.7 The center of the Brillouin zone showing the direct bandgap of GaAs drawn using the parabolic band approximation. The curvature is related to m^*. The band structure further from the bandgap (the Γ point) is ignored.

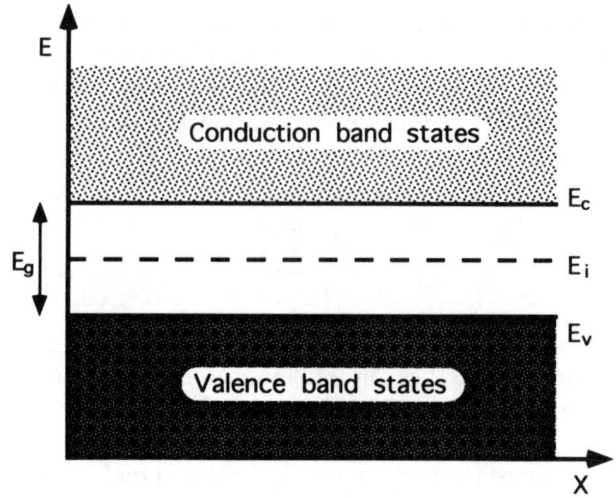

Figure 2.8 The schematic sketch of a generic semiconductor energy gap obtained by ignoring *k* and plotting distance in real space, **x**, along the ordinate.

gap is not a direct one and that a radiative transition is unlikely. The intrinsic level E_i is defined as lying at the energy gap center.

2.3 CARRIER CONCENTRATION

As we hinted above, the densities of both electrons in the conduction band and holes in the valence band are nonzero for a nonzero temperature. Since the carrier concentration, given as number of carriers per unit volume, is the basis for understanding the electrical and optical behavior of a semiconductor, we will now discuss how it is determined.

2.3.1 Density of States

Two factors contribute to the calculation of electron and hole densities: the number of available states for an electron and the probability that an electron finds itself in one of those states. The former property is known as the density of states, which we now determine.

Solving the Schrödinger equation in a given volume yields a uniform distribution of allowed values for k; we can imagine a sphere in k-space whose radius is the maximum momentum value. Because energy and momentum are related, in a particular region of the band, by Equation 2.5, we can determine the number of energy states for a given value of k. From there, we can calculate the density of states,[4] which is found for the conduction band, assuming that m^* is isotropic, to be

$$D_c(E) = \frac{\sqrt{2}}{\pi^2} \frac{m^{*3/2}}{\hbar^3} (E - E_c)^{1/2} \tag{2.6}$$

$D_c(E)$ [J^{-1} cm^{-3}] represents the density of available electron states per unit volume and energy for energy E above the conduction band E_c. We use precisely the same arguments to calculate the density of electron states in the valence band below E_v, namely

$$D_v(E) = \frac{\sqrt{2}}{\pi^2} \frac{m^{*3/2}}{\hbar^3} (E_v - E)^{1/2} \tag{2.7}$$

The square-root dependence of the density of states with energy is shown schematically in Figure 2.9(a).

2.3.2 Fermi-Dirac Distribution

An available electron state does not yet guarantee that it is occupied. Electrons are fermions, meaning they are subject to the Pauli exclusion principle, which requires that any available state for a particle can be occupied by at most two such particles, each with opposite spin. This is in contrast to bosons, such as photons, for which an arbitrary number of identical particles can occupy a given state.

If we apply the rules of statistical quantum mechanics to a system of Fermions, including the restriction that no more than two particles may occupy the same state, we can derive a probability distribution[5] that a state at a given energy is occupied

$$f(E) = \frac{1}{1 + \exp\left(\frac{E - E_F}{k_B T}\right)} \tag{2.8}$$

$f(E)$ [dimensionless] is known as the Fermi-Dirac distribution, the probability of finding an electron at energy E. k_B [J/K] is the Boltzmann constant and T [K] the temperature. E_F is known as the Fermi energy or Fermi level, and is defined as the

[4]For a derivation, see for example: Kittel, C. and H. Kroemer, *Thermal Physics*, 2nd edition, San Francisco: Freeman, 1980.

[5]See for example: Huang, K., *Statistical Mechanics*, Chapter 8.5, 2nd edition, New York: John Wiley & Sons, 1987.

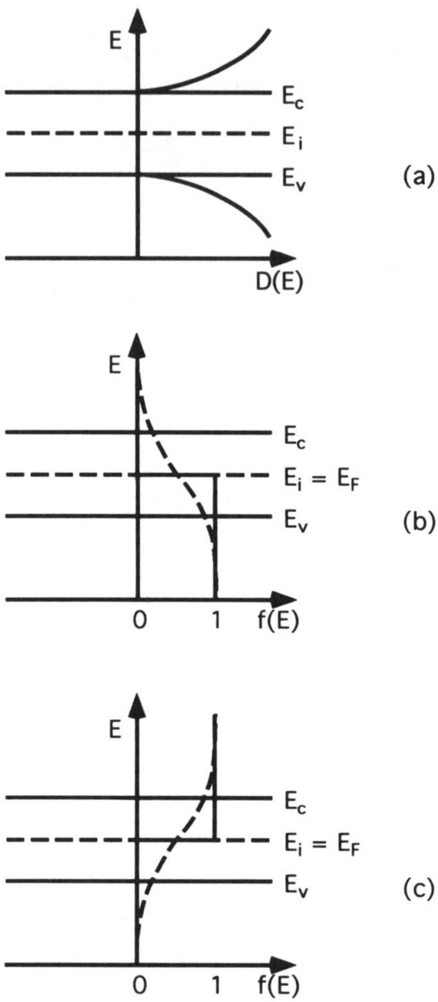

Figure 2.9 (a) The density of available electron states in the conduction band and hole states in the valence band varies as $(E)^{1/2}$. (b) The Fermi-Dirac distribution function $f(E)$ plotted schematically for $T = 0$ (solid line) and $T \neq 0$ (dashed line). The Fermi level E_F is defined as the point at which the distribution function has the value 0.5. (c) The function $1 - f(E)$ giving the occupation probability for holes in the valence band for $T = 0$ and $T \neq 0$.

energy at which, for zero temperature, all energy states below it are occupied and all above are empty; alternatively, at finite temperature, E_F is the energy at which the occupation probability is $f(E) = 0.5$. In Figure 2.9(b), we have sketched the Fermi-Dirac distribution for zero and nonzero temperature, assuming that the Fermi energy is at the middle of the gap, $E_F = E_i$. It is clear from Equation 2.8 that the occupation probability is a strong function of temperature.

For the valence band, the parameter of interest is not the probability that an electron is present but that one is absent: a missing electron is a hole with which an electron can recombine. The probability of vacancy in the conduction band is given by

$$1 - f(E) = \frac{\exp\left(\frac{E - E_F}{k_B T}\right)}{1 + \exp\left(\frac{E - E_F}{k_B T}\right)} = \frac{1}{1 + \exp\left(\frac{E_F - E}{k_B T}\right)} \tag{2.9}$$

which is shown schematically in Figure 2.9(c). The definition of the Fermi level remains the same.

The total carrier density, electrons and holes per unit volume, n and p [traditionally cm^{-3}], respectively, is then the product of the density of available states with the occupation probability, integrated over all relevant energies. These are expressed as

$$n = \int_{E_c}^{\infty} D_c(E) f(E) dE \tag{2.10}$$

for electrons and

$$p = \int_{-\infty}^{E_v} D_v(E) [1 - f(E)] dE \tag{2.11}$$

for holes.

The integrals of Equation 2.10 and 2.11 are in general not solvable analytically; we can, however, make an approximation to provide an analytic result. If the Fermi level is more than several $k_B T$ from either the valence or conduction band edge, which is certainly the case for an undoped semiconductor where $E_F \approx E_i$, we can approximate the Fermi-Dirac relation by the exponential Maxwell-Boltzmann distribution,

$$f(E) \approx \exp\left[-\frac{(E - E_F)}{k_B T}\right] \tag{2.12}$$

This probability distribution applies to bosons, but also to the high energy *tail* of a fermion distribution, where the interaction between electrons is negligible. Using this exponential probability distribution, Equation 2.10 and 2.11 are easy to integrate and yield

$$n = N_c \exp\left[-\frac{(E_c - E_F)}{k_B T}\right] \tag{2.13}$$

$$p = N_v \exp\left[-\frac{(E_F - E_v)}{k_B T}\right] \tag{2.14}$$

where

$$N_c = 2\left(\frac{2\pi m_c^* k_B T}{\mathrm{h}^2}\right)^{3/2} \tag{2.15}$$

$$N_v = 2\left(\frac{2\pi m_v^* k_B T}{\mathrm{h}^2}\right)^{3/2} \tag{2.16}$$

N_c and N_v [traditionally cm^{-3}] are known as the effective density of states, or the density of states at the band edge. In Figure 2.10 we see the distribution of electrons above the conduction band and holes below the valence band in energy. This distribution is a product of the density of states and the occupation probability.

Once we know the value of the Fermi level, we can then use Equation 2.13 and 2.14 to determine the density of electrons and holes in the semiconductor. We can simplify matters by writing

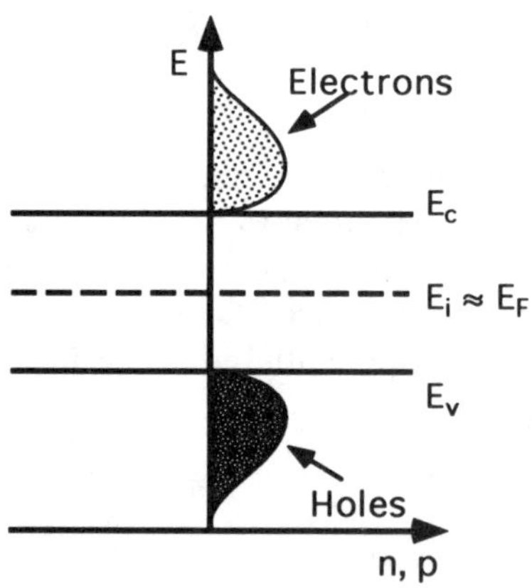

Figure 2.10 The distributions of electrons in the conduction band and holes in the valence band, derived from the product of the density of states and the Fermi-Dirac distribution function at nonzero temperature.

$$n = 5.39 \times 10^{15} T^{3/2} \exp\left[-\frac{(E_c - E_F)}{k_B T}\right] \quad \text{for Si}$$
$$p = 1.96 \times 10^{15} T^{3/2} \exp\left[-\frac{(E_F - E_v)}{k_B T}\right] \tag{2.17}$$

$$n = 9.05 \times 10^{13} T^{3/2} \exp\left[-\frac{(E_c - E_F)}{k_B T}\right] \quad \text{for GaAs}$$
$$p = 1.35 \times 10^{15} T^{3/2} \exp\left[-\frac{(E_F - E_v)}{k_B T}\right] \tag{2.18}$$

which retains the temperature dependence explicitly. We will return to how one calculates E_F below.

As a final point, we see that, for $n = p$, the Fermi level is almost, but not exactly, in the center of the gap. Strictly,

$$E_F = \frac{1}{2}E_g + \frac{3}{4}k_B T \ln\left(\frac{m_v^*}{m_c^*}\right) \tag{2.19}$$

Because the second term is only about −11 meV for Si and −44 meV for GaAs, we can usually safely take $E_F = 1/2E_g = E_i$ for undoped material.

2.3.3 Intrinsic Carrier Concentration

If we take the product of n and p, we see that

$$pn = N_C N_V \exp\left(\frac{E_v - E_c}{k_B T}\right) = N_C N_V \exp\left(\frac{-E_g}{k_B T}\right) \equiv n_i^2 \tag{2.20}$$

where

$$n_i = \sqrt{N_C N_V} \exp\left(\frac{-E_g}{2k_B T}\right) \tag{2.21}$$

defines the intrinsic carrier concentration n_i [cm^{-3}]. It is important to note that n_i is only a function of temperature and energy gap. The relationship between intrinsic carrier concentration and temperature is shown in Figure 2.11 from which we can see that, for Si and GaAs at room temperature, we have

$$\text{Si: } n_i = 1.45 \cdot 10^{10} \text{ cm}^{-3}$$

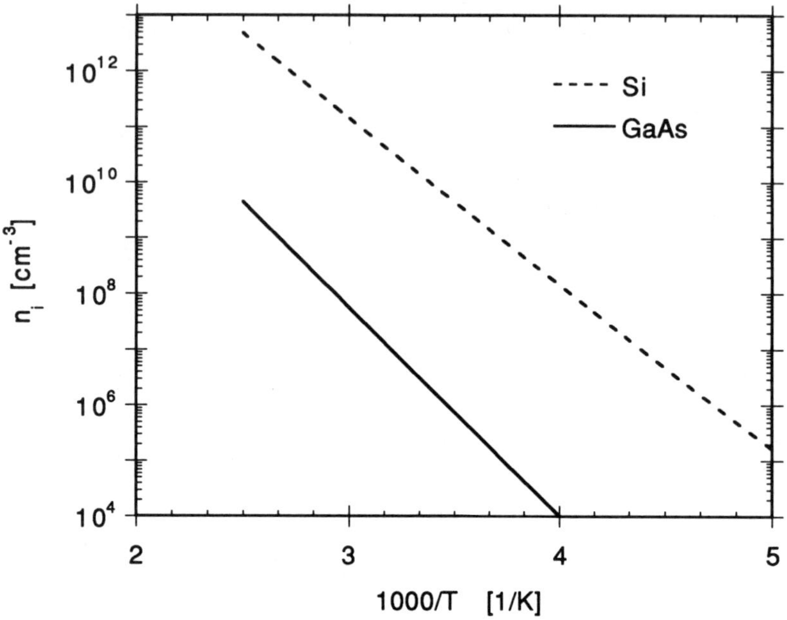

Figure 2.11 The intrinsic carrier concentration n_i plotted as a function of 1/T for Si and GaAs, showing the strong function of temperature.

$$\text{GaAs: } n_i = 1.79 \cdot 10^6 \text{ cm}^{-3}.$$

The product $pn = n_i^2$ is thus a constant at any given temperature, meaning that if the concentration of electrons increases in a piece of semiconductor, the concentration of holes decreases, and vice versa. This relationship is known as the law of mass-action but holds only for nondegenerate semiconductors; we discuss degeneracy in the next section.

For nonequilibrium conditions, such as under the injection or generation of carriers, electrons and holes have different Fermi levels, because the equilibrium condition $pn = n_i^2$ may not necessarily hold. The Fermi level for each is then known as a quasi-Fermi level, and is still a function of the local electron or hole concentration.

2.4 DOPING

Intrinsic semiconductors are interesting for many types of experiments and some applications, but most electronic and optical devices rely on controlled permanent changes in the concentration of electrons and holes in order to function. This process is known as doping and consists of replacing either silicon or, for GaAs, gallium or arsenic, atoms with certain other atomic species. These species, termed dopants, have either more or fewer electrons in their outer shell than those of the host lattice, and thus contribute either net negative or positive charge to the crystal. Atoms that contribute an electron are known as donors and form n-type material, with an excess of electrons. Those that have one electron too few attract electrons from the lattice and are known as acceptors; they produce p-type material, which has an excess of holes.

2.4.1 Donors and Acceptors

Donors and acceptors produce so-called donor and acceptor energy levels in the bandgap—the former close to the conduction band and the latter close to the valence band—as we can see in Figure 2.12. Many atomic species act as dopants in Si or GaAs; a few of the more popularly used ones are listed in Table 2.3 along with their distance in energy from the band edge, the activation energy [8].

Because the thermal energy k_BT = 26 meV at room temperature, we see that many of the dopant ions are ionized, meaning that they have given up or swallowed an electron and are electrically active. The extent of ionization is temperature dependent and can be determined through the use of Fermi-Dirac statistics. For donor doping concentration N_D [cm^{-3}], the ionized concentration N_D^+ may be written as [9]

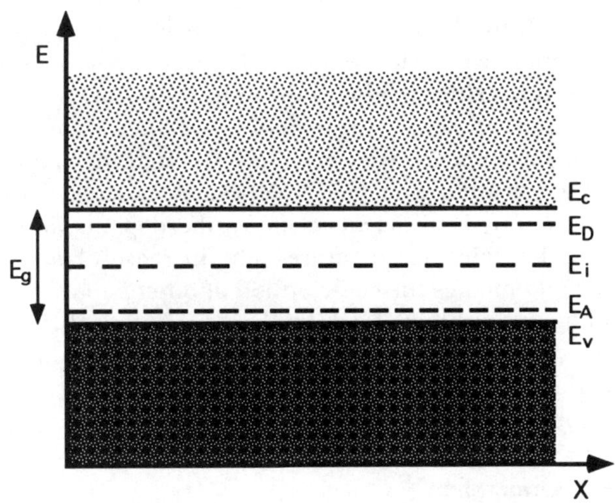

Figure 2.12 Donor, E_D, and acceptor, E_A, levels create states inside the energy gap that donate an electron or a hole to the semiconductor, respectively.

Table 2.3
Some Popular Donors and Acceptors in Si and GaAs

Element	*Si* Type	Activation energy [eV]	Element	*GaAs* Type	Activation energy [eV]
As	n	0.054	Si	n	0.006
P	n	0.045	C	p	0.026
B	p	0.045	Zn	p	0.031
			Mg	p	0.031

$$N_D^+ = N_D \frac{1}{1 + 2\exp\left(\frac{E_F - E_D}{k_B T}\right)} \tag{2.22}$$

while for acceptor concentration N_A, the ionized portion is N_A^-

$$N_A^- = N_A \frac{1}{1 + 4\exp\left(\frac{E_A - E_F}{k_B T}\right)} \tag{2.23}$$

The factors 2 and 4 in the denominator derive from the number of available states for each electron—2 for fermions in a single conduction band and 4 for fermions in the two valence bands that meet at the Γ point. We note that for both the n- and p-type doping shown here, the Fermi level E_F is still below E_D or above E_A. For this, termed the nondegenerate, case, the law of mass-action, $pn = n_i^2$, still holds.

Because temperature critically affects the ionization state of a dopant level, the concentrations of electrons and holes due to doping are likewise temperature dependent. As we can see in Figure 2.13, where we have plotted the carrier concentration in GaAs logarithmically as a function of $1/T$, there are three regimes of interest [10]. At low temperatures, $k_B T$ is insufficient to ionize the dopants and the carriers are subject to freeze-out. For high temperatures, the increase of n_i with temperature (from Equation 2.21) dominates the contribution of doping to carrier concentration, and there is a rapid increase with T. Luckily, there is a relatively large regime where the carrier concentration is relatively flat with temperature; room temperature falls into this range. In this flat region, most of the dopants are ionized, and N_A or N_D defines the carrier concentration.

2.4.2 Fermi Level Determination

The Fermi level for doped semiconductors can also be determined, although it is not as straightforward as is the case for undoped material. The total charge distribution

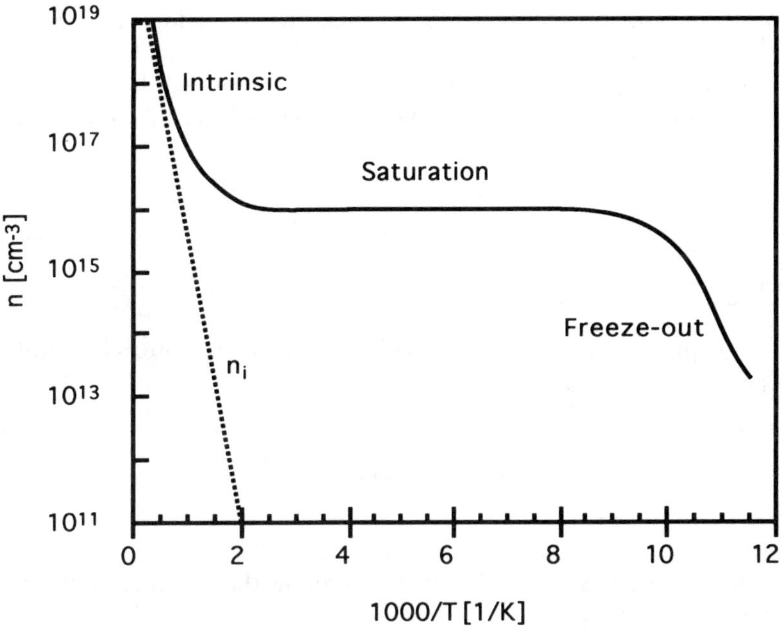

Figure 2.13 Three regimes are visible in the variation of carrier concentration with inverse temperature, drawn here for GaAs doped n-type with $N_D = 10^{16}$ cm^{-3}. At high T, the rapid increase is due to the rise in n_i seen in Figure 2.11; at low temperature, the carriers freeze-out when the thermal energy cannot ionize the dopant ions. The large flat region typically includes room temperature.

in the semiconductor is now the sum of free carriers and the ionized dopant cores, which have given or taken up an electron and are thus charged. Charge neutrality then requires

$$p + N_D^+ = n + N_A^- \tag{2.24}$$

We can expand this by adding the expressions for carrier density and ionized dopant concentration to yield

$$N_v \exp\left[\frac{-(E_F - E_v)}{k_B T}\right] + \frac{N_D}{1 + 2\exp\left(\frac{E_F - E_D}{k_B T}\right)} = N_c \exp\left[\frac{-(E_c - E_F)}{k_B T}\right] + \frac{N_A}{1 + 4\exp\left(\frac{E_A - E_F}{k_B T}\right)} \tag{2.25}$$

This expression must be solved for E_F, no mean feat, but for which Schockley had already derived an elegant graphical approach [11]. We can make an approximation, however, by assuming that for an n-type semiconductor, all the donors are ionized so that $N_D^+ = N_D = n$. We can then relate the Fermi level to the conduction band simply as

$$E_F = E_c - k_B T \ln\left(\frac{N_c}{N_D}\right) \tag{2.26}$$

The identical approximation can be made for a p-type substrate, where full ionization yields $N_A^- = N_A = p$, so that

$$E_F = E_v + k_B T \ln\left(\frac{N_v}{N_A}\right) \tag{2.27}$$

Thus, for many practical cases, we can determine the position of the Fermi level without too much trouble.

2.4.3 Degenerate Semiconductors

For very high doping levels, the Fermi level will move closer to and eventually beyond the conduction or valence bands, for n- and p-type semiconductors, respectively. When the Fermi level is above E_C or below E_V the semiconductor is known as degenerate, and many of the approximations we have just made are no longer valid. In particular, in a degenerate semiconductor, $pn = n_i^2$ no longer holds and the Fermi-Dirac distribution must be solved exactly to determine the carrier density.

Heavy doping also results in a transition from a distinct doping energy level to an impurity band, which widens as the doping density increases. As a result, the activation, or ionization, energy decreases and can eventually reach zero.

2.5 TRANSPORT

Once we have filled out a semiconductor with electrons or holes from donor or acceptor dopants, most applications will require that we make them move and do something useful. We can accomplish this in one of two ways: create a gradient in concentration and induce carrier movement by diffusion or apply an electric field and rely on drift.

2.5.1 Drift and Diffusion

Carrier diffusion is due to a gradient in concentration. Diffusion current density, J [A/cm^2], can be written separately for electrons and holes as

$$J_n = qD_n\nabla n \rightarrow qD_n\frac{\partial n}{\partial x} \tag{2.28}$$

$$J_p = -qD_p\nabla p \rightarrow -qD_p\frac{\partial p}{\partial x} \tag{2.29}$$

where the second terms after the arrow represent the 1-dimensional case, which is useful in many situations. D_n and D_p are the diffusion constants [cm^2/s], which are related to mobility and essentially define how fast the carriers move in response to a concentration gradient. Typical diffusion constant and mobility values for Si and GaAs are given in Table 2.4.[6]

The energy bands and Fermi level for the case of diffusion are shown in Figure 2.14(a); since the concentration of, in this case, electrons varies with position, so must the Fermi level. We note that the Fermi level must be tilted for current to flow; a flat Fermi level in system implies that currents are zero.

If we apply an electric field to the semiconductor, the electrons and holes, being charged carriers, will move, although they will be strongly affected by the crystal lattice in which they find themselves. The application of an electric field will tilt the energy bands as shown in Figure 2.14(b) and give rise to electron and hole drift currents given by

$$J_n = q\mu_n n\mathcal{E} \tag{2.30}$$

Table 2.4
Typical Values for Diffusion Constant, D, and Mobility, μ, for Si and GaAs

Parameter	*Si*	*GaAs*	*Units*
D_n	35	230	cm^2/s
D_p	12	7	cm^2/s
μ_n	1350	8600	cm^2/Vs
μ_p	480	250	cm^2/Vs

[6]For example: Grove, A. S., *Physics and Technology of Semiconductor Devices*, New York: John Wiley & Sons, 1967, p. 102.

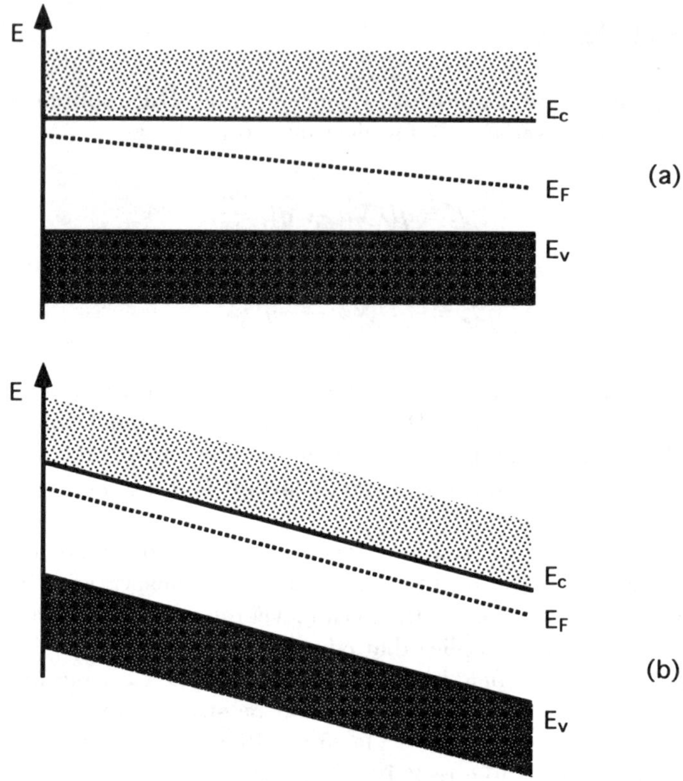

Figure 2.14 The energy bands of a semiconductor for the cases of (a) diffusion, with a larger carrier concentration to the left; and (b) drift, where the electric field points to the left. In both cases, electrons move to the right.

$$J_p = q\mu_p p\mathcal{E} \tag{2.31}$$

for electric field $\mathcal{E}$ [V/m] and mobilities μ_n and μ_p [cm^2/Vs]. Mobility is the proportionality between carrier velocity, v [m/s], and applied electric field, $v = \mu\mathcal{E}$, and thus contains the effects of the medium through which the carrier is moving; typical values are again given in Table 2.4.

For a given distribution of carriers and applied field in a semiconductor, the total electron and hole current densities are then the sum of the drift and diffusion components so that we have

$$J_n = q\mu_n n\mathcal{E} + qD_n\nabla n \tag{2.32}$$

$$J_p = q\mu_p p\mathcal{E} - qD_p\nabla p \tag{2.33}$$

Notice the hole diffusion term is of opposite sign due to the convention of current flow being that of *positive* charge carriers.

2.5.2 Contributions to Mobility

The mobility term collects all of the effects that the crystal lattice and the various bits of atomic rubbish that float about in it have on the moving charge carrier. There are several components, each a function of different material parameters. Lattice vibrations, due to nonzero temperature and represented by acoustic phonons, contribute a mobility factor that varies with temperature and effective mass as

$$\mu_l \propto \frac{T^{-3/2}}{m^{*5/2}} \tag{2.34}$$

We can imagine the atoms of the lattice jiggling in place due to their thermal energy, thus scattering charge carriers through elastic collisions. This component of mobility decreases with increasing temperature, since with increasing T such scattering increases. For GaAs, which is a polar semiconductor, optical phonons also contribute to lattice scattering, adding a further mobility component

$$\mu_p \propto \frac{T^{1/2}}{m^{*3/2}} \tag{2.35}$$

An ionized impurity, which is a charged site in the lattice, will also scatter a charge carrier due to the coulombic interaction. This contributes a term, for ionized impurity density N_i

$$\mu_i \propto \frac{T^{3/2}}{N_i m^{*1/2}} \tag{2.36}$$

which increases with increasing temperature. We can see why this is so by imagining that at elevated temperatures, the carriers move faster and thus have less time to be deflected by an ionized impurity; thus, the scattering is smaller and this component of the mobility is higher at higher temperatures.

The total, effective mobility that a carrier sees when moving through the lattice is given by the reciprocal sum of the components, namely

$$\frac{1}{\mu_T} = \frac{1}{\mu_l} + \frac{1}{\mu_i} \tag{2.37}$$

We note that for all of the contributions to μ_T we consider, mobility is inversely related to effective mass, m^*. Thus, lighter carriers will move faster with applied field, which usually means that the electrons move more quickly than holes.

We can often derive empirical relationships between μ_T and temperature, which accounts for all processes occurring in the lattice. The variations of mobility with temperature, doping, and other parameters are in general complex but have been studied in detail by numerous workers in the field [12,13]. Mobility and diffusion constants are related to one another by the Einstein relations, which state

$$\frac{D_{n,p}}{\mu_{n,p}} = \frac{k_B T}{q} \tag{2.38}$$

and are valid for nondegenerate semiconductors.

2.5.3 High Field Effects

When the applied electric fields reach large values, on the order of 10^4 to 10^5 V/cm, several nonlinear effects become apparent. The most prominent is velocity saturation: as the field increases, the carrier velocity does not increase in direct proportion but eventually saturates at a constant value. Through optical phonon emission, carriers lose additional energy imparted to them by the field and velocity does not increase beyond its saturation value.

The band structure of GaAs, with its two neighboring valleys at the X and L points of the Brillouin zone as we saw in Figure 2.6, also gives rise to interesting electrical effects at high applied electric fields. When electrons have gained sufficient energy from the field, they can transfer from the conduction band minimum at the Γ point to one of the neighboring valleys, typically the L-valley; this process is known as k-space transfer. Because the L-valley, due to its curvature in the E-k plane, has a large effective mass, $m^* = 1.2m_0$, electrons that transfer into it have a low mobility, on the order of 100 cm^2/Vs. These electrons virtually come screeching to a halt and give rise to a number of electrical phenomena, including negative differential resistance and the Gunn effect [14].

2.6 THE PN JUNCTION

We have seen in the last few pages how we can take a piece of semiconductor, add electrons or holes to it, and make these move about at our whim. Although this is interesting in and of itself, the combination of n and p doped regions, imaginatively called a pn junction, forms the basis for diodes and a host of further structures.

We can begin with a *step* or *abrupt* junction which, while perhaps not exactly physically realizable, allows us to simply calculate the behavior of a pn junction. If we take two separate semiconductors, one n-type and one p-type as shown in Figure

2.15, and bring them together, the conduction and valence bands will bend as shown in the lower half of the drawing because the Fermi level must be flat in the absence of current flow. When we form the junction, electrons will initially diffuse from the n side to the p side and holes in the opposite direction, due to the strong concentration gradients. The diffusing carriers will, however, uncover the dopant ion cores, positive donors and negative acceptors, and this will establish an electric field that will eventually hinder any further current flow. The flow of carriers will stop when the drift and diffusion currents become equal, which, for electrons, would imply

$$J_n = q\mu_n n\mathscr{E} + qD_n\frac{\partial n}{\partial x} = 0 \tag{2.39}$$

for the 1-dimensional case. Thus, without any external bias, the pn junction has a built-in field that balances the drift and diffusion currents so that total current is zero.

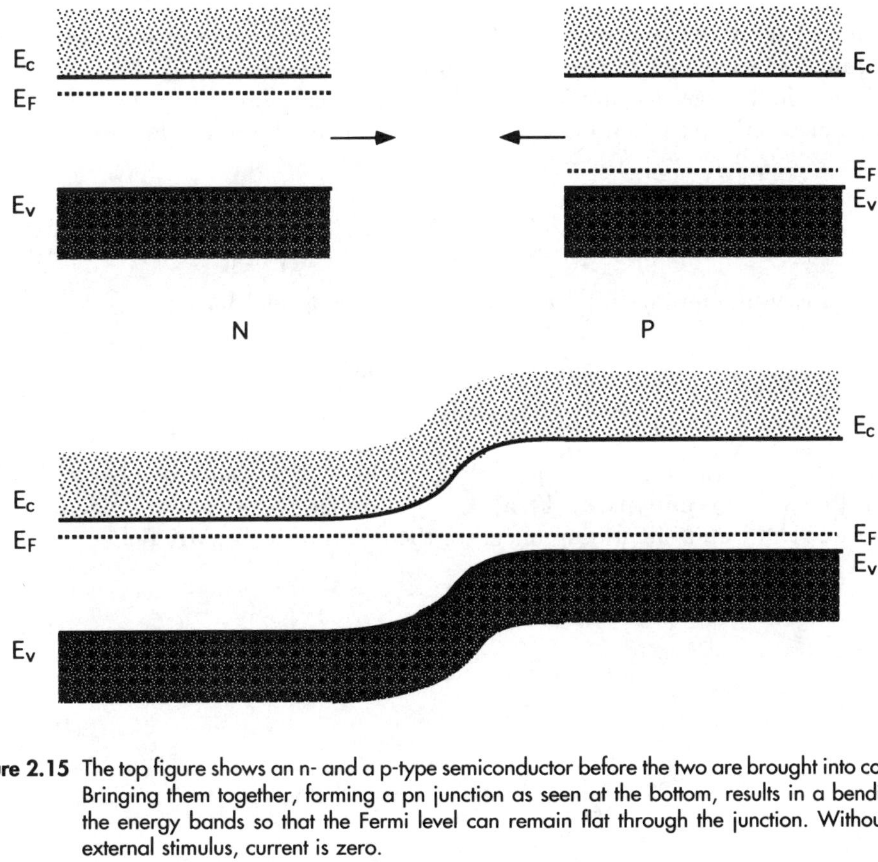

Figure 2.15 The top figure shows an n- and a p-type semiconductor before the two are brought into contact. Bringing them together, forming a pn junction as seen at the bottom, results in a bending of the energy bands so that the Fermi level can remain flat through the junction. Without any external stimulus, current is zero.

2.6.1 The Depletion Approximation

Near the junction, there is a region that is depleted of electrons and holes: these have diffused across the junction until a sufficiently large field is built up to inhibit further diffusion. In this depletion region, the dopant ion cores are no longer neutralized by mobile carriers so that they represent a region of fixed charge. By use of the depletion approximation, which assumes that we can define distinct regions that are fully depleted, with a sharp boundary to the neutral semiconductor, we can simplify our analysis of the pn junction. As shown in Figure 2.16(a); on the n side, we assume a positively charged N_D^+ extends to a distance $-W_n$ [m] from the junction, and on the p side, a negatively charged N_A^- extends to W_p [m] from the junction. Because the rest of the semiconductor is neutral (i.e., holes balance ionized acceptors and electrons balance ionized donors), charge neutrality then requires

$$W_n N_D^+ = W_p N_A^- \tag{2.40}$$

which means nothing more than that the two sheets of fixed charge at opposite sides of the junction are equal in magnitude and opposite in sign.

From the charge distribution, we can directly calculate the electric field and potential across the pn junction. We accomplish this by use of the Poisson equation

$$\nabla^2 V = -\nabla \mathcal{E} = -\frac{\rho}{\epsilon_m} \tag{2.41}$$

which is presently more useful for us in its one-dimensional form

$$\frac{\partial^2 V}{\partial x^2} = -\frac{\partial \mathcal{E}}{\partial x} = -\frac{\rho(x)}{\epsilon_m} \tag{2.42}$$

in which the potential V [V] is related to the field $\mathcal{E}$ [V/m] and the charge density ρ [C/m^3] plus the permitivity ϵ_m [F/m]. Using the charge distribution, $\rho(x)$ as given in Figure 2.16(a) and integrating once, we then obtain the electric field

$$\mathcal{E} = \frac{qN_D}{\epsilon_m}(x + W_n) \qquad \text{for } -W_n \leq x \leq 0 \tag{2.43}$$

and

$$\mathcal{E} = \frac{qN_A}{\epsilon_m}(W_p - x) \qquad \text{for } 0 \leq x \leq W_p \tag{2.44}$$

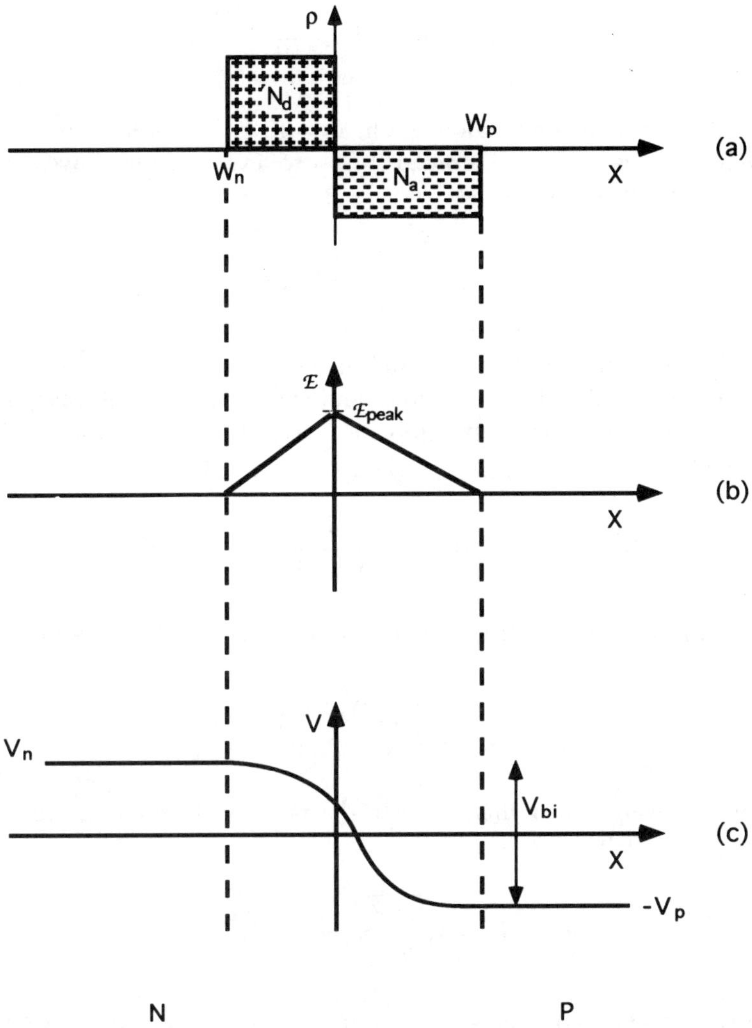

Figure 2.16 The (a) charge distribution, (b) electric field, and (c) potential across a pn junction in the depletion approximation. Both the pn junction and the depletion regions are assumed to be abrupt.

which is shown in Figure 2.16(b). Note that the magnitude of the peak field, at x = 0, is given by $\mathscr{E}_{\text{peak}} = qN_dW_n/\epsilon_m = qN_AW_p/\epsilon_m$ and that the electric field points to the right.

We can integrate the expression for electric field a second time to determine the potential as a function of position, namely

$$V(x) = \mathcal{E}_{peak}\left(x - \frac{x^2}{2W}\right) \tag{2.45}$$

where W is the total depletion region width, W = $W_n + W_p$ and the change in potential across the junction is seen in Figure 2.16(c). If we evaluate the above expression at $x = (-W_n, W_p)$, we then obtain

$$V_{total} = V_{bi} = \frac{1}{2}\mathcal{E}_{peak}W \tag{2.46}$$

where V_{bi} represents the *built-in* potential of the junction; this is the potential that exists due to the initial diffusion of holes and electrons that took place in order to flatten the Fermi level. We can now apply the charge neutrality condition of Equation 2.40 and write

$$W_n = W\frac{N_A}{N_A + N_D} \tag{2.47}$$

so that the expression for built-in voltage can be expanded into the form

$$V_{bi} = \frac{q}{2\epsilon_m}\left(\frac{N_A N_D}{N_A + N_D}\right)W^2 \tag{2.48}$$

We can alternatively invert this to write the total depletion layer thickness as a function of V_{bi}, namely

$$W = \sqrt{\frac{2\epsilon_m}{q}\left(\frac{1}{N_A} + \frac{1}{N_D}\right)V_{bi}} \tag{2.49}$$

The two previous equations still require that we know either W or V_{bi} in order to calculate the other. The junction potential can be found from the doping levels on either side by remembering, first, that energy and voltage are related by $V = E/q$ and, second, that the difference between the Fermi and intrinsic levels for a doped semiconductor is given by

$$V_F - V_i = \frac{k_B T}{q}\ln\left(\frac{N_D}{n_i}\right) \equiv V_n \text{ for n-type} \tag{2.50}$$

and

$$V_F - V_i = \frac{k_B T}{q}\ln\left(\frac{N_A}{n_i}\right) \equiv -V_p \text{ for p-type} \tag{2.51}$$

The built-in voltage is then, as is clear from Figure 2.16, $V_{bi} = V_n - V_p$ which yields

$$V_{bi} = \frac{k_B T}{q}\ln\left(\frac{N_A N_D}{n_i^2}\right) \tag{2.52}$$

If we then combine this last expression with Equation 2.49, we have a relationship between depletion region width and doping density, namely

$$W = \sqrt{\frac{2\epsilon_m k_B T}{q^2}\left(\frac{1}{N_A} + \frac{1}{N_D}\right)\ln\left(\frac{N_A N_D}{n_i^2}\right)} \tag{2.53}$$

With Equations 2.52 and 2.53, we have expressions for the built-in junction potential and the depletion region widths, which can be calculated knowing only the doping levels on both sides of the junction.

Most real pn junctions are more heavily doped on one side, so that either $N_A \gg N_D$ or vice versa. We see from Equation 2.49 that it is the more lightly doped side that determines W, meaning that the depletion region extends further into the side of the junction with lower doping. If we define N_S as the *smaller* of N_A or N_D we can simplify the expression for depletion region width as

$$W = \sqrt{\frac{2\epsilon_m}{qN_S}V_{bi}} \tag{2.54}$$

Such typical structures are known as one-sided junctions.

2.6.2 Debye Length

The picture of charge distribution in the depletion approximation we have just used, Figure 2.16(a), is indeed an approximation: the distribution does not start abruptly as we have drawn and assumed in our calculations. Due to majority carrier (electrons on the n side, holes on the p side) screening near the edges of the depletion region, the transition is characterized by the Debye length, which is given by [15]

$$L_D = \sqrt{\frac{\epsilon_m k_B T}{q^2 N}} \tag{2.55}$$

for doping level N. Thus, the depletion region starts and stops within a Debye length, but we see that the abrupt approximation is valid if $L_D \ll W$. The Debye length is a function of temperature, but the depletion approximation is usually a good one.

2.6.3 The PN Junction Under Bias

A pn junction sitting on the table without wires attached to it is not particularly exciting; however, when we apply a bias to it, it not only becomes interesting but shows the first signs of operating as a semiconductor device. An externally applied voltage is either added to or subtracted from the built in voltage of the junction, V_{bi}, depending on its sign. In forward bias, the positive voltage is applied to the p side of the junction, thereby decreasing the junction potential and reducing the potential barrier as is seen in Figure 2.17. The depletion region becomes narrower and it becomes easier for current to flow across the junction.

In reverse bias, the positive voltage is applied to the n side of the junction, causing an increase in the junction potential, barrier, and depletion region width, again as we can see from Figure 2.17. The effect of an applied external bias V_a on the potential and depletion regions can be seen directly from the relationships derived in the last section by replacing V_{bi} with $(V_{bi} - V_a)$ where V_a is positive for forward bias and negative for reverse bias.

Both forward and reverse bias are basically nonequilibrium situations, meaning that external inputs are supplying energy into the system. For that reason, the Fermi level is often ill-defined in the junction because the densities of holes and electrons are not necessarily at their equilibrium values. For that reason, we often define quasi-Fermi levels, one each for electrons and holes. In nonequilibrium, these quasi-Fermi levels then define the concentrations of electrons and holes separately because $pn = n_i^2$ no longer holds. In equilibrium, the two quasi-Fermi levels coalesce into the single Fermi level.

In the case of an applied forward bias, we see from Figure 2.17 that the potential barrier to carriers becomes smaller and, as a result, current flows through the junction. The relationship between applied bias and current is defined by the well-known Shockley equation, in reference to the person who first figured out how currents flow through a semiconductor junction:

$$J = J_0(e^{qV_a/k_BT} - 1) \tag{2.56}$$

The term J_0 [A/cm^2] is known as the reverse saturation current density, given by

$$J_0 = q\sqrt{\frac{D_p}{\tau_p}}\frac{n_i^2}{N_D} + q\sqrt{\frac{D_n}{\tau_n}}\frac{n_i^2}{N_A} \tag{2.57}$$

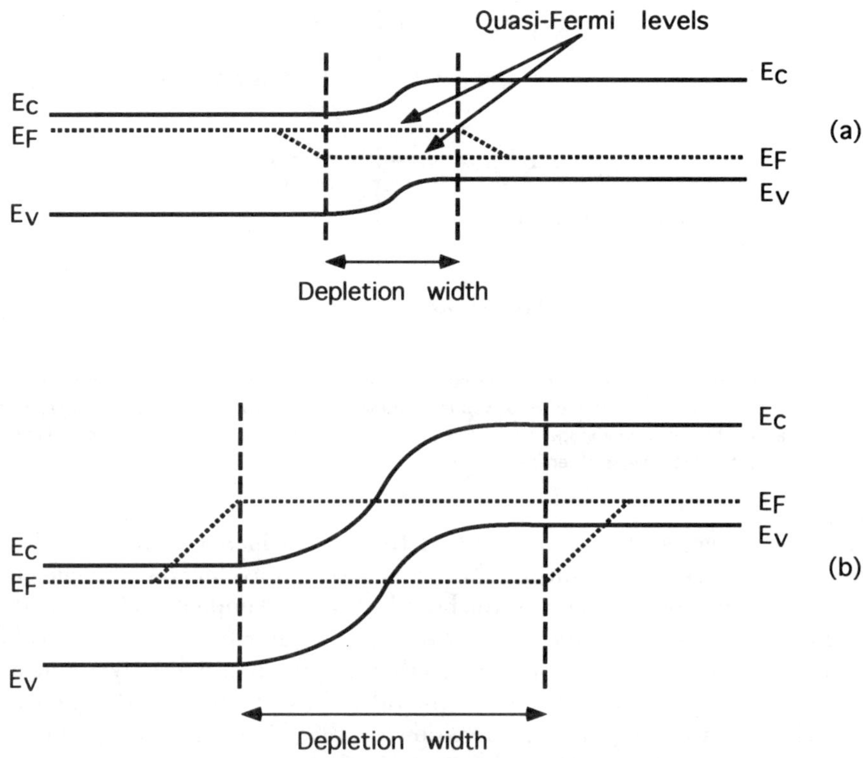

Figure 2.17 The pn junction under (a) forward, and (b) reverse bias. Forward bias results in a decrease of the junction barrier and a reduction in the depletion region width. In reverse bias, the junction barrier is increased and the depletion region widens. In the vicinity of the junction, nonequilibrium electron and hole concentrations exist, hence the quasi-Fermi levels for each carrier.

where the τ [s] terms are hole and electron lifetimes, meaning the average time that elapses before they recombine. J_0 is a strong function of temperature, mostly through variations in the energy gap and thus the intrinsic carrier density.

The current-voltage, or IV, characteristic of a pn junction is schematically sketched in Figure 2.18; current I [A] and current density J are simply related by the junction area, A [m^2], as $I = AJ$. It is clear that J_0 defines the magnitude of current flow in the reverse direction; because this is usually desired to be as small as possible, a long carrier lifetime is often required for good junction diodes.

2.6.4 Reverse Bias Breakdown

At large reverse biases, the electric fields in the pn junction become too large for the material to support and the junction breaks down; for an applied voltage above the

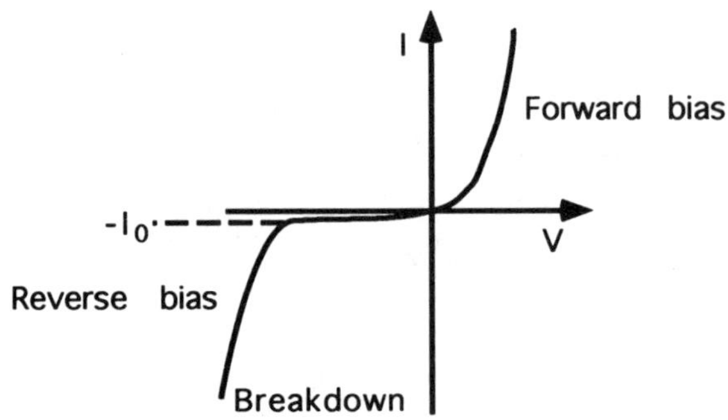

Figure 2.18 A schematic representation of the current-voltage characteristic of a pn diode. The reverse current is typically many of orders of magnitude smaller than usable forward currents until the reverse breakdown voltage is reached. Breakdown manifests itself as a very rapid increase in current in the reverse direction.

breakdown voltage, as shown in Figure 2.18, current increases very rapidly. Two mechanisms are usually responsible for breakdown: in avalanche, a single electron accelerated by the field generates further electrons by impact ionization, thereby causing a rapid increase in the number of high energy carriers. Zener breakdown results from tunneling of electrons through the energy gap and usually occurs at high doping densities. For either mechanism, provided the current levels that flow are not so high as to destroy the junction, breakdown is reversible and does not necessarily imply the demise of a device. This brief discussion belies the importance of breakdown in many device structures; we will see the phenomenon again when we discuss detectors in Chapter 10.

2.6.5 Junction Capacitance

The pn junction is responsible for a sizable contribution to the capacitance of a circuit, $C = \mathrm{d}Q/\mathrm{d}V$ [F], which can affect the high-speed switching performance of an electronic or optoelectronic device. Two components contribute: the depletion capacitance, dominant in reverse bias, is due to the parallel plate nature of the depleted pn junction and is given, for a one-sided junction, by

$$C_{\mathrm{depl}} = \frac{\epsilon_m}{W}A = \sqrt{\frac{qN_S\epsilon_m A^2}{2(V_{bi} - V_a)}} \tag{2.58}$$

where A [m^2] is the area of the junction. In forward bias, a diffusion component also comes into play, which results from charge storage in the junction as carriers traverse it. That component is again a function of carrier lifetime and junction bias

$$C_{\text{diff}} = \frac{qI_0}{k_B T}(\tau_p + \tau_n)e^{qV_a/k_BT} \tag{2.59}$$

We see that a long lifetime may indeed imply a lower reverse saturation current (Equation 2.57), but we pay for this with an augmented diffusion capacitance.

2.6.6 The P-I-N Diode

Placing an intrinsic (i), or very lightly doped (ν), region between the n and p sides of a junction can offer important advantages for certain device concepts. As we can see from Figure 2.19, when an undoped region is completely depleted of carriers, charge neutrality implies that the electric field is constant through it. Because the applied voltage derives from the integral of the electric field across the entire junction, $V = \int \mathcal{E} dx$, the increased width of the junction now implies that we can apply a larger voltage for a given maximum electric field. It is the peak electric field, not the voltage, that defines the onset of junction breakdown, so that a p-i-n or p-ν-n junction will support a higher breakdown voltage than a normal pn junction. Although we can idealize the center region as being truly undoped, and thus intrinsic, in practice this is rarely achieved, so we should strictly treat the intrinsic region as one that has very light, but nonzero, doping. In that case, the electric field does increase slightly across the center of the junction, as shown in Figure 2.19.

In addition to its high-voltage robustness, a p-i-n junction has a faster time response, due to lowered depletion capacitance (increased W) than a normal pn junction, a factor that becomes useful in the design of high-speed photodetectors. Finally, the i region supports a constant electric field: a slab of constant, and poten-

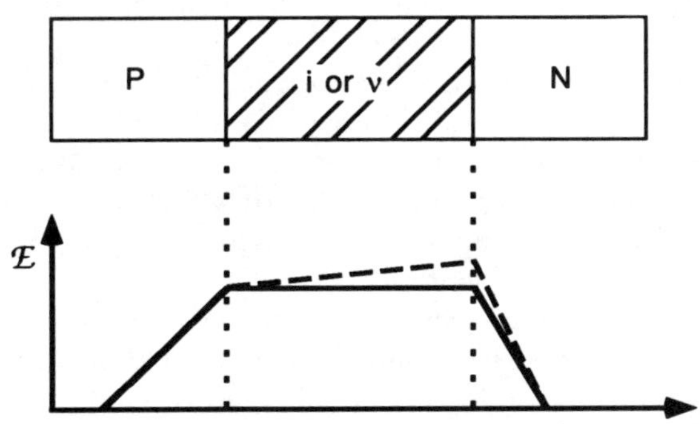

Figure 2.19 The p-i-n diode, structure, and electric field distribution. If the center region is lightly doped, as opposed to truly intrinsic, it is referred to as a p-ν-n diode and the field rises slightly as shown.

tially large, $\mathcal{E}$ is often useful for modulators and other devices that employ electro-optic effects.

2.7 HETEROJUNCTIONS AND QUANTUM WELLS

We have seen up to this point that doping density and bias variations across a junction give us a host of parameters to play with, even though we have considered only junctions made from a single material where the energy gap remained constant. Such junctions are generally referred to as homostructures to reflect that they are made only of Si or GaAs or any other single material. We now toss a further variable into the works by making use of the variations in energy gap that arise if we make sandwiches of compatible materials, which differ in their energy gap; such configurations are termed heterostructures. As we will see in Chapter 4, a large subset of the III-V materials have variable energy gap and optical properties, but identical or closely matched lattice constant, so that we can form single-crystal layered structures with an energy gap variation tuned to our needs. Such tuning is of central importance for the fabrication of many optoelectronic devices.

2.7.1 Single and Double Heterostructures

In Figure 2.20, the conduction and valence band variations of both a single and double heterostructure are shown. The former is a combination of two materials with differing energy gap, and the latter represents a layer structure in which a material with smaller energy gap is sandwiched between two regions of higher gap; hence two heterojunctions in that case. In these pictures, we have ignored any band bending that takes place as electrons diffuse into energetically more attractive regions, and we've drawn the heterojunction as an isotype, meaning the doping is the same on both sides of the junction.

An important parameter in a heterostructure is the energy gap difference between the two regions. The total difference is naturally equal to the sum of the valence and conduction band discontinuities, $\Delta E_g = \Delta E_c + \Delta E_v$, but ΔE_c and ΔE_v are not necessarily equal. For the important GaAs/AlGaAs system, which we will examine in detail in Chapter 4, the conduction band discontinuity represents about 60% of the gap difference; the valence band, 40%.

We can see from the figure why such heterostructures are useful. Electrons and holes generated anywhere near the junction will tend to "fall into" the low-gap region in the center. Thus, a wide bandgap material could very effectively be used to make an efficient electron emitter for, say, a bipolar transistor; the electrons would be efficiently injected into the low bandgap base, but they need considerable thermal energy to be injected in the other direction. For that reason, wide bandgap emitters have been used for bipolar transistor applications.

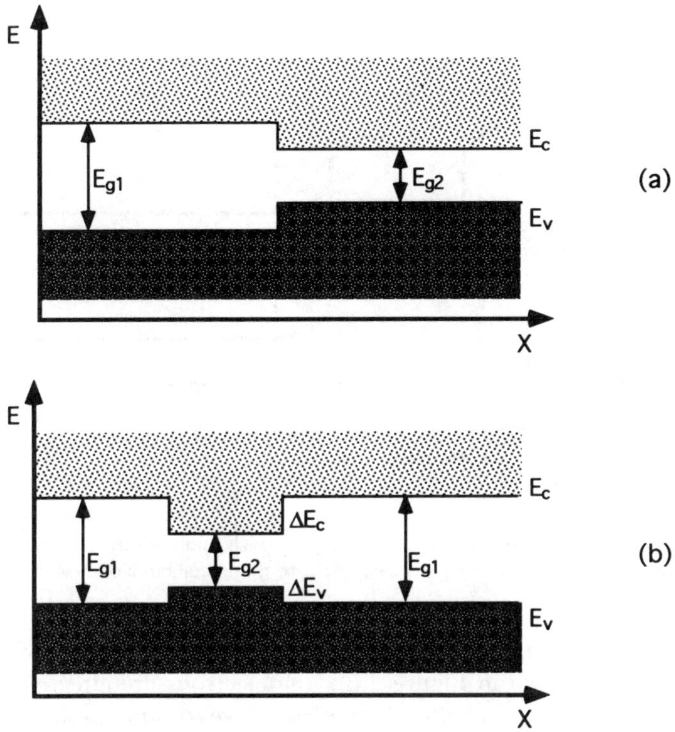

Figure 2.20 Energy band diagrams for (a) single, and (b) double heterostructures. Doping and carrier diffusion effects have been ignored in this schematic representation.

For optoelectronic devices, the *double* heterostructure has proven to be very useful; a semiconductor laser might make use of the doped double heterostructure shown in Figure 2.21. Injected electrons and holes again collect in the narrow bandgap center region and can only leap back out with difficulty. For that reason, electron hole recombination is much more likely in the narrow gap region enhancing the efficiency of the light emitter. An additional attraction of this configuration is that the low bandgap center region typically has a higher refractive index than the outer layers; thus, emitted light is guided in this layer as well. These considerations will become clearer in the discussions of Chapters 7 and 9; an excellent discussion of heterostructures of many types can be found in [16].

2.7.2 Quantum Wells

As the width of a low bandgap center region of a double heterostructure decreases to the order of the electron wavelength, quantum effects become apparent and eventually

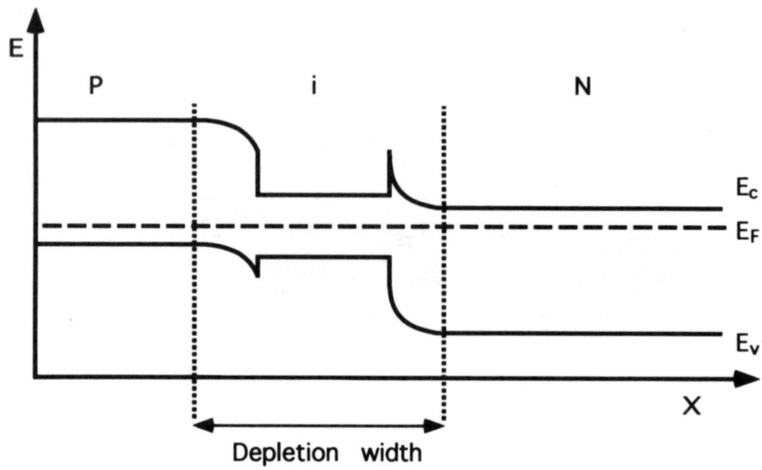

Figure 2.21 A doped double heterojunction, such as that which might be used for a laser. Carriers are attracted to the low bandgap center region, so that recombination is enhanced there.

dominate the electrical and optical behavior. For center layer thicknesses below t = 10 to 20 nm, as shown in Figure 2.22, the electrons are effectively confined in one dimension (that normal to the plane of the layers), and we have to determine their energies by solving the quantum mechanical problem of a particle in a potential well. This thin double heterostructure is called a quantum well and its characteristics require that we solve in two dimensions much of which we solved in three dimensions previously.

In Section 2.2.4, we had used the approximation that electron energy varies parabolically with k above the conduction band (Equation 2.5), so that in three dimensions the electron energy E_e is given by

$$E_e = E_c + \frac{\hbar^2 k^2}{2m^*} \tag{2.60}$$

A particle in a potential well has discrete energy levels, no longer a continuum of allowed energies,[7] so that electron and hole energies become, respectively

$$E_e = E_c + \frac{i^2}{2m_c^*}\left(\frac{\hbar\pi}{t}\right)^2 + \frac{\hbar^2 k^2}{2m_c^*} \tag{2.61}$$

[7]See virtually any quantum mechanics text; for example, Cohen-Tannoudji, C., B. Diu, and F. Laloë, *Quantum Mechanics*, Paris: Hermann, 1977, pp. 67 ff.

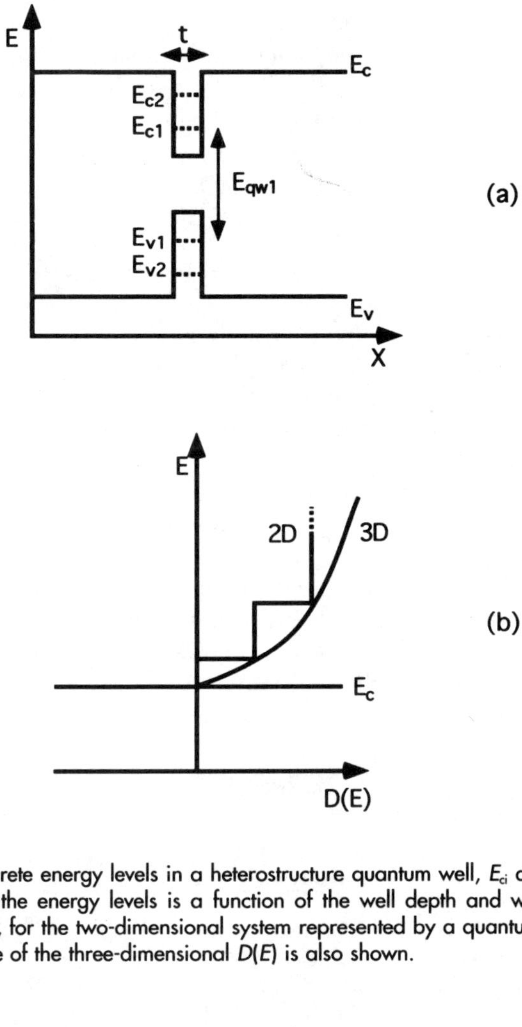

Figure 2.22 (a) The discrete energy levels in a heterostructure quantum well, E_{ci} and E_{vi}. The number and position of the energy levels is a function of the well depth and width; (b) The density of states, $D(E)$, for the two-dimensional system represented by a quantum well. The square-root dependence of the three-dimensional $D(E)$ is also shown.

$$E_h = E_v + \frac{i^2}{2m_v^*}\left(\frac{\hbar\pi}{t}\right)^2 + \frac{\hbar^2 k^2}{2m_v^*} \tag{2.62}$$

for well thickness t [m] and index $i = 1,2,3, \ldots$ Important to note is that for Equation 2.62, hole energy is measured *downward*. Thus, we see that the electron and hole energies are discrete, a function of well width and limited in number, the latter being a function of well depth. We have labeled these E_{c1}, E_{c2}, and E_{v1}, E_{v2}, in the figure. The total transition energy, E_{qwi}, is then equal to $E_{ci} + E_{vi} + E_g$ for index $i = 1,2,3, \ldots$

Finding the electron density of states is now also a two-dimensional problem. Rather than the smooth square-root energy dependence we found for the three-dimensional case (Equations 2.6 and 2.7), the two-dimensional density of states, D_{eqw} [$J^{-1}m^{-3}$] is a stair-shaped function [17]

$$D_{eqw}(E) = \frac{im^*}{\pi\hbar^2 t} \tag{2.63}$$

where $i = 1,2,3, \ldots$. The variation is shown in Figure 2.22(b), on which the three-dimensional density of states is also sketched. Thus, the density of available electron states in a quantum well is zero for energies $E < E_c + E_{c1}$, then becomes $(m^*/\pi\hbar^2 t)$ for $E_c + E_{c1} < E < E_c + E_{c2}$, then $(2m^*/\pi\hbar^2 t)$ for $E_c + E_{c2} < E < E_c + E_{c3}$ and so forth. Whereas at the bottom of the quantum well energy level, the density of states is larger than that for the three-dimensional density of states at the bottom of its energy band (where for E_c, the density of states is actually zero), the total density of states is smaller in two dimensions. It is often useful to place more than one quantum well in the active region of, say, a laser. A collection of quantum wells then represents what is called, a multiquantum well structure. One can also fabricate structures with many quantum wells, which are so close together that tunneling through the barriers becomes probable and the energy states broaden into bands. In this case, the wells cannot be analyzed separately. Such a configuration is termed a superlattice and has numerous applications for physical material studies, electro-optic devices, and improvements in material quality during semiconductor growth.

REFERENCES

[1] Hoddeson, L., E. Braun, J. Teichmann, and S. Weart, *Out of the Crystal Maze*, New York: Oxford University Press, 1992, p. 171.

[2] Schockley, W., "The Theory of P-N Junctions in Semiconductors and P-N Junction Transistors," *Bell System Technical Journal*, Vol. 28, 1949, pp. 435–489.

[3] Goldschmidt, V. M., *Transactions of the Faraday Society*, Vol. 25, 1929, pp. 253–283.

[4] Nathan, M. I., W. P. Dumke, G. Burns, F. H. Dills, and G. Lasher, "Stimulated Emission of Radiation from GaAs PN Junctions,"*Applied Physics Letters*, Vol. 1, 1962, pp. 62–64; and Hall, R. N., G. E. Fenner, J. D. Kingsley, T. J. Soltys, and R. O. Carlson, "Coherent Light Emission from GaAs Junctions," *Physical Review Letters*, Vol. 9, 1962, pp. 366–368.

[5] Sze, S. M., *Physics of Semiconductor Devices*, 2nd edition, New York: John Wiley & Sons, 1981, p. 848.

[6] Ziman, J. M., *Principles of the Theory of Solids*, 2nd edition, Cambridge, England: Cambridge University Press, 1972.

[7] Kittel, C., *Introduction to Solid State Physics*, 5th edition, New York: John Wiley & Sons, 1976.

[8] Sze, S. M. and J. C. Irwin, "Resistivity, Mobility and Impurity Levels in GaAs, Ge and Si at 300°K," *Solid State Electronics*, Vol. 11, 1968, pp. 599–602.

[9] Ghandi, S. K., *The Theory and Practice of Microelectronics*, New York: John Wiley & Sons, 1968, p. 212 ff.

[10] Smith, R. A., *Semiconductors*, 2nd edition, Cambridge, England: Cambridge University Press, 1978, p. 95.
[11] Schockley, W., *Electrons and Holes in Semiconductors*, Princeton, NJ: D. Van Nostrand, 1950, pp. 467–471.
[12] Jacobini, C., C. Canali, G. Ottaviani, and A. Alberigi Quaranta, "A Review of Some Charge Transport Properties of Silicon," *Solid State Electronics*, Vol. 20, 1977, pp. 77–89.
[13] Sze, S. M., *Physics of Semiconductor Devices*, 2nd edition, New York: John Wiley & Sons, 1981, p. 29.
[14] Shur, M., *GaAs Devices and Circuits*, Chapter 4, New York: Plenum Press, 1987.
[15] Muller, R. S. and T. I. Kamins, *Device Electronics for Integrated Circuits*, New York: John Wiley & Sons, 1977, p. 118.
[16] Casey, H. C., and M. B. Panish, *Heterostructure Lasers A*, Chapter 4, New York: Academic Press, 1978.
[17] G. Bastard, *Wave Mechanics Applied to Semiconductor Heterostructures*, Les Ulis, Les Editions de Physique, 1988.

Chapter 3

Optical Properties of Semiconductors

La luce è sempre uguale ad altra luce.[1]

Pier Paolo Pasolini, 1961 [1]

We saw during our scratch onto the surface of semiconductors in the last chapter that they have interesting electrical properties. Since our intent is to learn about the behavior of optical devices, we now need to turn to the optical characteristics of semiconductors and see what happens when light propagates through a crystal. In a material, the dielectric constant may no longer be unity and conductivity may not be zero; these two factors affect the propagation of an optical field. Thus, when we go from vacuum to a material, light is not equal to light.

The effects of a medium on light are usually lumped into the refractive index, n, which is in general complex and simply related to the propagation constant, k. The propagation constant in a material is related to the propagation constant in a vacuum, the k_0 we saw in the first chapter, by the dielectric constant, which is also in general complex. Crudely stated, the dielectric constant defines how much the light is attenuated and how fast it goes in a medium; in this chapter, we will look at the origin of these two effects on an optical field.

We will defer the consideration of specific material parameters for particular species of semiconductors to Chapter 4 and consider now the general behavior of electromagnetic waves in a semiconducting medium.

3.1 REFRACTIVE INDEX

As a beam of light propagates through a semiconducting material, it can be subject to resonant and nonresonant scattering; the former is absorption due to the complete transfer of energy from a photon to an electron, and the latter results from an induced polarization of the electron clouds and the polar molecules. The extent of wave/

[1] "All light is equal to all other light."

material interaction is summarized by the dielectric constant or, directly related to it, the index of refraction; these parameters define the contributions of both absorption and polarization of a material.

When a wave no longer travels in free space, the permitivity ϵ_0 [F/m] and permeability μ_0 [H/m] become

$$\epsilon_0 \rightarrow \epsilon_m = \epsilon \cdot \epsilon_0 \tag{3.1}$$

$$\mu_0 \rightarrow \mu_m = \mu \cdot \mu_0 \tag{3.2}$$

where the subscript m denotes the value of permitivity or permeability in a material as opposed to a vacuum. The constants ϵ and μ are dimensionless factors that describe the strength of material interaction with a wave. The former, ϵ, is the dielectric constant; thus a material with $\epsilon = 2$ is one in which the dielectric constant is twice that in a vacuum. We will assume from here that $\mu = 1$, which is usually the case in semiconductors.

We recall from Chapter 1 that the wave velocity is given by

$$v = \frac{1}{\sqrt{\epsilon_m \mu_m}} \tag{3.3}$$

which in a vacuum implies

$$c = \frac{1}{\sqrt{\epsilon_0 \mu_0}} \tag{3.4}$$

We can now define the index of refraction, n, as

$$\mathrm{n} = \frac{c}{v} = \sqrt{\frac{\epsilon\epsilon_0\mu\mu_0}{\epsilon_0\mu_0}} = \sqrt{\epsilon\mu} \tag{3.5}$$

Since $\mu \approx 1$ as we restricted above, $\mu_m = \mu_0$. This implies that we can write

$$\mathrm{n} = \sqrt{\epsilon} \tag{3.6}$$

Since $\mathrm{n} > 1$, this of course implies that $v < c$ as we would expect. For our material systems of interest, n typically varies from 1 for a vacuum (and roughly clean air) to about 4 for Ge.

It will become clear presently that the index of refraction is one of the most important parameters involved in the discussion of wave propagation through materials. Since we have hinted that the dielectric constant is typically complex, n is also usually complex and we will see what that implies for the relations above in a moment.

3.2 ABSORPTION

We found in Chapter 2 that the band structure of semiconductor materials leads to their varied and interesting electrical properties. For optical applications, the interactions of semiconductors with light become the relevant issues and again the band structure plays a dominant role when we consider the phenomenon of absorption. In this and the next section, we will look at mechanisms of photon absorption and generation to which we had already alluded in Section 2.2; in Section 3.4, we will see how the former is treated mathematically in the dielectric constant.

In semiconductors, absorption of photons generates an electron/hole pair, and recombination of electrons and holes leads to the creation of photons. Both of these processes result from transitions of electrons from the valence to conduction bands, or vice versa, and are known as band-to-band transitions.

3.2.1 Direct Bandgap

The band diagram of Figure 3.1, a variation of Figure 2.7, represents a direct bandgap semiconductor such as GaAs. Absorption occurs when an incident photon of energy hν [eV] greater than the bandgap at a given k, $E_c - E_v$, promotes an electron from the valence to the conduction band. Both energy, E, and momentum, k, are conserved in this process: $\Delta E = E_c - E_v = h\nu$ and $\Delta k = 0$. These transitions are likely since photons can have considerable energy (several eV) but little momentum.

Absorption can occur at any energy difference, as shown by three possible transitions in the figure, as long as the photon has sufficient energy to overcome the band-to-band energy difference and there is an available state for an electron to enter. The lowest possible photon energy that still leads to absorption, called the absorption edge, is at the bandgap minimum, which, for the semiconductor shown in Figure 3.1, is at the center of the Brillouin zone where $k = 0$. This is an example of a direct bandgap semiconductor, meaning that the conduction band minimum is at the same k value as the valence band maximum; this has important consequences, as we shall see.

Using the parabolic approximation of Section 2.2.4 for the band shape near the bandgap minimum, we can write allowed electron energies from the conduction band edge, E_{c0}, as

$$E_c = E_{c0} + \frac{\hbar^2 k^2}{2m_c^{*2}} \tag{3.7}$$

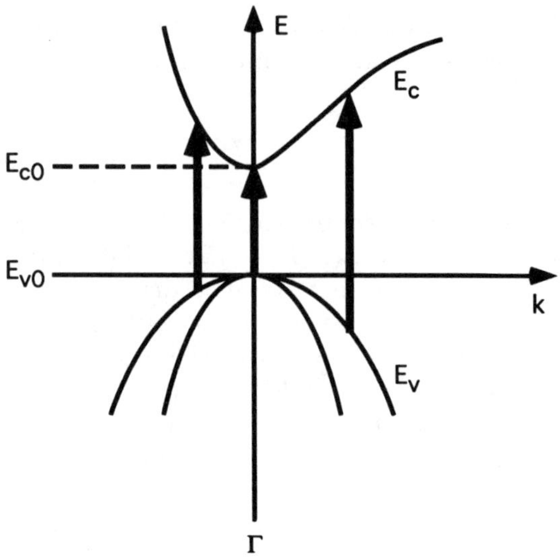

Figure 3.1 The energy bands for GaAs showing three possible transitions for band-to-band absorption with $\Delta k = 0$. The lowest absorption energy, the absorption edge, is at the Brillouin zone center, the Γ point where $k = 0$.

and the hole energies in the valence band, measured downward from the band edge, E_{v0}, as

$$E_v = E_{v0} - \frac{\hbar^2 k^2}{2m_v^{*2}} \tag{3.8}$$

In these last two expressions, the first term represents the potential energy, and the second, the kinetic energy of an electron. Thus the allowed photon energies for absorption near the Brillouin zone center are given by

$$\mathrm{h}\nu = (E_{c0} - E_{v0}) + \frac{\hbar^2 k^2}{2}\left(\frac{1}{m_c^{*2}} + \frac{1}{m_v^{*2}}\right) \tag{3.9}$$

For these direct transitions, we can then define an absorption coefficient, α [m^{-1}] which is a function of the photon energy and is also proportional to the density of electron states in the conduction band. Recalling that the density of states increases as the square root of energy above E_{c0} (Equation 2.6), we can write

$$\alpha(\mathrm{h}\nu) \propto \sqrt{(\mathrm{h}\nu - E_g)} \tag{3.10}$$

The absorption coefficient α defines how fast the optical field, with photon energy above the bandgap energy, decays with distance. For distance x [m] into a semiconductor, the power or intensity, S [W/m^2] falls off as

$$S = S_0 \exp(-\alpha x) \tag{3.11}$$

where S_0 is the power density at the surface. We will derive this last expression below.

3.2.2 Indirect Bandgap

If we look at the band diagram of Figure 3.2, which is similar to that of Si, we see that for this case the valence band maximum and the conduction band minimum do not coincide at the same k value; this is an indirect bandgap semiconductor. Because photons have negligible momentum, or k, a transition from the valence band maximum to the conduction band minimum can only be accomplished through the mediation of phonons. Phonons typically have k values on the order of the Brillouin zone width, so that an interaction between an electron, a photon, and one or more phonons is required for the transition shown.

The absorption coefficient in an indirect bandgap semiconductor is a function of both phonon density and electron density of states, as all three suspects (the

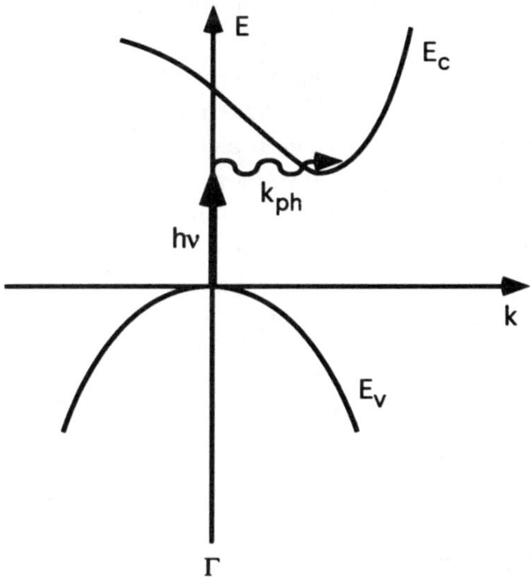

Figure 3.2 Absorption of a photon with energy $h\nu$ in an indirect bandgap semiconductor, such as Si. A phonon with momentum k_{ph} is required for momentum conservation; the energy contribution of the phonon is ignored in this figure.

electron, the photon, and the phonon) must be present for the transition to occur. For phonon energy E_{ph} [eV], the phonon density N_{ph} [cm^{-3}] follows a Bose-Einstein distribution,

$$N_{ph} \propto \frac{1}{\exp\left(\frac{E_{ph}}{k_b T}\right) - 1} \quad (3.12)$$

such that α is a strong function of temperature [2].

Of course, the direct transition from the valence band maximum into the conduction band is also still possible in an indirect material (as are all the other direct transitions), but such a transition requires a photon of considerably higher energy. In Si, for example, the direct transition at the Brillouin zone center requires an energy of about 3.5 eV, implying a photon wavelength of roughly 350 nm, in the ultraviolet; the indirect, phonon assisted transition occurs at a photon wavelength of 1.1 μm, in the near infrared.

3.2.3 Band Tails

Photon absorption in semiconductors as we have discussed it would appear to begin abruptly at the energy of the lowest bandgap and increase with density of available electron states for higher photon energies. In reality, the energy bands are not abrupt but extend into the energy gap. These extensions are known as band tails and result from crystal imperfections and impurities, which lead to local variations in the energy bands. As a result, the absorption coefficient as a function of photon energy does not have the discontinuous increase at the bandgap, as shown by the solid line in Figure 3.3, but rather extends exponentially into the bandgap as shown by the dotted line. This exponential behavior, known as Urbach's Rule and given mathematically as

$$\alpha \propto \exp\left(\frac{h\nu}{k_B T}\right) \quad (3.13)$$

for photon energy $h\nu$ below the bandgap energy, explains the soft absorption edge which is seen experimentally [3].

3.2.4 Free Carrier Absorption

In addition to the interband transitions we have just considered, intraband transitions involving the transfer of an electron to a higher energy within the same band can also play an important role in optical behavior. In Figure 3.4 we see that an electron

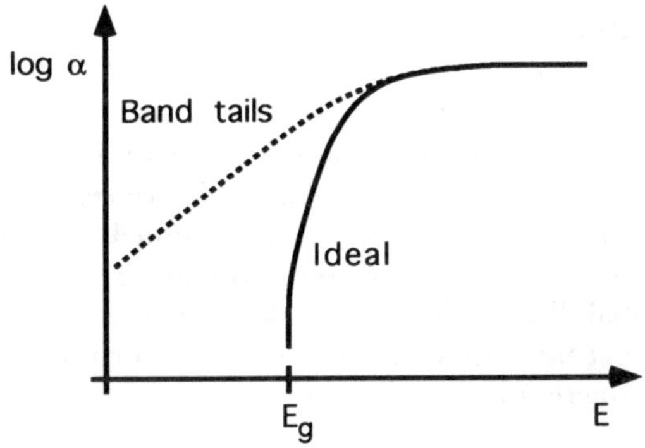

Figure 3.3 Photon absorption as a function of energy for the ideal case of a sharp band edge (solid line) and the more realistic situation of band tails (Urbach tails) into the bandgap (dotted line).

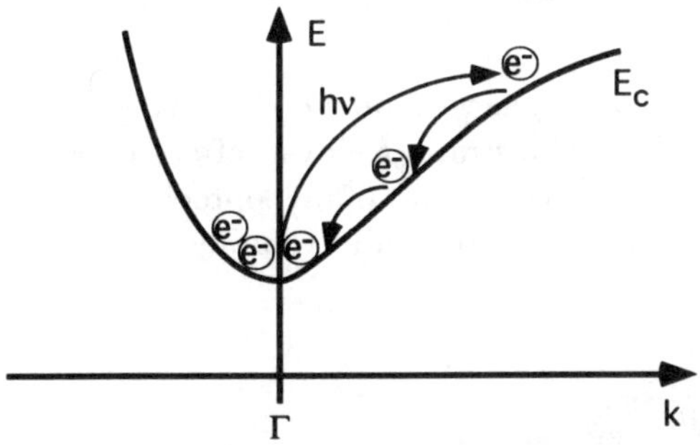

Figure 3.4 Free carrier absorption in the conduction band of a semiconductor. Low photon energies result in the transfer of electrons to higher energies in the conduction band; they return to the band minimum by thermalization, giving off their energy to phonons.

can absorb the energy from a low-energy, long-wavelength photon, in order to move higher into the conduction band. Since even low energies can be absorbed, a broad absorption spectrum is found provided the conduction band is sufficiently populated with free carriers; this effect is known as free carrier absorption.

The absorption coefficient α, due to free carrier absorption, generally increases with wavelength λ and carrier density n. Depending on which mechanisms mediate the intraband transitions, the variation with wavelength is given by λ^2 for the Drude model for electrons, $\lambda^{1.5}$ for acoustic phonons, $\lambda^{2.5}$ for optical phonons, and λ^3 for ionized impurities [4]. Once excited to higher parts of the band, the electrons tend to quickly drop back to the conduction band minimum, giving off their energy to phonons, a process known as thermalization.

3.2.5 Franz-Keldysh Effect

An externally applied electric field gives rise to a shift of the absorption spectrum with energy and results in the absorption of photons with lower energies than would be the case in the absence of a field. In Figure 3.5, we see that this phenomenon, known as the Franz-Keldysh effect [5,6], is due to tunneling of the electron into the conduction band. The electric field leads to a tilting of the energy bands (see Figure 2.14) so that the probability of finding an electron, with energy E_g', as shown in Figure 3.5, in the gap becomes appreciable. Due to a finite probability of the electron tunneling into the conduction band, absorption occurs for excitation energies below the bandgap.

The change in absorption edge is a function of the band shape in *k*-space, and can be written as [7]

$$\Delta E = -\frac{3}{2}(m^*)^{-1/3}(q\hbar\mathcal{E})^{2/3} \tag{3.14}$$

which shows the shift to lower energies of the effective energy gap as a function of the applied electric field $\mathcal{E}$. The Franz-Keldysh effect is one of numerous electro-optical effects that are very useful for many optoelectronic devices, particularly modulators, a subject to which we return in Chapter 11.

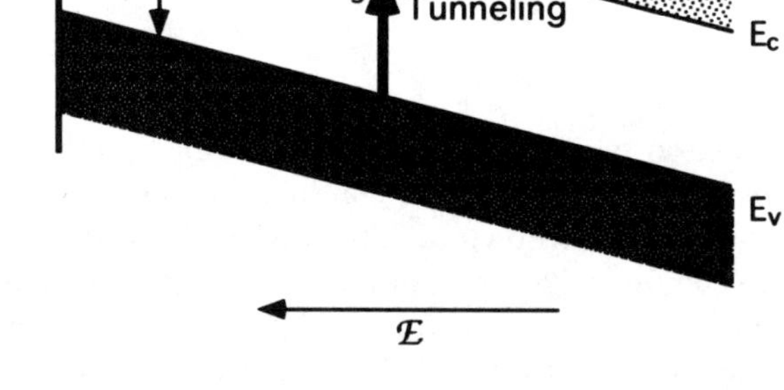

Figure 3.5 The Franz-Keldysh effect demonstrating the absorption of photons with energy below the bandgap energy due to an applied electric field. The electron tunnels into the conduction band, leading to absorption although it has only absorbed a photon with energy $E_g' < E_g$. Note that the abscissa is real space, not *k* space.

3.3 RECOMBINATION

Just as electron-hole pairs are generated by the absorption of a photon, electrons and holes can recombine and generate new photons. Recombination and photon emission are important issues for lasers; we will treat these in detail in Chapter 9 and only take a brief glance at them here.

Recombination transitions are just the inverse of absorption, as shown in Figure 3.6. A salient difference is that due to thermalization, which implies that most electrons quickly pile up at the bottom of the conduction band, recombination predominantly takes place from the conduction band minimum to the valence band maximum. Recombination transitions such as the one shown at the right of the bandgap minimum are allowed, just less likely.

3.3.1 Radiative Transitions

Recombination that gives rise to the emission of photons results from radiative transitions, as opposed to nonradiative transitions in which the electron gives its energy to a succession of phonons, and no light is emitted [8]. In the absence of an electromagnetic field that can induce a transition, the emission is spontaneous, meaning the electron decides at some time to leap from the conduction band and emit a photon,

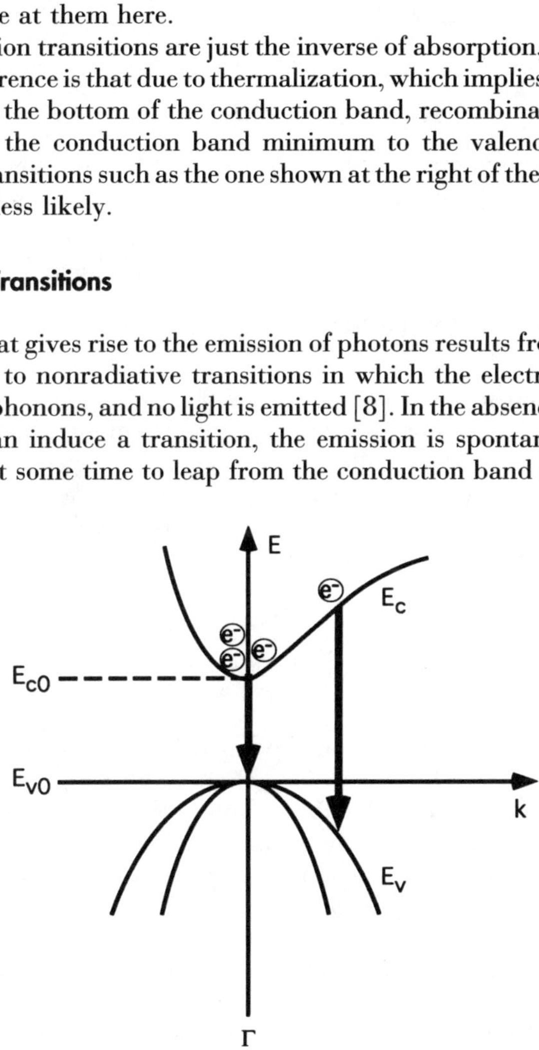

Figure 3.6 Recombination of electrons and holes leads to the generation of photons. Since electrons thermalize to the conduction band minimum, most transitions take place from there. The transition shown at the right is unlikely.

without any external stimulus. The average time between the generation of an electron and its spontaneous radiative recombination is known as the spontaneous lifetime, τ_{sp} [s], where values of τ_{sp} can range from ns to seconds, depending strongly on the material and the presence of impurities and crystalline imperfections.

From the lifetime, we can define a radiative transition rate, W_{sp} [s^{-1}] simply as

$$W_{sp} = \frac{1}{\tau_{sp}} \tag{3.15}$$

giving the average number of photons emitted per atom and per unit time due to radiative recombination. Transition rates will again become relevant when we discuss lasers.

3.3.2 Direct and Indirect Transitions

The wavelength spectrum of the photons emitted from a piece of semiconductor directly reflects the distribution of electrons that recombine radiatively. Thus, the spectrum of a direct bandgap semiconductor has a large peak near the bandgap energy, with additional wavelength components at higher energies due to transitions from higher up in the conduction band. The higher energy portions of the spectrum often decay exponentially with energy, namely

$$S = S_0 \exp\left(-\frac{E}{k_B T_e}\right) \tag{3.16}$$

for emitted power S, peak power S_0, and energy E above the conduction band minimum. T_e [K] is in this case the effective temperature of the excited electrons, which can be considerably higher than the temperature of the semiconductor crystal. Because the Fermi-Dirac distribution for the electrons excited high into the conduction band can be approximated by the exponential Boltzmann distribution of Equation 3.12 [9], this decay is often called the Boltzmann tail and can be used to characterize the electron temperature, or energy, in the conduction band.

Indirect semiconductors suffer from the same problem in recombination as they face in absorption: one or more phonon interactions are required in order to permit the transition of an electron from the conduction band minimum to the valence band maximum. As a result, radiative recombination in indirect bandgap semiconductors is very unlikely, and electrons and holes will recombine predominantly through nonradiative channels. This is the reason that an efficient light emitter based on, for example, bulk Si, has not been demonstrated.

3.3.3 Nonradiative Recombination

There are numerous pathways for an electron and hole to recombine nonradiatively, that is without the generation of a photon. Electron energy is given up to phonons in most of these cases. The mechanisms involved include transitions via defect levels in the bandgap, resulting in lower energy photon emission or phonon generation, or surface recombination resulting from the numerous states in the bandgap that appear at the physical boundaries of the semiconductor due to the local destruction of crystal symmetry there. Finally, Auger recombination, in which a recombining electron gives its energy to a second electron, which then loses that energy to phonons, is an important nonradiative recombination mechanism, particularly in longer wavelength semiconductors such as InGaAsP.

3.4 POLARIZATION

A photon happily zipping through a semiconductor crystal may be absorbed and thus disappear, as we saw in the last sections, but it may also interact with the electrons and atoms of the material, thereby changing its velocity without being absorbed. The photon/material interaction is characterized by the susceptibility, which essentially defines how the material can be polarized by an incident photon and leads to the definition of the dielectric constant.

3.4.1 Susceptibility

In the derivations of Chapter 1, we ignored the effect of polarization P in the wave equation. We will now remedy this omission in order to define the susceptibility [10].

We recall the electric flux density D was defined with the aid of the material permitivity, ϵ_m, as

$$D = \epsilon_m \mathcal{E} = \epsilon_0 \mathcal{E} + P \tag{3.17}$$

such that we can include any departure of the permitivity from its vacuum value ϵ_0 in the polarization term P. Writing this explicitly,

$$P = (\epsilon_m - \epsilon_0)\mathcal{E} = (\epsilon\epsilon_0 - \epsilon_0)\mathcal{E} = \epsilon_0(\epsilon - 1)\mathcal{E} = \epsilon_0 \chi \mathcal{E} \tag{3.18}$$

where χ, the susceptibility, is defined as

$$\chi = \epsilon - 1 \tag{3.19}$$

where we remember that ϵ is the dielectric constant, a dimensionless term. The susceptibility is in general both complex and frequency dependent, such that we can write it as the sum of its real and imaginary parts, or

$$\chi = \mathrm{Re}(\chi) + \mathrm{Im}(\chi) = \chi_R + j\chi_I. \tag{3.20}$$

In a vacuum, $\chi = 0$, so it seems clear that χ represents the deviation of the dielectric constant from 1, and thus the deviation from the vacuum value. In anticipation of what we shall find below, the real part of the susceptibility defines the extent of polarization induced by an optical field and the imaginary part, its absorption.

3.4.2 Polarization Model

Microscopically, the real part of the susceptibility χ is derived from the dipole response of atoms and electrons in the material to an electromagnetic wave. Since this is the basis of polarization, it is fun to examine the mechanism for polarization in more detail by means of a simple model. At low electromagnetic frequencies, the atoms and electrons can follow the stimulation of an oscillating externally applied field, at certain frequencies we will find resonances, and at sufficiently high frequencies, we reach a point where the elements of the medium (electrons and polar molecules) can no longer respond and the strength of the polarization interaction decreases. For electrons, this is on the order of 10^{14} Hz. Thus, polarization, susceptibility, dielectric constant, and index of refraction are all frequency-dependent.

We can now derive a crude but useful model for polarization by representing the material system as a set of masses connected by springs [11], such that the exciting wave causes the whole mass to oscillate like a blob of jello. We can consider atomic (lattice) vibrations and the oscillations of electron distributions separately. The force-balance equation representing this system is then in general [12]

$$\frac{\mathrm{d}^2 r}{\mathrm{d}t^2} + \gamma\frac{\mathrm{d}r}{\mathrm{d}t} + \frac{K}{m}r = \frac{q}{m}\mathcal{E} \tag{3.21}$$

for displacement r [m]. In Equation 3.21, γ [s^{-1}] is a damping term, K [N/m] a spring constant, q [$1.6 \cdot 10^{-19}$ C] is the electronic charge, and $\mathcal{E}$ [V/m] is the applied electric field. Using our model, the atom or electron of mass m [kg] is attached to a spring of spring constant K such that the restoring force of the spring, F [N], is given by

$$F = -Kr \tag{3.22}$$

for displacement r; the system thus has a resonance at frequency ω_0 [rad/s] given by

$$\omega_0 = \sqrt{\frac{K}{m}} \tag{3.23}$$

If the incident wave has the usual form $\mathcal{E} = \mathcal{E}_0 \cdot \exp(j\omega t)$, and assuming the same time dependence for $r = r_0 \exp(j\omega t)$, Equation 3.21 becomes

$$-\omega^2 r + j\omega\gamma r + \frac{Kr}{m} = \frac{q}{m}\mathcal{E} \tag{3.24}$$

which can easily be shown to have a solution

$$r = \frac{q\mathcal{E}}{m\left(\frac{K}{m} - \omega^2 + j\omega\gamma\right)} \tag{3.25}$$

Equation 3.25 can be written as the sum of real and imaginary parts, which, when we include the definition of ω_0 from above, becomes

$$r = \frac{(\omega_0^2 - \omega^2)q\mathcal{E}}{m[(\omega_0^2 - \omega^2)^2 + \omega^2\gamma^2]} - j\frac{q\mathcal{E}\omega\gamma}{m[(\omega_0^2 - \omega^2)^2 + \omega^2\gamma^2]} \tag{3.26}$$

We can now use this general solution for displacement to derive polarizability due to atoms and electrons. Let us first examine the atoms of the crystal lattice, which begins to vibrate when an optical field is incident on it; we take mass $m \rightarrow M$ and ignore any damping; i.e., $\gamma = 0$. We then have a solution

$$r_{\text{atom}} = \frac{q\mathcal{E}}{M(\omega_0^2 - \omega^2)} \tag{3.27}$$

Because the polarization is given by the dipole moment times the charge, $P = qn_{\text{atom}}r$, for an atomic density of n_{atom} [cm^{-3}], and recalling that the susceptibility is defined as (Equation 3.18),

$$\chi = \frac{P}{\epsilon_0\mathcal{E}} \tag{3.28}$$

we may write the atomic contribution to susceptibility as

$$\chi_{\text{atom}} = \frac{q^2 n_{\text{atom}}}{\epsilon_0 M(\omega_0^2 - \omega^2)} \tag{3.29}$$

Let's now repeat this exercise and analyze the contributions of electrons to the susceptibility. In this case, the deformation of the charge distribution is absorbed in the damping term, γ, and the spring constant is ignored, so $K = 0$; m is now electron mass. The solution for r is in this case complex and of the form

$$r_e = \frac{q\mathcal{E}}{m(\omega^2 + \gamma^2)} - j\frac{q\mathcal{E}\gamma}{m\omega(\omega^2 + \gamma^2)} \tag{3.30}$$

A value for the damping constant, γ, can be determined by solving the steady-state version of Equation 3.21, namely with $\mathrm{d}^2r/\mathrm{d}t^2 = 0$. In that case

$$\gamma\frac{\mathrm{d}r}{\mathrm{d}t} = \gamma\mu\mathcal{E} = \frac{q}{m}\mathcal{E} \tag{3.31}$$

where we have also used the electron velocity, given by $\mathrm{d}r/\mathrm{d}t = \mu\mathcal{E}$ for mobility μ [cm^2/Vs]. Thus, $\gamma = q/\mu m$ and we can write the electronic contribution to susceptibility as the sum of its real and imaginary parts,

$$\chi_{eR} + j\chi_{eI} = -\frac{q^2 n_e}{\epsilon_0 m(\omega^2 + \gamma^2)} - j\frac{q^3 n_e}{\mu\epsilon_0 m^2\omega(\omega^2 + \gamma^2)} \tag{3.32}$$

for electron density n_e [cm^{-3}].

From our simple jello model, we see the contributions of heavy, relatively fixed atomic cores and the more mobile electrons to the susceptibility and therefore polarizability of the medium. The atoms give rise to a real contribution, whereas the electrons add both real and imaginary components. Heuristically, the effect of $\mathrm{n} \neq 1$ is that the velocity of the wave is less than c; we can imagine, then, that although the photons are not absorbed, these polarization effects do slow them down. In the next section, we will see directly how these define the dielectric constant of the material.

Equation 3.29 shows that the susceptibility is constant when it is far from the resonance frequency, ω_0, but it varies rapidly when it is near and has a singularity at ω_0. For polar ionic materials, the large value of M yields resonances in the visible and infrared optical ranges such that the susceptibility (and thus ϵ and n, as we will see below) can be strongly frequency-dependent in the wavelength range of interest. It is important to note that when we specify "ϵ" or "n" in future chapters, we mean a static value valid for the stimulation wavelength being used.

3.5 COMPLEX MATERIAL CONSTANTS

Having had a taste of the physical origins of polarization through our simple model, we can now use the susceptibility as a tool to derive the dielectric and propagation constants and the index of refraction in a medium.

3.5.1 Complex Dielectric Constant

We remember that for a sinusoidal electromagnetic field, we have as solutions to the wave equation, for electric field and polarization,

$$\mathcal{E} = \mathcal{E}_0 e^{j(kz-\omega t)}, \; P = P_0 e^{j(kz-\omega t)} \tag{3.33}$$

assuming propagation in the **z** direction; k [m^{-1}] is the propagation constant in the material and ω [rad/s] the angular frequency. By plugging these solutions back into the wave equation, Equation 1.15, we derive a characteristic equation

$$-k^2\mu_0\mathcal{E} + j\omega\sigma\mu_0\mathcal{E} + \omega^2\epsilon_0\mu_0\mathcal{E} = -\omega^2\mu_0 P \tag{3.34}$$

Note that σ has been included in order to see the effect of nonzero conductivity and also that $\mu_m = \mu_0$. Recall that the free-space propagation constant is defined as $k_0 = \omega_0/c$. The polarization and electric field are related by Equation 3.18 as

$$P = \epsilon_0\chi\mathcal{E} = \epsilon_0(\chi_{\mathrm{atom}} + \chi_{eR} + j\chi_{eI})\mathcal{E} \tag{3.35}$$

using the susceptibilities determined in the previous section. Equation 3.34 can thus be rewritten as

$$k_0^2(1 + \chi_{\mathrm{atom}} + \chi_{eR}) + jk_0^2\left(\frac{\sigma}{\omega\epsilon_0} + \chi_{eI}\right) = k^2 \tag{3.36}$$

which relates k and k_0 to real and imaginary χ, ω and σ. In this last equation, it is important to note again the distinction between the two propagation constants k_0 and k. k_0 is the propagation constant of the wave *in free space*, and k is the *effective* propagation constant *in the material*.

We now define the complex dielectric constant, ϵ', as

$$\epsilon' = \epsilon_R + j\epsilon_I \tag{3.37}$$

where ϵ_R and ϵ_I are the real and imaginary parts, respectively. Equation 3.36 then takes on a much simpler form, namely

$$k_0^2\epsilon' = k^2 \tag{3.38}$$

This last relationship clearly shows that ϵ' relates the propagation constant in the material, k, to the free-space value, k_0. Comparing with Equation 3.36 then reveals that

$$\epsilon' = (1 + \chi_{\text{atom}} + \chi_{eR}) + j\left(\frac{\sigma}{\omega\epsilon_0} + \chi_{eI}\right) \tag{3.39}$$

implying that

$$\epsilon_R = 1 + \chi_{\text{atom}} + \chi_{eR} \tag{3.40}$$

and

$$\epsilon_I = \frac{\sigma}{\omega\epsilon_0} + \chi_{eI} \tag{3.41}$$

Thus we may easily relate the complex susceptibility to the complex dielectric constant. As we see from Equation 3.38, a complex dielectric constant implies a complex propagation constant; we will see the implications of this in the next section.

3.5.2 Complex Refractive Index

The refractive index will often prove to be a useful parameter in the discussion of material effects on transmitted waves, so we can also derive a complex n′ from ϵ'. Complex n′ is given by

$$n' = n_R + jn_I \tag{3.42}$$

which relates to the propagation constants as

$$k = k_0\sqrt{\epsilon'} = k_0 n' \tag{3.43}$$

n_R is the real component of the index of refraction; often, when we say "refractive index," we mean just this real term. n_I corresponds to the imaginary component, which defines absorption.

The real and imaginary parts of Equation 3.37 and the square of Equation 3.42 can be equated to yield

$$\epsilon_R = n_R^2 - n_I^2 \tag{3.44}$$

and

$$\epsilon_I = 2n_R n_I \tag{3.45}$$

Conversely, the components of refractive index are written as a function of the dielectric constant as [13]

$$n_R^2 = \frac{1}{2}\epsilon_R\left[\sqrt{1 + \left(\frac{\epsilon_I}{\epsilon_R}\right)^2} + 1\right] \tag{3.46}$$

and

$$n_I^2 = \frac{1}{2}\epsilon_R\left[\sqrt{1 + \left(\frac{\epsilon_I}{\epsilon_R}\right)^2} - 1\right] \tag{3.47}$$

Thus, in summary, the refractive index is in general complex, with a real part based on the susceptibility derived from the atomic lattice and free electrons and an imaginary component based only on the behavior of the electrons.

3.5.3 Complex Propagation Constant

We can now relate the refractive index to the propagation constant and finally see what all of this has to do with a wave propagating in a semiconductor. Considering an electromagnetic wave of the form of Equation 3.33, and remembering that the propagation constant k is given by Equation 3.43, we can include the complex refractive index, n′, to yield

$$\mathcal{E} = \mathcal{E}_0 e^{j(kz-\omega t)} = \mathcal{E}_0 e^{jk_0 n' z} e^{-j\omega t} \tag{3.48}$$

$$= \mathcal{E}_0 e^{-k_0 n_I z} \cdot e^{jk_0 n_R z} \cdot e^{-j\omega t} \tag{3.49}$$

In Equation 3.49, the three exponential terms correspond (left to right) to absorption, propagation, and time dependence of a wave in a material. Absorption takes the form of an exponential decay, propagation as a sinusoidal space dependence, and the last term as a sinusoidal time dependence. The significance of the real and imaginary components should now be clear: the real part defines the propagation of a wave, its velocity, and the imaginary part defines its change in intensity, or absorption.

The time dependence is not usually of interest because we assume that the wave is oscillating happily at a frequency ω and that this does not change with time. For that reason, the $\exp(j\omega t)$ term is often ignored and, in leaving it off, it will be understood that the waves have a sinusoidal time dependence. We often write the field in the form, thus ignoring the time dependence,

$$\mathcal{E} = \mathcal{E}_0 e^{-\frac{1}{2}\alpha z} e^{jk_z z} \tag{3.50}$$

where k_z is the propagation constant in the **z** direction and is often replaced by β, particularly in a waveguide, as we shall see in Chapter 7. The α term is of course the loss coefficient. The factor 1/2 is employed since, by convention, α is an *intensity* loss coefficient, such that the intensity $\mathcal{I}$ decays with propagation through a material as

$$\mathcal{I} = \mathcal{I}_0 e^{-\alpha z} \tag{3.51}$$

this expression is exactly equivalent to Equation 3.11.

With Equation 3.50, we can leave the theory behind and begin to form a relationship with measurable properties. By measuring what fraction of incident power is transmitted through a piece of semiconductor of known thickness, we can directly determine α. Through a measurement of the speed of a wave, usually done indirectly by, for example, ellipsometry, k_z can be found.

3.5.4 Summary of Complexities

Let's sift through our salad bowl of constants and terms and pick out those that will be most important later:

1. The effect of a material on a wave traveling through it is summarized in either the dielectric constant, ϵ, or the index of refraction, n, where the two are simply related by $\epsilon = \mathrm{n}^2$. n is generally complex, in which case we write it as n′.
2. The propagation constant of a wave traveling through free space is k_0, which is given by $k_0 = \omega/c$.
3. The complex refractive index has a real part that corresponds to the polarizability of the material and determines the velocity of a wave in a material, and an imaginary part that defines the absorption. This leads to a complex propagation constant of a wave traveling in a material, k, where $k = k_0 \cdot \mathrm{n}' = k_0(\mathrm{n_R} + j\mathrm{n_I})$.
4. The magnitude, phase, and time dependence of an electric field propagating in the **z** direction in a material is then given by $\mathcal{E} = \mathcal{E}_0 \cdot \exp(-\alpha z/2) \cdot \exp(jk_z z) \cdot \exp(j\omega t)$.
5. For the case where n is purely real (i.e., no absorption) we can define an effective propagation constant, wavelength and velocity in a material as: $k = \mathrm{n}k_0$, $\lambda = \lambda_0/\mathrm{n}$, and $v = c/\mathrm{n}$. These are often useful quantities for waves that remain in a particular material, such as a waveguide.

3.6 DIELECTRIC TENSOR

In our discussions of dielectric constant or indices of refraction until now, we have tacitly assumed that these parameters were independent of the direction in which the wave was moving through the material. We chose the **z** axis for convenience, but the results would not have changed had we chosen another. For many materials, this is sufficient.

In general, however, it is not.[2] Whereas we have written the definition of the electric flux density, D, to be

$$D = \epsilon_m \mathcal{E} \tag{3.52}$$

without specifying the directions of either $\mathcal{E}$ or D, ϵ is strictly a second rank tensor so that the relation of electric field to flux density must be written as

$$\begin{bmatrix} D_x \\ D_y \\ D_z \end{bmatrix} = \begin{bmatrix} \epsilon_{11} & \epsilon_{12} & \epsilon_{13} \\ \epsilon_{21} & \epsilon_{22} & \epsilon_{23} \\ \epsilon_{31} & \epsilon_{32} & \epsilon_{33} \end{bmatrix} \begin{bmatrix} \mathcal{E}_x \\ \mathcal{E}_y \\ \mathcal{E}_z \end{bmatrix} \tag{3.53}$$

The dielectric constant, and thus the conditions of propagation, are dependent on both the *direction* of wave propagation and the *polarization*. By clever choice of axes, it is always possible to rid ourselves of the off-diagonal elements, such that Equation 3.53 simplifies to

$$\begin{bmatrix} D_x \\ D_y \\ D_z \end{bmatrix} = \begin{bmatrix} \epsilon_{11} & 0 & 0 \\ 0 & \epsilon_{22} & 0 \\ 0 & 0 & \epsilon_{33} \end{bmatrix} \begin{bmatrix} \mathcal{E}_x \\ \mathcal{E}_y \\ \mathcal{E}_z \end{bmatrix} \tag{3.54}$$

which may be more compactly written as

$$D_{x,y,z} = \epsilon_{11,22,33} \mathcal{E}_{x,y,z} \tag{3.55}$$

For isotropic media, all of the dielectric constants are equal and we are left with the familiar Equation 3.52. For birefringent materials, the two differing polarizations for a wave traveling along an axis of the crystal will experience different refractive indices, thus affecting the propagation characteristics of the transmitted wave differently in different directions. In that case, Equation 3.54 or 3.55 must be used.

3.7 KRAMERS-KRONIG RELATIONS

It is a fortuitous property of linear passive systems that exhibit a response of the form

$$\xi(\omega) = \xi_R(\omega) + j\xi_I(\omega) \tag{3.56}$$

[2]See, for example, Born, M. and E. Wolf, *Principles of Optics*, Chapter XIV, 6th edition, Oxford, England: Pergamon, 1980.

that the real and imaginary components, ξ_R and ξ_I, are interrelated [14]. For our complex index of refraction, n′, the two components n_R and n_I are interrelated by

$$n_R(\omega) = \frac{2}{\pi} P \int_0^\infty \frac{s n_I(\omega)}{s^2 - \omega^2} ds \tag{3.57}$$

and

$$n_I(\omega) = -\frac{2\omega}{\pi} P \int_0^\infty \frac{n_R(\omega)}{s^2 - \omega^2} ds \tag{3.58}$$

which are known as the Kramers-Kronig relations. In these two equations, P denotes the Cauchy principal part of the integral, namely

$$P\int_0^\infty \equiv \lim_{\delta \to 0} \left(\int_0^{\omega - \delta} + \int_{\omega+\delta}^0 \right) \tag{3.59}$$

Thus, knowledge of either the real or imaginary part of the dielectric constant allows us to calculate the other [15]. The Kramers-Kronig relations are powerful tools in the evaluation of dielectric constant and absorption coefficient, for example, from reflectance spectra of materials. Alternatively, if we know the absorption spectrum, $\alpha(\omega)$, of a sample (a relatively easy thing to measure), we can determine, using Equation 3.52, the real part of the refractive index, $n_R(\omega)$ (a relatively difficult thing to measure). In practice, determining *absolute* values of n_R from n_I is difficult, since n_I would have to be known for all ω, $0 \le \omega \le \infty$. Although typically the absorption data is not known for all ω, but only over a limited range, reasonable assumptions may be made for α outside of the regions of interest.

REFERENCES

[1] Pasolini, P., "La Resistenza e la Sua Luce," *Roman Poems,* San Francisco: City Lights Books, 1986.
[2] Pankove, J., *Optical Processes in Semiconductors,* New York: Dover, 1971, p. 38.
[3] Urbach, F., "The Long-Wavelength Edge of Photographic Sensitivity and of the Electronic Absorption of Solids," *Physical Review,* Vol. 92, 1953, p. 1,324.
[4] Pankove, J., op.cit., p. 75.
[5] Franz, W., "Einfluβ Eines Elektrischen Feldes auf Eine Optische Absorptionskante," *Zeitschrift für Naturforschung,* Vol. 13a, 1958, pp. 484–489 (in German).
[6] Keldysh, L. V., "The Effect of a Strong Electric Field on the Optical Properties of Insulating Crystals," *Soviet Physics JETP,* Vol. 34, 1958, pp. 788–790.
[7] Hunsperger, R. G., *Integrated Optics: Theory and Technology,* 3rd edition, Berlin: Springer-Verlag, 1991, p. 269.
[8] Ridley, B. K., *Quantum Processes in Solids,* Chapters 5 and 6, 2nd edition, Oxford, England: Clarendon, 1988.

[9] Yourgrau, W., A. van de Merwe, and G. Raw, *Treatise on Irreversible and Statistical Thermophysics,* New York: Dover, 1982, p. 139.
[10] Ziman, J. M., *Principles of the Theory of Solids,* Chapter 8, 2nd edition, Cambridge, England: Cambridge University Press, 1972.
[11] Ashcroft, N. W. and N. D. Mermin, *Solid State Physics,* Chapter 27, Philadelphia: Saunders, 1976.
[12] Yariv, A., *Optical Electronics,* 4th edition, Philadelphia: Saunders, 1991, p. 161.
[13] Moss, T. S., *Optical Properties of Semi-Conductors,* Chapter 1, London: Butterworth-Heinemann, 1959.
[14] Papoulis, A., *The Fourier Integral and its Applications,* New York: McGraw Hill, 1962, p. 198.
[15] Madelung, O., *Introduction to Solid State Theory,* Berlin: Springer-Verlag, 1978, p. 261.

Chapter 4

Optical Semiconductor Materials

Through most of history, matter has been a concern of metaphysics more than physics, and materials of neither.

Cyril Stanley Smith, 1968 [1]

Aside from an occasional reference to GaAs or Si as representative of two classes of potentially useful semiconductors, we have up to this point refrained from looking at the nitty gritty of these materials. In order to avoid landing in the regime of metaphysics or worse, we will now turn to the specifics and discuss some of the properties of the materials from which semiconductor integrated optical structures can be made.

The III-V system is the dominant one for optical applications, and benefits (or suffers) from a very large variety of compounds with distinct characteristics. The advantage of the III-Vs over Si stems from the formers' ability to efficiently emit light; thus, this material family is suitable for the fabrication of optically active devices such as lasers. Silicon has proven itself to be useful for many passive optical applications, so we will not fail to study a few of the many capabilities of this material at the end of the chapter.

4.1 III-V MATERIALS

The III-Vs, as their name suggests, are compound semiconductors formed from a combination of elements from columns 3 and 5 of the periodic table. The binary compounds gallium arsenide (GaAs) and indium phosphide (InP) are the two grandparents of these materials, and indeed most of the more complex materials we are about to discuss are grown on wafers of either GaAs or InP.

4.1.1 The III-V Extended Family

Stoichiometric GaAs or InP have one gallium atom per one arsenic atom, or one indium per one phosphorus. The move to ternary or quaternary materials opens a route to considerably more flexibility. These are crystalline semiconductors composed of three or four elements, the relative concentrations of which can be varied, within limits, independently. AlGaAs is a well-developed material system, which we discuss further below; InGaAs, GaAsP, and GaAsSb, each containing two anion elements, are also popular materials.

As an example, the compound $Al_xGa_{1-x}As$ is a ternary alloy formed by varying the Al concentration between x = 0 (GaAs) and x = 1 (AlAs); Al and Ga occupy the same positions in the lattice for this compound. The lattice constant of $Al_xGa_{1-x}As$ varies linearly [2] with x, namely as $a_{AlGaAs} = a_{GaAs} + 0.00078x$ nm, from which we can see that the lattice mismatch between GaAs (0.5654 nm) to AlAs (0.5662 nm) does not exceed 0.15%. This close lattice matching allows relatively unproblematic epitaxial growth, such that high-quality layers of $Al_xGa_{1-x}As$ of any x can be grown lattice matched onto GaAs. This lattice-matching property is essential for the growth of good crystalline layers and is one reason that the GaAs/AlGaAs system has enjoyed considerable popularity. In addition, by varying the mole fraction of aluminum, we can take advantage of the concomitant variation of the electrical and optical properties of the material.

We can continue this game and add a further element to produce a quaternary alloy, adding an additional degree of freedom and complexity. For example, in $In_{1-x}Ga_xAs_{1-y}P_y$, the ratios of In to Ga and As to P can be varied independently; the material is lattice matched to InP for $0 \leq x \leq 0.47$ [3]. For the $(Al_xGa_{1-x})_yIn_{1-y}P$ system, the material is lattice matched to GaAs for y = 0.5 and a variation of its properties can be achieved with an independent change of x. These are just two examples; dozens of further variations are possible and in use for research or applications.

Varying the composition of these alloys, by changing x or y, gives rise to a change in the energy gap and the refractive index. Thus, emission or absorption wavelength as well as refractive optical properties may be tuned. Most of these III-V materials possess a large electro-optic coefficient, such that the fabrication of electro-optic modulators is facilitated. In Figure 4.1, we see an outline of some important III-V ternary materials with their energy gaps and lattice parameters as a function of alloy composition.

Every silver lining has a cloud, however, and the material features outlined above come with a few limiting factors. The III-Vs, in comparison to Si, are difficult to process. This is in part because of their very brittle nature, which requires great care in handling. One of the fundamental steps in silicon processing, the growth of a thermal SiO_2 passivation layer and its easy, selective removal in HF, has no analog in III-Vs. In addition, most of these materials can withstand only limited high-temperature processing, usually to the level of 400–500°C, without degradation due to the evaporation of one alloy component from the surface. Finally, and not insignificantly, III-V semiconductor wafers are considerably more expensive than their Si counterparts.

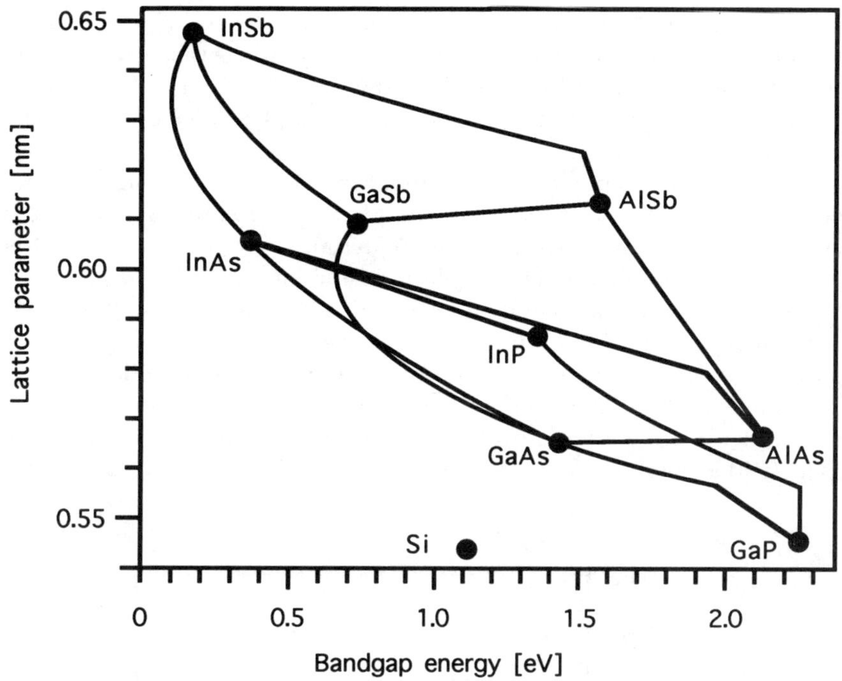

Figure 4.1 A selection of ternary III-V compounds plotted as a function of energy gap and lattice constant. The alloy, such as AlGaAs for example, is found on the solid line connecting the two binaries, GaAs and AlAs in this case. Si is also plotted for reference.

4.1.2 Regimes of Application

As we have suggested, the menagerie of III-V materials is an extensive and varied one. Which of the materials one would chose depends on numerous factors, but the first order is the wavelength of the intended application. The GaAs/AlGaAs system is perhaps the best known and most mature; it has been under development since the 1960s for high-speed electronics as well as optics and optoelectronics. Lasers fabricated from these materials typically emit in the wavelength range $780 < \lambda < 850$ nm and cheap, packaged GaAs-based lasers are now as common and unremarkable as disposable LCD wristwatches.

Advances in telecommunications employing optical fiber have provided an impetus for the development of longer wavelength lasers and the use of the InGaAsP system; this quaternary material can be grown, with almost perfect lattice matching, on InP substrates. Subject to research since the 1970s, InGaAsP lasers will emit in the range $1.3 < \lambda < 1.6$ μm. This range covers the two wavelengths of interest for fiber communications: 1.3 μm, at which point there is no dispersion along the fiber, and 1.55 μm, at which point fiber loss is minimum. Due to the degrees of freedom in alloy composition allowed by a quaternary compound, the range of lattice constants and energy gaps is larger for InGaAsP than, say, AlGaAs. Figure 4.2 shows the values achievable, and [4] presents the classic, more complete plot of all possible alloy compositions for this material.

Also of interest among the III-V materials is InGaAs; as a result of the relatively large lattice mismatch (GaAs: 0.565 nm and InAs: 0.606 nm), it can be grown as a strained layer on GaAs substrates. This material is attractive for high-frequency transistors and lasers, due to its high electron mobility. AlGaInP is under extensive development for lasers that operate at visible wavelengths, 600 to 700 nm. A wavelength summary of some important III-V ternary and quaternary compounds is given in Figure 4.3, along with a few values for the short-wavelength II-VI materials, which will be discussed in Section 4.5. This brief mention of the few materials and material systems available provides a hint of the extent of research and development in this area. The field is not only comprehensive, it is also subject to rapid development and change.

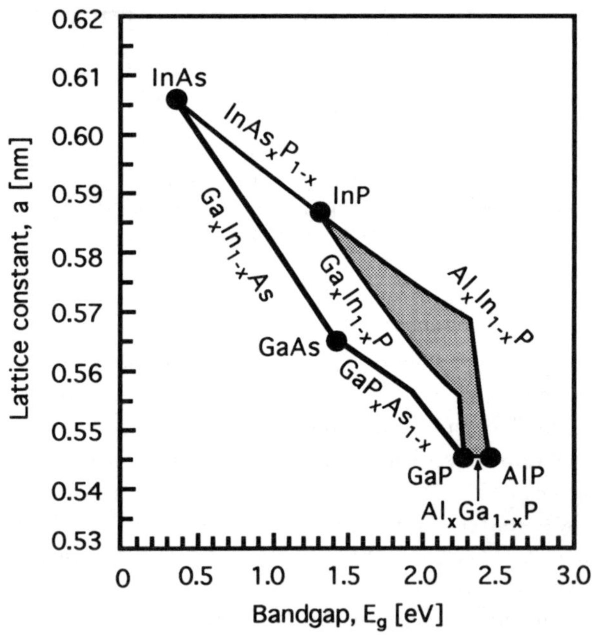

Figure 4.2 The allowed lattice constants and bandgap energies for the quaternary materials $In_{1-x}Ga_xAs_{1-y}P_y$ and $(Al_xGa_{1-x})_yIn_{1-y}P$ for varying alloy compositions. The former exists in the clear region and the latter in the dotted region. This figure contains data from [5], and is used with permission.

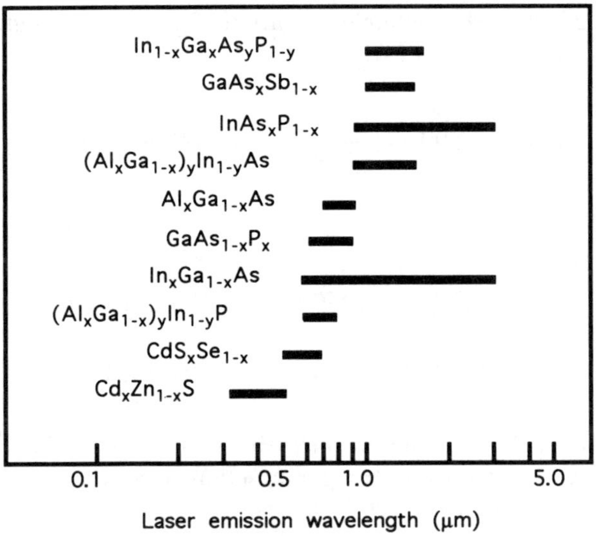

Figure 4.3 A plot of the wavelength ranges of several important III-V ternary and quaternary materials. The II-VI materials CdSSe and CdZnSe are also included for comparison. This figure contains data from [6]; © 1986 American Telephone and Telegraph Company, reprinted by permission.

4.2 COMPOSITION AND OPTICAL PROPERTIES

As mentioned above, changing the material composition of a III-V alloy changes the energy gap and the refractive index. The ability to tune these values at will provides a great deal of flexibility in the design of a semiconductor optical device. We can provide here a few relationships to facilitate the calculation of energy gap and refractive index for AlGaAs, a common material system.

4.2.1 Refractive Index

Refractive index is an important parameter for the design of waveguiding structures, as we will see in Chapter 7. An attractive feature of the AlGaAs system is that index varies with aluminum mole fraction (the value of x in $Al_xGa_{1-x}As$) so that layers of varying index can be grown, lattice matched, into a waveguide sandwich. The variations of index have been studied in depth[1] but a useful mathematical expression for

[1]One classic reference is: Blakemore, J. S., "Semiconducting and Other Major Properties of Gallium Arsenide," *Journal of Applied Physics*, Vol. 53, 1982, pp. R123–R181. Another classic reference: Adachi, S., "GaAs, AlAs, and $Al_xGa_{1-x}As$: Material Parameters for Use in Research and Device Applications," *Journal of Applied Physics*, Vol. 58, 1985, pp. R1–R29.

n of a material is given by the Sellmeier equation [7]. For wavelength λ [in μm] and x, the aluminum mole fraction in $Al_xGa_{1-x}As$,

$$n^2 = 10.906 - 2.92x + \frac{0.97501}{\lambda^2 + C} - 0.002467(1.41x + 1)\lambda^2 \qquad (4.1a)$$

where

$$C = (0.52886 - 0.735x)^2 \qquad \text{for } x \leq 0.36 \qquad (4.1b)$$

$$C = (0.30386 - 0.105x)^2 \qquad \text{for } x \geq 0.36 \qquad (4.1c)$$

We have plotted the variation of refractive index predicted by Equation 4.1 in Figure 4.4; the index becomes a stronger function of wavelength near the energy gap, and its absolute value decreases as we move from GaAs to AlAs. The refractive index also varies with temperature. The Sellmeier equation constants for refractive index have also been determined for SiO_2 [8], InGaAsP [9], and InP [10].

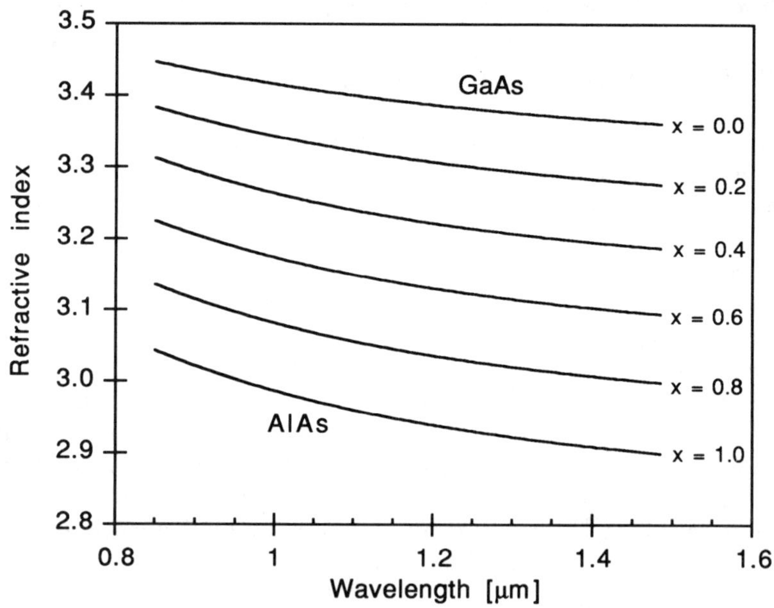

Figure 4.4 The variation of refractive index of $Al_xGa_{1-x}As$ as a function of aluminum mole fraction, x, and wavelength, λ, at T=300 K as predicted by the Sellmeier equation, Equation 4.1.

4.2.2 Energy Gap

The energy gap, the closest approach between the conduction and valence bands, is a function of material composition, which for the case of $Al_xGa_{1-x}As$ implies the possibility of a continuous variation as the Al mole fraction is varied. The variation of the energy gap at the Γ point as a function of x is given empirically as [11]

$$E_g = 1.424 + 1.247x \qquad \text{for } 0 \le x \le 0.45 \tag{4.2a}$$

$$E_g = 1.424 + 1.247x + 1.147(x - 0.45)^2 \qquad \text{for } 0.45 \le x \le 1 \tag{4.2b}$$

and for the X point by

$$E_g = 1.9 + 0.125x + 0.143x^2 \tag{4.3}$$

The energy gaps as predicted by Equations 4.2 and 4.3 are plotted in Figure 4.5; we see immediately that $Al_xGa_{1-x}As$ becomes an indirect bandgap semiconductor

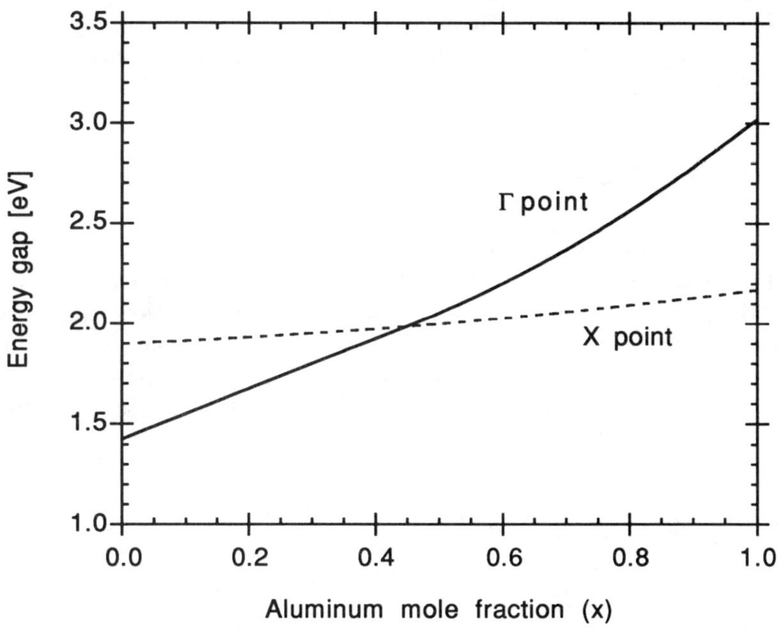

Figure 4.5 Variation of the energy gaps for the Γ and X points in $Al_xGa_{1-x}As$ as a function of Al mole fraction. For $x \ge 0.45$, AlGaAs becomes indirect. This figure contains data from [12] and is used with permission.

for $x \geq 0.45$ since the X point has a lower gap energy than the Γ point. Thus, in addition to the magnitude of the energy gap, its direct or indirect nature is of importance.

For a given semiconductor type or composition, the energy gap is also critically a function of temperature. Thermal effects in the crystal lattice cause E_g to decrease with temperature. It has been determined [13,14] that the relationship between gap and temperature for Si and GaAs can be expressed as

$$E_g = E_0 - \frac{\alpha T^2}{T + \beta} \tag{4.4}$$

where E_g [eV] is the energy gap, E_0 [eV] the energy gap at T = 0 K, T [K] the temperature in Kelvin, and α [eV/K] and β [K] empirical constants defined in Table 4.1. Figure 4.6 presents this relationship between energy gap and temperature for Si and GaAs graphically.

A summary of some material parameters for the semiconductors and dielectrics commonly used in integrated optical structures is given in Table 4.2; values for n are for infrared wavelengths, where Figure 4.4 gives the exact relation. The term λ_g is the absorption edge wavelength corresponding to the energy gap and can be calculated from $\lambda_g \cdot E_g = hc/q$. For E_g in eV and λ_g in μm, the right-hand side (hc/q) has a constant value 1.23985 [eV μm], a useful number to remember.

4.3 BULK CRYSTAL GROWTH

Despite the differences in their structure, crystals of Si and III-V semiconductors are fabricated in much the same way. Two industrially mature methods are of primary interest, both of which generate large ingots of semiconductor. These ingots are sliced into wafers, and the polished wafers are typically the starting point for any integrated optical fabrication process, such as those we will discuss in Chapter 5. We will focus here in the growth of GaAs [15], realizing that Si is made by much the same means.

Table 4.1
Parameters for the Calculation of Energy Gap as a Function of Temperature Using Equation 4.4 for Si and GaAs

Parameter	*[units]*	*Si*	*GaAs*
E_0	eV	1.17	1.519
α	eV/K	$4.73 \cdot 10^{-4}$	$5.405 \cdot 10^{-4}$
β	K	636	204

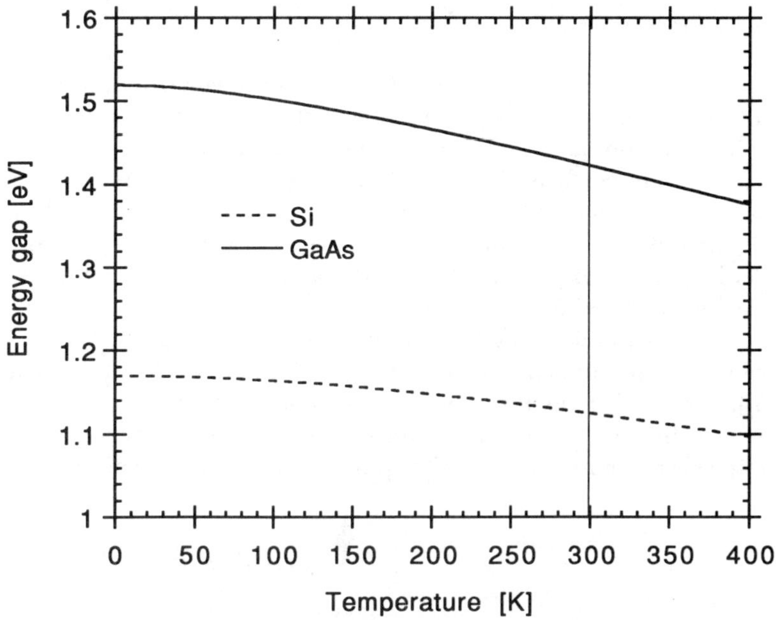

Figure 4.6 The variation of the intrinsic energy gap as a function of temperature for Si and GaAs, as calculated by the relation of Equation 4.4.

Table 4.2
Energy Gap, E_g, Absorption Edge Wavelength, λ_g, Lattice Constant, a, Infrared Refractive Index, n, Thermal Conductivity, σ_{th}, Breakdown Field, $\mathcal{E}_{bd}$, and Thermal Coefficient of Expansion, $\Delta L/L\Delta T$, for Several Commonly Used Semiconductors and Dielectrics

Parameter	*[units]*	*GaAs*	*Ge*	*InP*	*Si*	Si_3N_4	SiO_2
E_g	eV	1.42	0.66	1.35	1.12	~5	~9
λ_g	μm	0.87	1.88	0.919	1.11	~0.248	~0.138
a	nm	0.5654	0.5657	0.5869	0.5431	-	-
n		3.36	4.00	3.45	3.44	2.05	1.46
σ_{th}	W/cm K	0.46	0.6	0.68	1.45		0.014
$\Delta L/L\Delta T$	K^{-1}	$5.7 \cdot 10^{-6}$	$5.8 \cdot 10^{-6}$	$4.6 \cdot 10^{-6}$	$2.6 \cdot 10^{-6}$		$0.5 \cdot 10^{-6}$
$\mathcal{E}_{bd}$	V/cm	$4 \cdot 10^5$	$1 \cdot 10^5$	$2 \cdot 10^5$	$3 \cdot 10^5$	$1 \cdot 10^7$	$1 \cdot 10^7$

4.3.1 Liquid-Encapsulated Czochralski

The liquid-encapsulated Czochralski (LEC) technique involves pulling a single crystal ingot from a melt of gallium arsenide. As shown schematically in Figure 4.7, a crucible of SiO_2 or pyrolytic boron nitride (pBN) is heated by a radio frequency (RF) heater to temperatures where Ga and As react in a process known as compounding, an exothermal reaction. Since As has a high vapor pressure, the melt must be encapsulated by a material in which both elements are insoluble and the reactions must take place under high inert gas pressure ($3–4 \cdot 10^6$ Pa) in order to prevent As evaporation; B_2O_3 is usually employed for encapsulation.

Once the crucible is heated to the melting temperature of GaAs, 1238°C, a crystal seed is dipped into the melt and slowly withdrawn by a rotating arm; the crucible rotates as well (in the opposite direction), which contributes to uniform crystal growth. In this manner, the semiconductor ingot is pulled slowly from the melt. The speeds of rotation and vertical pulling vary, as considerations of remelting

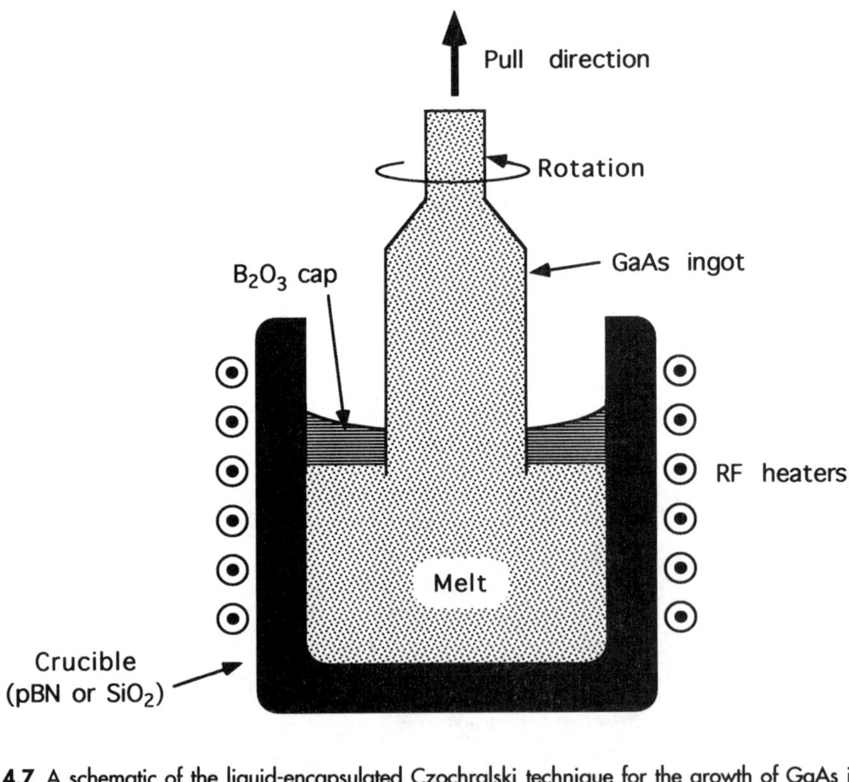

Figure 4.7 A schematic of the liquid-encapsulated Czochralski technique for the growth of GaAs ingots. The ingot is pulled from a melt of gallium and arsenic by a rotating arm. The crucible rotates in the opposite direction.

and crystal quality define the minimum and maximum growth rates. Vertical pull rate is usually on the order of 10 mm/hr and decreases with increasing ingot diameter.

Doping is performed either intentionally, by dissolving the dopant in the melt, or unintentionally as, for example, by the incorporation of Si dissolved from the sides of the SiO_2 crucible. GaAs may be grown in semi-insulating form, a misnomer that implies a very high sheet resistance. A high-resistance substrate ($\rho \sim 5 \cdot 10^6$ to $5 \cdot 10^8$ Ω-cm) is achieved, not by high purity, but through compensation in which an n-type dopant cancels out a p-type dopant. For GaAs, Cr can be added to the melt to compensate Si, a donor, since that element produces deep acceptor traps. In addition, the EL2 defect, a deep-level trap due to the Ga/As antisite defect (in which an arsenic atom occupies a gallium site in the lattice) is used to compensate carbon, a shallow acceptor. Si substrates, grown by the same technique, can reach considerably higher levels of purity but not the same high resistivity values. The thermal gradients during pulling and cooling are often large and give rise to the relatively high dislocation densities in LEC-grown GaAs material.

4.3.2 Horizontal Bridgman and Vertical Gradient Freeze

An alternative technique keeps the ingot fixed in a crucible but moves the heater to generate a moving temperature gradient along the growth direction. The horizontal Bridgman (HB) process, shown schematically in Figure 4.8, employs an enclosed ambient with As overpressure to keep arsenic from leaving the heated GaAs surface.

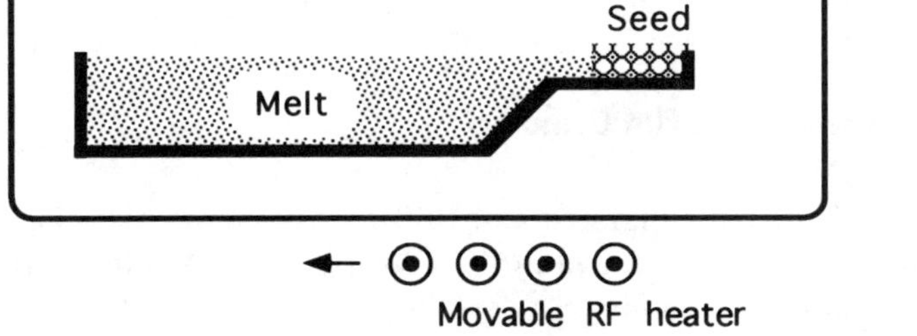

Figure 4.8 A schematic representation of the horizontal Bridgman technique for crystal growth. A seed initiates crystalline growth from the right and an RF heater is slowly pulled along the melt to the left.

On the right, a seed at the solidification temperature provides an orientation reference for the crystal that grows as the heater is pulled to the left. The heater generates a known, gradual temperature profile, employing up to 50 independent heaters; this arrangement gradually cools the ingot after solidification at a controlled rate.

Ingots grown by HB are typically D-shaped, which derives from the shape of the boat in which the GaAs charge is crystallized. The possible inconvenience due to the odd shape or material loss during rounding is compensated for by the considerably reduced dislocation densities with respect to LEC material. An additional advantage is that the reproducibility from run to run is more easily achieved with this crystal growth technique [16], and Si doping is facilitated since the B_2O_3 cap, which pulls silicon out of the LEC melt, is not required.

Variations on this approach have been employed as alternative crystal growth or purification methods. Vertical Bridgman is conceptually identical to the horizontal variant, but uses a vertically oriented chamber. The low temperature gradients generated by the movable heater again allow the fabrication of materials with very low dislocation densities. Vertical or horizontal gradient freeze processes use chambers similar to that shown in Figure 4.8, but with both heater and crucible fixed. The temperature is slowly ramped downward, allowing slow cooling of the solidified crystal, again with both vertical and horizontal arrangements.

In float zone processes, a crystal ingot is held vertically and a movable RF heater is pulled from bottom to top. A thin molten floating zone of material is thus pulled the entire length of the ingot, which can either be crystallized if an appropriate seed is provided at the bottom or, if already a single crystal, it can be purified, as impurities can be pushed ahead of the molten region. The lack of a crucible to hold the ingot removes one possible source of contamination during refining. Alternatively, doping can be accurately defined if a controlled quantity of dopant is present as the molten zone is passed through the ingot. Multiple passes of this technique can be used to reach high levels of purity in a process known as zone refining; the technique is very useful for Si purification but impossible to apply to GaAs except under 0 G conditions (i.e., outer space) due to instabilities in the molten zone and poor surface-tension properties.

4.3.3 Wafer Sawing and Flat Grinding

Si crystal ingots can have diameters of 200 mm or more and weigh from several kilograms to several tens of kilograms; GaAs ingots are smaller. After growth, the ingots are first ground circular to a specified diameter and then flattened, also by grinding, along two sides to produce the wafer flats that allow the end user to determine the crystalline direction on the wafer surface. As is shown in Figure 4.9, sliced and polished wafers are not round but have flat sections cut into the edge. The relative positions of the major (long) and minor (short) flats provide information about the wafer surface orientation (both (100) for the examples shown) and, for

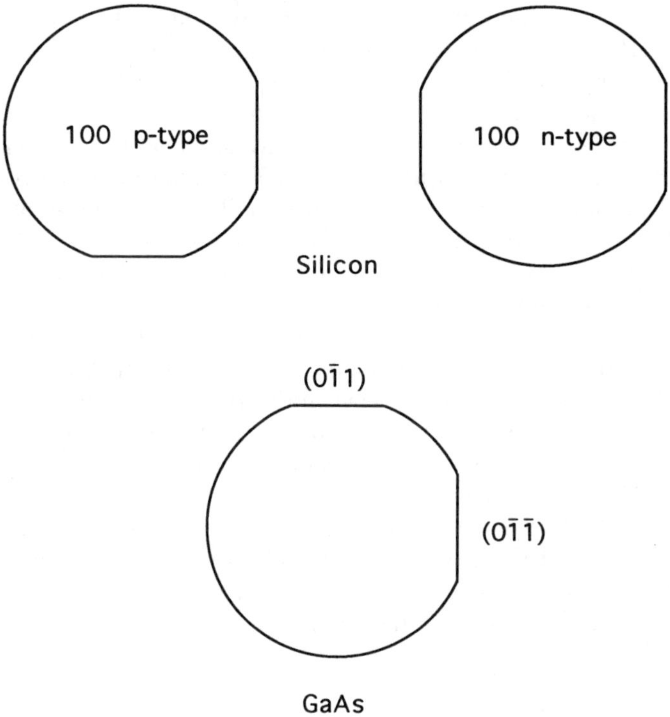

Figure 4.9 Orientation of the polished flats for (100) silicon wafers and (100) GaAs wafers. For silicon, the orientation is different for wafers with a (111) surface.

the case of Si, the substrate doping type. Although the flat is usually accurately aligned (by X-ray diffraction), exceptionally precise alignment, such as for laser mirrors, usually requires a cleave before processing to assure adequate precision in orientation.

After grinding, the ingots, like large salamis, must be sliced into wafers using the technological cousin of the salami slicer. This rather abusive mechanical treatment is followed by lapping and a contouring of the edges plus a wet etch, which removes most of the damage from the surface. The wafers, typically 300 to 500 μm thick, are then mechanically and chemically polished in a slurry in order to yield the brightly reflecting disks with which the process technician begins work. A nice discussion of the technology involved is found in [17], which looks at the fabrication of Si wafers.

4.4 EPITAXIAL LAYER GROWTH

A great deal of flexibility in device structures is gained when layers of different semiconducting materials are "grown" on bulk GaAs or Si substrates. If these layers are grown epitaxially, meaning as a single crystal, and with precisely controlled thickness and composition, we can take advantage of the variability in refractive index and energy gap in order to fabricate interesting vertical structures.

A III-V-based waveguide, for example, could begin with a wafer of GaAs on which a thin layer of $Al_xGa_{1-x}As$ is grown, with, say, x = 0.4. This would be followed by another layer of $Al_xGa_{1-x}As$, with a slightly lower value for x, perhaps 0.2 such that the refractive index is slightly higher, and finally another layer of $Al_{0.4}Ga_{0.6}As$. The central AlGaAs layer, by virtue of its higher refractive index, will guide light, and we will thus have grown a two-dimensional slab waveguide structure.

Typical thicknesses for such waveguide layers vary from several tens of nanometers to several microns; quantum wells are several nanometers in thickness, which generally corresponds to only a few monolayers of material. Thickness control is thus important for the successful growth of waveguide layers. In addition, optical quality layers must have an accurately defined composition and a very low density of defects. $Al_xGa_{1-x}As$, for example, is plagued by DX centers, deep donor levels due to various impurities. The DX center concentration, which becomes more of a problem for $x \geq 0.2$, can be diminished by reducing the background of undesired dopants in the grown layers. We see that the growth of epitaxial layers is a demanding but central technological concern for the fabrication of most semiconductor-based optical devices, and we thus turn to an overview of the primary techniques involved.

4.4.1 Liquid Phase Epitaxy

In liquid phase epitaxy (LPE), the substrate wafer is essentially dunked into a melt of the material that either is to be deposited or has a molten zone passed over it. The technique has a distinguished history and has been used for several decades, although its popularity has diminished somewhat in recent years. Although it is inexpensive and relatively simple when compared to the other approaches we will discuss, difficulties in achieving good uniformity and the reduced ability to generate thin layers have proven to be essential limitations.

4.4.2 Molecular Beam Epitaxy

Under high-vacuum (UHV) conditions, using a controlled stream of the constituent atoms that make up a film, highly pure layers of well-defined thickness can be grown with a technique known as molecular beam epitaxy (MBE) [18]. An MBE chamber is presented schematically in Figure 4.10, which shows the substrate on which the layers will be grown at the center of the ultrahigh vacuum chamber. The pressures commonly employed, 10^{-8} to 10^{-9} Pa, imply that a very clean environment, relatively free from the impurities that might otherwise be incorporated into the films, is given.

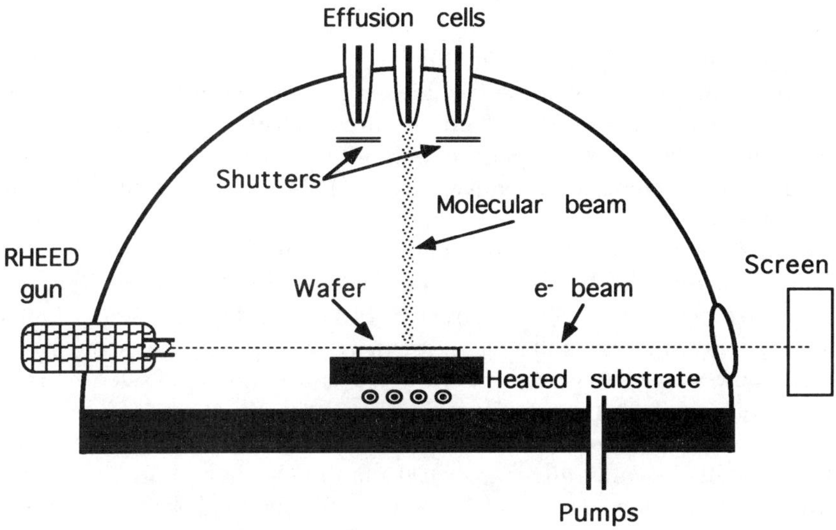

Figure 4.10 Schematic cross-section of an ultrahigh vacuum MBE apparatus. The solid-source effusion cells emit a molecular beam, controlled by the shutters and cell temperature. The atoms adhere to the wafer where epitaxial layers are grown with monolayer precision. The RHEED gun permits monitoring of the layers during growth.

The low pressure also permits the accurate control of epitaxial film thickness. Material sources for crystal growth are high-temperature effusion cells, in which either the semiconductor components (Ga, As, ...) or the dopants (Si, Be, ...) are evaporated—in the absence of scattering at very low pressures, evaporation gives rise to a molecular beam that impinges on the wafer surface. Computer-controlled shutters regulate the atomic flow and the flux may be adjusted by setting the cell temperature. This process is far from thermal equilibrium and the kinetics of the atoms that adhere to the substrate surface are important in defining the nature of the grown film.

Typical MBE growth rates are about one monolayer per second, with substrate temperatures of 500–700°C depending on the material and the desired growth conditions. Purity of the layers is primarily a function of the quality of the metal sources in the effusion cells—sources of 99.99999% purity are routinely employed. The mechanical shutters allow a rapid switching of the molecular beams, such that both accurate epitaxial layer thicknesses and precise interfaces may be achieved. The crystal grower is assisted in this process by the ability to monitor *in situ* the growth of the layers; this is accomplished by a reflection high-energy electron diffraction (RHEED) gun, which uses electrons reflected from the surface to generate oscillating patterns with each grown monolayer on a phosphorescent screen. The low growth rate and the ability to measure grown film thickness in real time allow the reproducible generation of layers with subnanometer accuracy.

Several variations on the MBE technique have been developed, notably by moving to gas sources as replacements for the effusion cells; this change simplifies the apparatus somewhat and can result in higher growth rates and films with reduced defect densities. Metalorganic sources, liquids or gases with high vapor pressures that may thus be employed at low temperatures, are the basis for chemical beam epitaxy (CBE) or metalorganic molecular beam epitaxy (MO-MBE), two names for essentially the same technique.

MBE may be used to grow epitaxial films of numerous materials; the most popular are probably the III-Vs, based on GaAs or InP substrates. The growth technique is also popular for silicon, either Si/Si homoepitaxy, SiGe on Si, or GaAs on Si. The latter, which we will discuss further, generally requires the growth of intermediate buffer layers to generate films of reasonable quality. Other semiconductor systems, such as HgCdTe, may be grown by MBE as well.

Film thickness accuracy, interface quality, and purity achievable by MBE are probably unmatched by any other growth technique—very high mobility and very long lifetime structures are typically grown by this method for both device and research applications. Such performance has a price, however—relatively low growth rates and limited throughput, factors which translate into a truly pecuniary price as well.

4.4.3 Metalorganic Chemical Vapor Deposition

Back in the days of yore, GaAs was grown from the vapor phase using $AsCl_3$ and Ga as sources. Such vapor phase epitaxy (VPE) has been refined—particularly by the use of metalorganics as sources for the atomic species and the move to lower pressures—into a technique known as metalorganic chemical vapor deposition(MOCVD) or metalorganic vapor phase epitaxy (MOVPE). This advanced form of chemical vapor deposition, in which materials are deposited from tightly controlled vapor phase, is popular due to its flexibility and relatively high growth rates [19], for both electronic and optical [20] applications.

As we may see from the diagram in Figure 4.11, the epitaxial layers are again grown in a vacuum chamber, although the UHV conditions used in MBE are not reached. Pressures range from about 10^5 (atmospheric) to 10^4 Pa for low-pressure MOCVD and further to 10^{-2} to 10^{-4} Pa for so-called vacuum MOCVD. The substrates are placed on a susceptor, which is RF heated to the temperatures suitable for growth (500–750°C) and often spins at high speed to improve uniformity. A laminar flow of gas-phase precursors is then directed over the wafers, of which the hydrodynamics is important for film growth.

Metalorganic precursors for the group III atoms are typically alkyls, either triethyl ($(C_2H_5)_3$) gallium, TEGa, or tri-methyl ($(CH_3)_3$) gallium, TMGa; TEAl is the source for aluminum. The group V component precursors are the hydrides, AsH_3

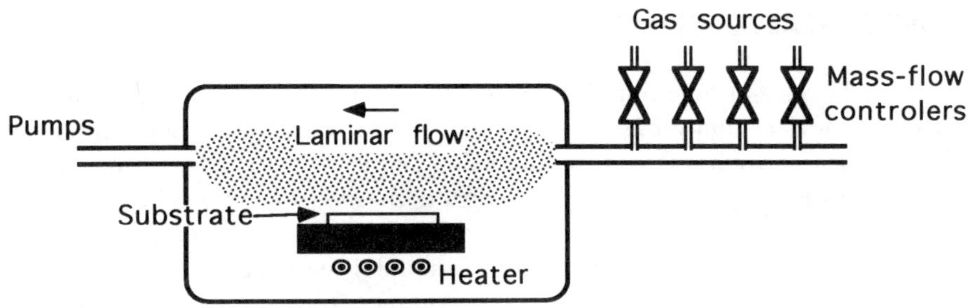

Figure 4.11 Schematic of an MOCVD reactor. The wafers are RF heated and placed in the laminar flow of gas precursors for the materials to be grown. The flow rates of the metalorganic sources are precisely controlled by the mass-flow controllers to the right. At the heated wafer surface, the gas dissociates into the atomic species for epitaxial layer growth.

and PH_3 for arsenic and phosphorus, respectively. These gas species are introduced into the chamber using hydrogen as a carrier gas at precisely regulated flow rates through mass flow controllers. The high temperatures near the wafer surface then lead to a typical reaction, for GaAs from TMGa and arsine for example, such as

$$Ga(CH_3)_3 + AsH_3 \rightarrow GaAs + 3CH_4 \tag{4.5}$$

The group III alkyls and group V hydrides are cracked at the hot wafer surface, meaning that the atom required for growth becomes free to bind with the crystal surface. This process is close to thermal equilibrium, and the growth rate is usually limited by diffusion of the atomic species to the substrate surface. Growth rates are typically 10–30 nm/min, and *in situ* growth monitoring is possible by monitoring the reflection of a laser beam incident on the surface [21].

A positive feature of MOCVD is the relatively wide spectrum of materials that can be grown epitaxially with it; many of the precursors for the III-V and II-VI compounds can be found in gas form. The antimonides, such as GaSb, and indium-based materials use TMSb and TEIn as sources and the nitrides, GaN or AlN, simply employ NH_3. The II-VI compound semiconductors, CdTe, ZnSe, and the like, use DEZn, DECd, and DMTe, with H_2S and H_2Se for the sulfides and selenides. Lead-salt structures, such as PbSSe, can further use TMPb as a precursor.

Limiting for MOCVD, with respect to MBE, is film purity—carbon incorporation into the films, resulting from the organic nature of the precursors, is often unavoidable. In addition, the toxic nature of the gaseous precursors is clearly not a positive attribute. The relatively high growth rates, with thickness control suitable and uniform for the growth of high-quality quantum wells, plus material flexibility are, however, distinct advantages.

4.5 MORE OPTICAL MATERIALS

For integrated optics applications that require the monolithic integration of various optoelectronic devices including light emitters, the III-V materials dominate at present. The ease with which one can make a III-V semiconductor laser, combined with an advanced fabrication technology due in part to the well-developed III-V electronics industry, make the GaAs and InP systems lucrative for the manufacture of complete semiconductor optical circuits. Other semiconductor systems, however, have features that have provided an impetus to their development for optical applications: Si is cheap, robust, manufacturable, and has an even more advanced fabrication technology, and numerous further compound semiconductor systems have proven to be useful for detection and imaging in certain wavelength ranges. Much of the work in this area has focused on discrete semiconductor optical devices, mostly photon detectors, and passive applications, where no light source is integrated.

4.5.1 Silicon

Unmatched in its development for electronics, the application of silicon to optics is limited by the indirect bandgap, which makes light emission inefficient, and the lack of electro-optic effects. Nevertheless, Si has numerous attractive properties [22]. Were the fabrication of Si lasers a possibility in the past, the III-V compound semiconductors would likely still have an edge due to their ability to tune energy gap and refractive index with composition. The question of whether the III-Vs would have reached their current state of development had that been the case is a religious one, which the Si and GaAs tribes can battle over while reaching for their bludgeons.

Silicon is an indirect bandgap semiconductor (see Figure 2.6) which, due to its symmetry, also does not exhibit the Pockels, or linear electro-optic, effect. This latter limitation can be circumvented to some extent by the use of carrier injection effects to change the refractive index and fabricate an electro-optic modulator—heavy doping or carrier injection result in a decrease of n, as we will see in more detail in Chapter 11. Thus, doping is an important parameter for the fabrication of passive optical components from Si. Combining Si with SiO_2 and Si_3N_4 dielectrics or Ge, as we shall see, improves the flexibility in device structuring.

Bulk Si crystals, with a room temperature bandgap of 1.12 eV and a temperature dependence as outlined in Figure 4.6, are predominantly grown by the same type of Czochralski puller as is GaAs. The crucible is SiO_2, in which the Si is molten, held by a graphite susceptor; nuclear-grade, high-purity materials are required for industrial production. Pull rates of 2 mm/min are typical and ingots of 200 mm in diameter are routinely produced in industrial quantities; 300-mm diameters are in preparation. The purity of Si is exceptional and it is doubtful that any other material is produced with such quality at such magnitude. Wafer grinding, slicing, and lapping are essentially the same as described for GaAs.

Epitaxial layer growth for Si has focused more on electronic applications than optical ones. Silicon homoepitaxy (growing Si on Si) has allowed the growth of high-resistivity material on low-resistivity substrates, which is useful for improving the performance of bipolar transistors. Epitaxy on dielectric materials (SiO_2 or Al_2O_3) generates silicon-on-insulator structures with enhanced electrical isolation. Epitaxial growth is accomplished in high-temperature reactors, in the range 1200°C. These reactors are also employed for the recrystallization of CVD-deposited polysilicon films or by MBE at somewhat lower temperatures of about 950°C.

Despite the large lattice mismatch between Si and GaAs (about 4%) or Si and InP (8%), the heteroepitaxy of GaAs or InP on Si, by both MBE and MOCVD, is an area subject to research by crystal growers [23]. The successful epitaxial growth of optical quality III-V films on a Si substrate could combine the best of both crystalline worlds: active optical devices (lasers) and the advanced electronics and micromachining capabilities of silicon. The lattice mismatch of the materials, however, causes the generation of dislocations at the interface, which migrate into the optical layers and contribute to nonradiative recombination, reducing laser quality. This problem may be defused somewhat by the growth of intermediate buffer layers or superlattices, or by using off-axis substrates, all of which increase complexity and hence cost. An additional limitation is the mismatch of thermal coefficients of expansion among the different materials. Nevertheless, a number of active optical elements have been fabricated from III-V materials grown on a silicon substrate [24].

4.5.2 Silicon: Modifications and Accessories

Since no direct analog to the useful lattice-matched GaAs/AlGaAs system, with its mole-fraction-dependent energy gap and refractive index, exists in Si, numerous auxiliary materials can be employed to make simple optical structures. Most silicon-based waveguide structures involve the use of dielectric sandwiches or doping variations; the former employ SiO_2 and Si_3N_4 as common dielectrics and the latter make use of carrier-induced changes in refractive index. Because the refractive indices of the materials are usually fixed (for a given stoichiometry), a creative combination of dielectrics is usually necessary to obtain a useful layer structure. SiO_2 is usually used as a buffer (due to its low index—see Table 4.2) with waveguiding layers deposited on top. ZnO, Nb_2O_5, Ta_2O_5, or TiO_2 are further materials, usually formed by sputtering and subsequent annealing, which have proven useful for the fabrication of Si-based optical structures [25]. We will examine these in more detail in Chapter 8.

Silicon may also be modified in its bulk state to alter its optical characteristics—light emission can be affected by the incorporation of rare earth dopants, predominantly Er [26]. Alternatively, isoelectronic impurities that bind excitons, such as Be, can enhance the luminescence efficiency of silicon and may lead to improved light-emission properties [27]. Finally, an area of research at the time of writing is so-called porous silicon, bulk Si substrates that have been microstructured by electrochemical etching or spark-erosion techniques. Samples abused in this manner appear to luminesce with reasonable efficiency in the visible (orange) wavelength range [28]. The mechanism is as yet not understood, and it is too early to decide whether this is a flash in the dark or a flash in the pan.

4.5.3 Silicon/Germanium

The development of silicon technology for optical applications has also focused on the combination of Si with Ge [29]. The growth of silicon/germanium layers, $Si_{1-x}Ge_x$, primarily by MBE or MOCVD, is possible due to the relatively close lattice matching of Si and Ge; see Table 4.2. Since the refractive indices of Si (3.45) and Ge (4.00) are sufficiently distinct, and the energy gaps are different as well, some flexibility in the tuning of optical properties is expected.

The values of refractive index and energy gap are a function of both the mole fraction of Ge in the alloy and the nature of the strain in the grown film. SiGe is compressively strained when grown on Si; the nature of the strain defines both the value and nature of the energy gap. The refractive index of SiGe is higher than that of Si, varying as a function of germanium content for $Si_{1-x}Ge_x$ at a wavelength of 1.3 μm as [30]

$$n = 3.5 + 0.38x \tag{4.6}$$

The bandgap energy decreases with increasing Ge content, so that heterostructures can be designed and grown in this system.

In addition to the strained layers of $Si_{1-x}Ge_x$ on Si, alternating layers of pure Ge and Si can be grown by epitaxial techniques, as can SiGe quantum wells in Si and Si_nGe_m superlattices. Much of the work on this material system has focused on electrical applications for high-speed modulation-doped transistors (analogous to the III-V HEMT); optically, the combination of silicon and germanium is interesting for applications in the 1.3- to 1.6-μm wavelength range, especially for detectors [31]. Active light emitters suffer from low emission efficiencies (on the order of 10^{-4}) since the indirect bandgap haunts SiGe as it does Si. The efficient, silicon-based light emitter is as yet an elusive goal.

4.5.4 Other Compound Semiconductors

Numerous other compound semiconductor materials have seen extensive development, primarily for discrete devices at particular wavelengths. The antimonides, such as GaSb or InAsSb, are III-V compounds useful for long-wavelength applications around $\lambda = 5$ μm. The lead-salt devices, which are IV-VI semiconductors (PbSnTe,

PbSnSe or PbSSe), extend the wavelength range to 50 μm or longer and have proven useful for long-wavelength tunable lasers [32].

The well advanced II-VI system, $Hg_{1-x}Cd_xTe$, has been subject to significant work primarily for imaging in the mid-infrared range (around 5 to 15 μm), since the energy gap can be varied virtually continuously from 0 to 1.5 eV with Cd mole fraction. CdTe or GaAs substrates are usually used as the basis for MBE or MOCVD growth of these semiconductors. All of these materials, and most others in this wavelength range, are only useful at low temperatures and must often be operated at liquid nitrogen temperatures (77 K) or lower. For short-wavelength applications, the II-VI semiconductors (ZnSe, CdZnSe, CdTE, or ZnTe) are also enjoying popularity, figuring predominantly in the quest for a blue/green semiconductor laser.[2]

REFERENCES

[1] Smith, C. S., "Matter vs. Materials: a Historical View," *A Search for Structure*, Cambridge, MA: MIT Press, 1981, p. 115.

[2] *Properties of Gallium Arsenide*, 2nd Ed., London: INSPEC, 1990.

[3] Pearsall, T. P., *GaInAsP Alloy Semiconductors*, Chichester, England: Wiley, 1982.

[4] Moon, R. L., G. A. Antypas, and L. W. James, "Bandgap and Lattice Constant of GaInAsP as a Function of Alloy Composition," *Journal of Electronic Materials*, Vol. 3, 1974, pp. 635–644

[5] Casey, H. C., and M. B. Panish, *Heterostructure Lasers*, Part B, San Diego: Academic Press, 1978, p. 3.

[6] Agrawal, G. P., and N. K. Dutta, *Long Wavelength Semiconductor Lasers*, New York: Van Nostrand Reinhold, 1986, p. 6.

[7] Hunsperger, R. G., *Integrated Optics: Theory and Technology*, 3rd Ed., Berlin: Springer-Verlag, 1991, p. 60.

[8] Malitson, I. H., "Interspecimen Comparison of the Refractive Index of Fused Silica," *Journal of the Optical Society of America*, Vol. 55, 1965, pp. 1,205–1,209.

[9] Fiedler, F., and A. Schlachetzki, "Optical Parameters of InP-Based Waveguides," *Solid-State Electronics*, Vol. 30, 1987, pp. 73–83.

[10] Pettit, G. D., and W. J. Turner, "Refractive Index of InP," *Journal of Applied Physics*, Vol. 36, 1965, p. 2,081.

[11] Casey and Panish, op.cit., p. 16.

[12] Casey, H. C.. and M. B. Panish, *Heterostructure Lasers*, Part A, San Diego: Academic Press, 1978, p. 193.

[13] Varshni, Y. P., "Temperature Dependence of the Energy Gap in Semiconductors," *Physica*, Vol. 34, 1967, pp. 149–154.

[14] Thurmond, C. D., "The Standard Thermodynamic Functions for the Formation of Electrons and Holes in Ge, Si, GaAs and GaP," *Journal of the Electrochemical Society*, Vol. 122, 1975, pp. 1,133–1,141.

[15] Ferry, D. K. (ed.), *Gallium Arsenide Technology*, Indianapolis, IN: Howard W. Sams, 1985, Chapter 2.

[16] Williams, R. E., *Gallium Arsenide Processing Techniques*, Dedham, MA: Artech House, 1984, Chapter 2.

[2]See the special issue on wide bandgap II-VI semiconductors in the *Journal of Electronic Materials*, Vol. 23, March 1994.

[17] Pearce, C. W., "Crystal Growth and Wafer Preparation," in Sze, S.M., *VLSI Technology*, New York: McGraw Hill, 1983.
[18] Herman, M. A. and H. Sitter, *Molecular Beam Epitaxy*, Berlin: Springer Verlag, 1989.
[19] Ludowise, M. J., "Metalorganic Chemical Vapor Deposition of III-V Semiconductors," *Journal of Applied Physics*, Vol. 58, 1985, pp. R31–R55.
[20] Dupuis, R. D., "AlGaAs-GaAs Lasers Grown by Metalorganic Chemical Vapor Deposition—A Review," *Journal of Crystal Growth*, Vol. 55, 1981, pp. 213–221.
[21] Epler, J. E., and H. P.Schweizer, "Evolution of Monolayer Terrace Topography During Metalorganic Chemical Vapor Deposition of (100) GaAs," *Applied Physics Letters*, Vol. 63, 1993, pp. 1,228–1,230.
[22] Wolf, H. F., *Silicon Semiconductor Data*, Oxford, England: Pergamon Press, 1969.
[23] Fang, S. F., K. Adoni, S. Iyer, H. Morkoç, H. Zabel, C. Choi, and N. Otsuka, "Gallium Arsenide and Other Compound Semiconductors on Silicon," *Journal of Applied Physics*, Vol. 68, 1990, pp. R31–R58.
[24] Razeghi, M., *The MOCVD Challenge*, Vol. 1, Bristol, England: IOP, 1989, Chapter 6.5.
[25] Boyd, J. T., R. W. Wu, D. E. Zelmon, A. Naumaan, and H. A. Timlin, "Guided Wave Optical Structures Utilizing Silicon," *Optical Engineering*, Vol. 24, 1985, pp. 230–234.
[26] Ren, F., J. Michel, Q. Sun-Paduano, B. Zheng, H. Kitagawa, D. Jacobson, J. Poate, and L. Kimerling, "IC Compatible Processing of Si:Er for Optoelectronics," *MRS Proceedings*, Vol. 298, 1993, pp. 415–423.
[27] Brown, T. J., P. L. Bradfield, D. G. Hall, and R. A. Soref, "Optical Emission from Impurities within an Epitaxial-Silicon Optical Waveguide," *Optics Letters*, Vol. 12, 1987, pp. 753–755.
[28] Koshida, N. and H. Koyama, "Visible Electroluminescence from Porous Silicon," *Applied Physics Letters*, Vol. 60, 1992, pp. 347–349.
[29] König, U., "Challenges for a Si/Ge Heterodevice Technology," *Microelectronic Engineering*, Vol. 23, 1994, pp. 3–13.
[30] Splett, A., T. Zinke, K. Petermann, E. Kasper, H. Kibbel, H.-J. Herzog, and H. Presting, "Integration of Waveguides and Photodetectors in SiGe for 1.3μm Operation," *IEEE Photonics Technology Letters*, Vol. 6, 1994, pp. 59–61.
[31] Pearsall, T. P., H. Temkin, J. C. Bean, and S. Luryi, "Avalanche Gain in Ge_xSi_{1-x}/Si Infrared Waveguide Detectors," *Electron Device Letters*, Vol. 7, 1986, pp. 330–332.
[32] Shani, Y., R. Rosman, A. Katzir, P. Norton, M. Tacke, and H. M. Preier, "Distributed Bragg Reflector $Pb_{1-x}Sn_xSe/Pb_{1-x-y}Eu_ySn_xSe$ Diode Lasers with a Broad Single-Mode Tuning Range," *Journal of Applied Physics*, Vol. 63, 1988, pp. 5,603–5,606.

Chapter 5

Basics of Semiconductor Processing

La découverte d'un mets nouveau fait plus pour la bonheur pour du genre humain que la découverte d'une étoile.[1]

Brillat-Savarin, 1828 [1]

With our raw ingredients in hand, we can now cook up a semiconductor integrated optical device. Semiconductor processing techniques, initially developed and highly refined for the fabrication of microelectronics, allow us to structure our crystalline substrate into a useful optical element or circuit. Two features of semiconductor processing become prominent upon a cursory examination: the fabricated structures are small (microns or less) and we batch fabricate many at a time (thousands or more). While these aspects are generally a positive feature, they also stipulate that considerable effort be placed into microfabrication technology—extreme cleanliness, highly reproducible process steps, and machinery that permits the definition of structures of 100-nm dimensions are essential. For these reasons, newly developed processes, while perhaps doing less for humankind than a new recipe for the kitchen, are generally jealously guarded secrets.

There remains a vestige of art in the semiconductor fabrication sciences: invisible demons often wreak havoc with semiconductor processes, causing them to mysteriously fail after having worked for ages. In this chapter, however, we will peruse the hard technical side of semiconductor processing, as the more artistic aspects of successfully fabricating a semiconductor micro-optical device can only really be learned by actually doing it. We will focus on the fabrication technology important for integrated optics applications, treating some areas that are cornerstones of microelectronic processing only parenthetically. The chapter is intended as an introduction for readers who have limited experience with microelectronic fabrication, although it may serve as a quick review for those who already know how to get into a cleanroom suit.

[1] "The discovery of a new recipe [dish] does more for the happiness of mankind than the discovery of a star."

5.1 A TYPICAL OPTOELECTRONIC PROCESS

Most semiconductor-based optics processing can be roughly divided into two categories: one yields a waveguide-based structure and the other is useful for making discrete, often surface-emitting, devices. The former is usually implied under the rubric *integrated optics* and is the most widely applied, particularly for optical integrated circuits. The latter, nonetheless, has a particularly rich variety of clever structures and associated fabrication processes. For optoelectronics, which includes electrical as well as optical devices, the fabrication of transistors and their ilk must also be considered, which is a field onto itself and will not be a primary focus here; in any case, many of the basics are the same. We will use a simple waveguide process in this section as a means of illustration to present many of the microfabrication processes required by all types of devices.

5.1.1 The Basics

A typical semiconductor waveguide-based integrated optic fabrication procedure is outlined schematically in Figure 5.1. This example treats one possible way to fabricate a waveguide electro-optic modulator in GaAs/AlGaAs technology. Although many variations on the processing steps required to produce such a device are possible, it is instructive to study this example to appreciate the value of the processing techniques presented in the following sections.

As we will see in more detail in Chapter 11, a waveguide electro-optic modulator employs an electric field to change the phase or intensity of an optical field propagating in a waveguide. In this processing example, the electric field will occur across a pn-junction and the waveguide will be formed by an etched ridge; that latter structure will be dissected in Chapter 8.

We begin with an n-type, silicon-doped GaAs wafer on which a set of waveguide layers has been grown by MBE or MOCVD (Figure 5.1a); the design of such waveguiding layers is the subject of subsequent chapters. Suffice it to say here that we might use 1 μm of n-type AlGaAs, 200 nm of undoped GaAs and a final 1 μm of p-type AlGaAs. Our pn-junction is then a thin 1-μm p-region near the surface with the remainder of the wafer being n-type; the center GaAs layer forms the actual waveguide core.

A layer of light-sensitive photoresist is deposited on the surface by spinning, and a pattern is exposed onto the wafer by means of a mask aligner and a mask. The pattern is developed, leaving a stripe that defines the waveguide (Figure 5.1b). In a wet chemical or dry plasma etch, the waveguiding layers are etched away everywhere except under this photoresist stripe, leaving an etched ridge in the surface (Figure 5.1c); this is our waveguide. The remaining photoresist is now stripped, the wafer is cleaned and prepared for the deposition of a thin (200-nm) layer of Si_3N_4

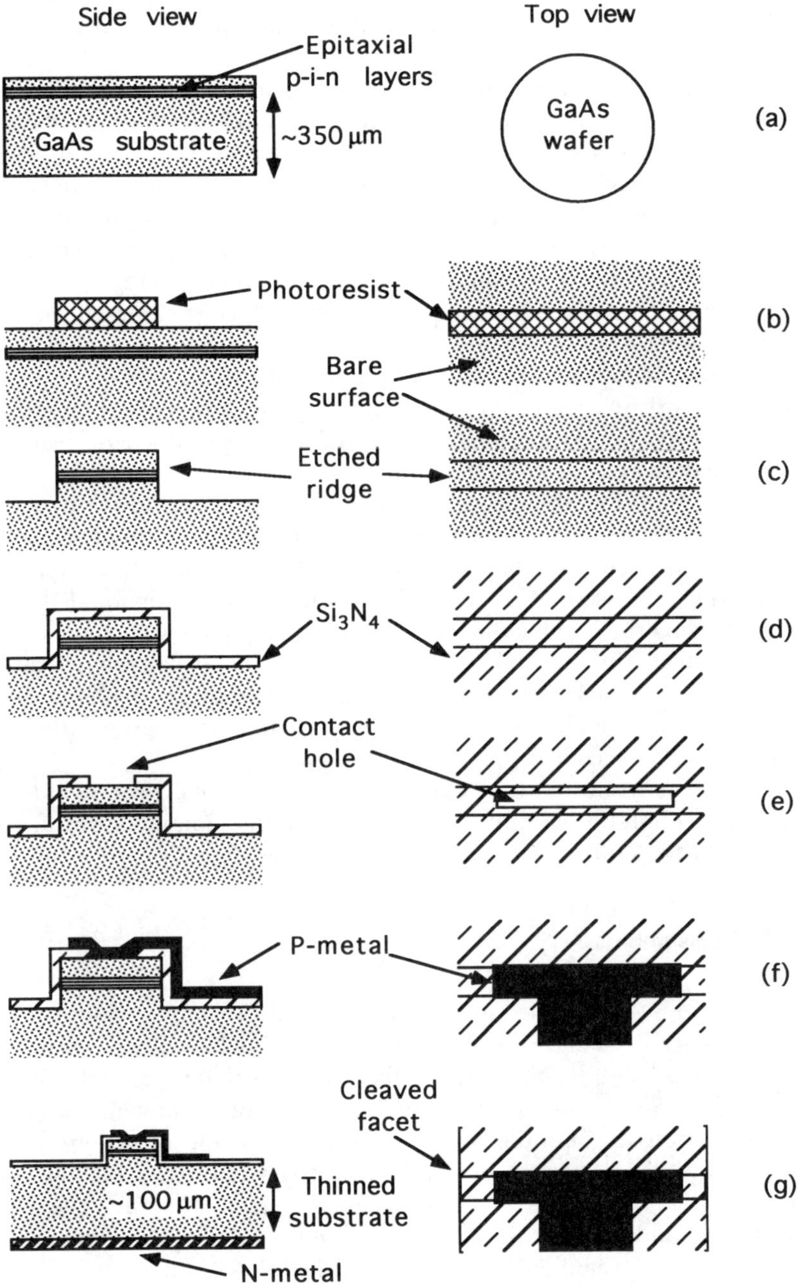

Figure 5.1 A schematic outline of a simple semiconductor process to make a waveguide-based electro-optic modulator in III-V semiconductors; both cross-sectional (left) and top (right) views are shown.

(Figure 5.1d). This nitride acts as an electrical insulator and protects the surface to some extent.

We now repeat the entire photolithography process, spinning on resist and exposing a new pattern, this time to define a contact hole in the Si_3N_4 layer. This step does not employ a stripe of photoresist after development, but rather its inverse, an open region in the resist over the waveguide. The wafer is placed into an etcher, and the nitride is removed only in the hole over the waveguide, leaving the bare GaAs surface (Figure 5.1e). After another photoresist strip and wafer clean, the entire photolithography sequence is repeated once more, this time leaving an open region in the photoresist where a metal contact pad is desired. A series of metalization layers, typically involving Ti, Pt, and Au for a p-type contact, is deposited over the entire surface, a step that is followed by a final stripping of the photoresist. In a process step known as lift-off, the dissolved resist takes any metal deposited on the surface with it, so that we are left with a metal pad only in the regions where the resist was initially developed away (Figure 5.1f).

The wafer is then often thinned to a thickness of roughly 100 μm by a mechanical lapping procedure to facilitate the subsequent cleaving into individual devices. A final metal layer is deposited on the backside of the thinned wafer, this n-type contact typically requiring the use of Ge, Ni, and Au. Finally, the individual devices are separated by cleaving the semiconductor, yielding a completed waveguide modulator as shown schematically in Figure 5.1g. A voltage applied between the p-type surface contact and the n-type backside contact generates a field in the depletion region of the pn-junction, and the etched ridge acts as a waveguide for light coupled into the structure through the cleaved facet.

This example uses most of the important steps in microfabrication at least once and some, such as photolithography, several times. In the remainder of this chapter, following a brief diversion, we discuss these processing steps in more detail.

5.1.2 The Embellishments

The rudimentary process we used as an example could certainly yield a functional modulator, but current semiconductor processing technology has a cornucopia of ingenuous variations, complications, and embellishments that permits the fabrication of more complex and more highly functional devices. For example, selective wet or dry etching is possible through the choice of certain etch-stop layers, grown in with the waveguide layers, and the proper etch chemistries [2]. After etching, one can regrow further layers by MBE or MOCVD on the patterned substrate; the use of Fe-doped InP, for example, allows one to grow a semi-insulating, crystalline isolation layer over an etched waveguide [3] in the InP material system.

Electrical isolation between devices, which we ignored altogether in our example, can be accomplished by the selective etching of doped layers or through implantation to produce locally defined regions of high resistance [4]. Ion implantation or spin-on glass dopants can also be employed to define doped and undoped regions independently of the initial epitaxial layer growth. The mirror facets for lasers, often formed by the final cleave step, can also be manufactured by dry etching [5], the same technology useful for fabricating angled waveguide mirrors [6] or lenses. Had a Si-based structure been fabricated, the deposition of dielectric waveguiding layers [7], or the use of implantation to define regions of higher refractive index, might have been used. The addition of electrical devices extends the horizons even further, involving the formation of gates, recesses, multilayer metals, and numerous other fabrication steps.

5.1.3 The Environment

Many of the demands on semiconductor processing derive from the sizes of the structures fabricated: typical dimensions are in the micrometer range and many types of structures are measured in tens of nanometers. On these length scales, the average dust particle is relatively large, such that cleanliness is one of the primary considerations. Fabrication takes place in cleanrooms, with extremely well-controlled temperature and humidity, plus filtered air handling with very low particle densities [8]. Cleanrooms are usually classified as Class N, where N represents the average number of particles with diameters greater than 0.5 μm in a cubic foot of air; class 1 to 10 is usually the minimum required.

Certain process steps, particularly photolithography, require even more stringent controls of environment: temperature control of 0.1°C is often necessary in the subenvironments established for this equipment. All surfaces must be free from dust, oils, and other contaminants. Extra attention must be paid to the removal of metals and sources of atomic species, which act as dopants in the materials being processed. Despite gloves, bunny suits, and face guards, the dirtiest components of any cleanroom are still the people in it.

Much of the fabrication and cleanroom technology was developed through the demands of silicon microelectronics [9]. Despite some essential differences, III-V and Si-based microfabrication techniques share numerous similarities, such that many of the discussions to follow are, to first order, material independent.

5.2 PHOTOLITHOGRAPHY

Lithography is, true to the Greek roots of its name, essentially writing on rocks. This is the method by which we create the patterns (the writing) on the wafer (the rock) from which the devices are formed; lithography defines where to etch, implant, leave metal, and so forth and represents a significant fraction of any semiconductor process.[2]

[2]See the very comprehensive: Moreau, W. M.., *Semiconductor Lithography: Principles, Practices and Materials*, New York: Plenum Press, 1988.

Most patterning is currently done by photolithography, implying that light is used for exposing the photoresist, but electron-beam lithography is important for very small critical dimensions. Lithography using focused ion beams and X-rays is an active research area.

As we glossed over briefly in our example process, *photolithography* is actually a collection of processing steps. It begins with a spin coating of the photoresist onto the wafer, which moves from the spinner to the mask aligner, where the mask pattern is aligned to previously fabricated structures on the wafer surface. Once aligned, the wafer is exposed, through the mask, and the photoresist is subsequently developed. The mask patterns in photoresist are left on the semiconductor surface. At this point, the actual processing step of interest takes place: an etch, an ion implantation, a metalization, etc. Finally, the photoresist is stripped, the wafer is cleaned, and processing continues. We now consider several aspects of photolithography in greater detail.

5.2.1 Masks

Electron or ion beams can be focused and directed in patterns onto the wafer by applied electric or magnetic fields; light cannot be controlled in this manner, so a mask is required to transfer a pattern onto the wafer. The mask is a patterned plate with transparent and opaque regions; a flood exposure by a light source transfers this pattern onto the photoresist coating the wafer. Structurally, the mask is usually a high-quality glass or quartz plate. One side is coated with a photo-sensitive emulsion or chrome and covered with photoresist. Chrome masks provide higher resolution than emulsion. Using a pattern generator or electron-beam writer, a pattern is written into this photo-sensitive layer, which is subsequently developed and transferred into the chrome.

Depending on the application and the nature of the lithography equipment used, the mask pattern is either actual size, denoted 1×, or 5× (for which the mask patterns are five times the final structure size, requiring reduction during exposure). In Figure 5.2, we have sketched the mask patterns that would have been required to fabricate our example modulator, assuming positive photoresist. Masks can be used repeatedly, although the lifetime of the glass plate for contact printers is lower than that for projection printing.

5.2.2 Photoresist

The primary task of photolithography is to transfer the pattern from the mask to the wafer. On the wafer surface, photoresist undertakes this function. Photoresist is a class of light-sensitive polymers, of which there are two basic kinds, positive and negative. Both kinds of resists have distinct advantages: the popularity of positive

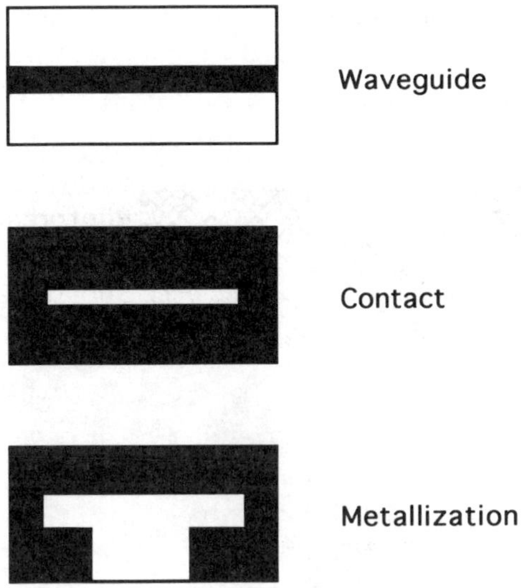

Figure 5.2 Mask patterns required to fabricate the example modulator structure of Figure 5.1; the polarity assumes positive photoresist.

resist is due to its higher resolution and its lack of swelling during development, which bedevils negative resists. Negative resists, on the other hand, are generally more sensitive and require less exposure time. The resist is uniformly coated on the wafer to a thickness of one to several microns. Coating takes place by spinning at a spin speed of 2,000–7,000 rpm for 30–60s, in an environment where cleanliness and dryness are important. Wafers are often subject to a prebake before resist spinning, for example, at 130°C for 30 minutes, to remove any residual moisture from the surface.

Exposure is done at the shortest optical wavelength feasible; the source is usually a Hg-arc lamp with the mask between the wafer and lamp. The mercury emission spectrum has several strong lines: g-line (436 nm) is now commonly used and i-line (365 nm) is considered state of the art. Depending on the intensity of the source, exposure time is in the range of one to several seconds; the total time-integrated power (energy) is the parameter of importance. The mechanics of exposure are revealed in the differing characteristics of positive and negative resist. For positive resist, exposure to light destroys the photosensitizers suspended in the base resin so that the dissolution rate during development increases. In negative resists, the activated photosensitizers promote a cross-linkage of the base polymer, so that it does not dissolve. As a result, the exposed parts of a positive resist are developed away and the unexposed parts of a negative resist are removed, as summarized in Figure 5.3.

Exposure is often followed by a postbake, before development, which improves uniformity and linewidth. The temperature and time required for this bake, as with

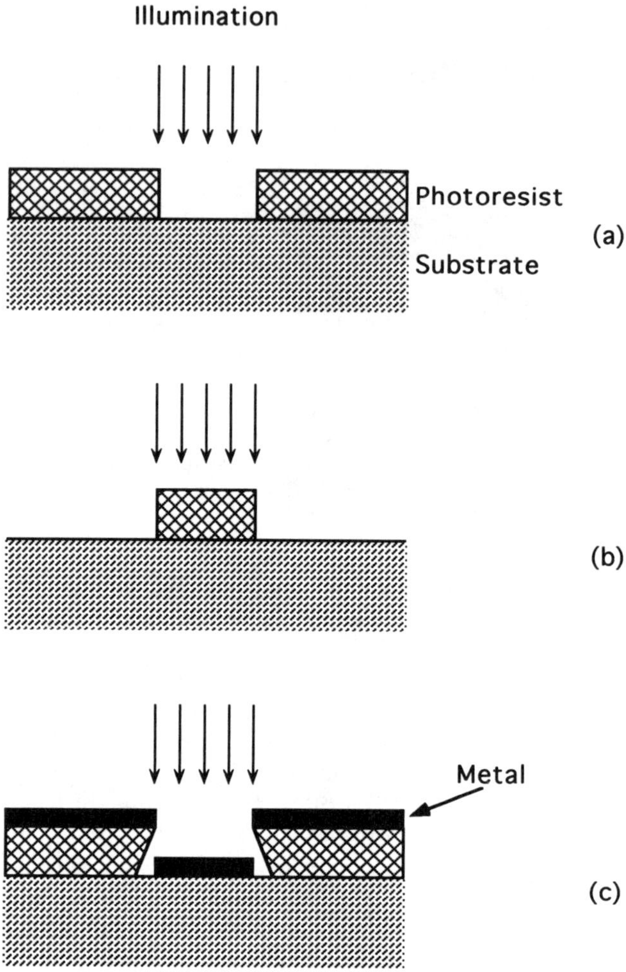

Figure 5.3 Positive (a) and negative (b) resists after exposure and development; the lift-off profile (c) is generated by a positive resist subject to a soak in chlorobenzene. A deposited metal would not deposit on the resist sidewall and could thus be stripped with the photoresist away from all unwanted areas.

most of the others, is a function of resist type. Following exposure and bake, the photoresist layers are developed in KOH-based solutions. Although the patterns are usually clean after development, the removal of photoresist residue from small structures is often necessary. This removal, known as descum, takes place in a low-power (50W) oxygen plasma. This can be followed by another bake, the hardbake, which toughens the resist in preparation for its following dunk into a chemical bath or plasma, a step that can also be used as a means to improve the resist sidewall smoothness by controlled reflow [10].

The wafer is now ready for the actual processing step for which the pattern had been generated, possibly an etch, a deposition, or implantation. The photoresist "resists" the action of the process so that covered regions of the wafer are protected from any effects. The waveguide region in Figure 5.1c, for example, is not attacked by the etchant. After that step, the resist must be stripped in preparation for another photolithography sequence. Stripping can be accomplished either in a solvent bath or in a high-power (300W) oxygen plasma. In both cases, the wafers are usually subject to a thorough cleaning after stripping.

5.2.3 Lift-Off and Image Reversal

A number of variations on the photoresist theme are available. A useful technique is the generation of a lift-off profile, as also shown in Figure 5.3c. This profile has a strong undercut, so that a material, often metal, deposited on the surface does not coat the resist sidewalls. Stripping the photoresist will concomitantly remove the metal on its surface, leaving only metal in the exposed regions; this technique is an easy way to pattern metals without resorting to a metal etch and is important for many III-V-based technologies.

The lift-off profile can be generated by soaking the positive photoresist in an aromatic solvent, such as chlorobenzene, for several minutes prior to exposure [11]. This hardens the outer layer of the resist so that upon development, the upper crust develops more slowly, giving the undercut shown in Figure 5.3c. Such a hardened upper layer can also be generated, to some degree, by doing a higher temperature postbake.

Image reversal is a further photolithography variation accomplished by the use of special photoresists that allow one to generate the negative image of the mask pattern. After initial exposure of the pattern, the wafer is subject to a flood exposure (no mask) and another bake; this converts a positive resist into a negative one. The profile generated also has an undercut, to an extent, but it is not as strong as the one obtainable by performing the lift-off process.

5.2.4 Mask Aligners

Exposure of the photoresist using a patterned mask requires a short-wavelength light source with a shutter, but also a means to align the pattern on the mask with that on the wafer, if one exists from a previous processing step. Thus, in the example of Figure 5.1, the waveguide could have been placed anywhere on the wafer surface, but the contact opening must be aligned with the etched waveguide and the tolerances must be sufficiently tight to ensure that the contact ends up in the correct place.

Useful for this procedure are alignment marks, patterns on the wafer whose only function is to facilitate this step; these often take the form of transparent and opaque crosshairs.

Alignment and the following exposure are done by means of mask aligners, of which two popular kinds are illustrated schematically in Figure 5.4. On the left is a contact printer with which a 1× mask is brought in direct contact with the photoresist-coated wafer. During alignment, the wafer and mask are separated by a gap of several microns, so that each may be moved independently of the other. Once aligned, the two are brought into contact for exposure.

This form of alignment and exposure has the advantage of being simple, being relatively inexpensive, and providing good resolution, and contact printers are still in use for many research and some industrial applications. On the other hand, both wafer and mask surfaces can be damaged by their physical contact, reducing the lifetime of the masks. In addition, the depth of focus is limited, implying that exposure over complex surface topology, such as the deeply etched waveguide of our example, is subject to reduced resolution.

Projection printing, in the form of the wafer stepper illustrated in the right of Figure 5.4, overcomes some of these limitations, albeit at the cost of increased complexity, price, and demands on machine subenvironment. The stepper employs a 5× mask and reduces the pattern to the correct size on the wafer by high-quality optics. The masks benefit in two regards: lack of contact with the wafer surface implies an

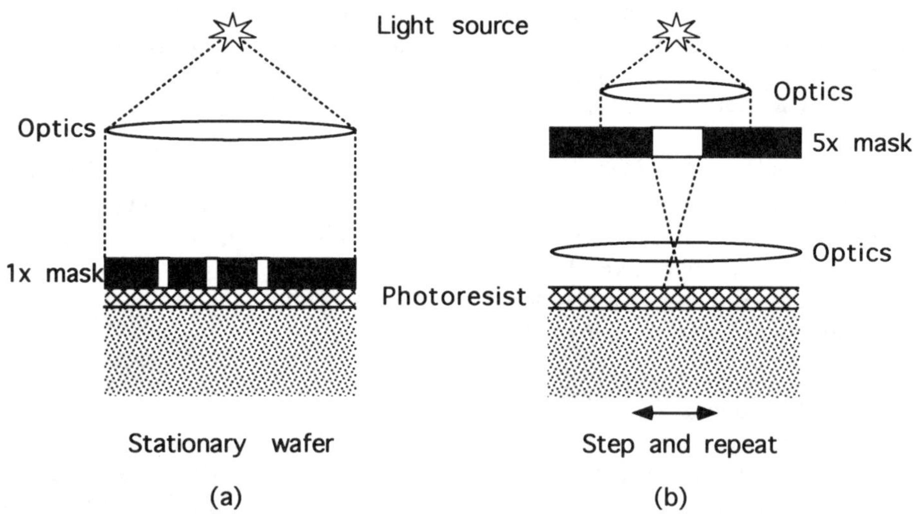

Figure 5.4 Schematic views of (a) a contact printer and (b) a wafer stepper. The former uses a 1× mask in direct contact with the wafer, and the latter uses a 5× mask projected onto the surface through an optical system.

indefinitely long lifetime, and the mask pattern generation is easier since structures can be written five times the required size. High-quality optics results in an improvement of the depth of focus with respect to that achievable by contact printing.

In practice, both the mask and the wafer are aligned to the wafer stepper and thus to each other; the mask pattern is *stepped* over the wafer surface by repeated wafer positioning and exposure, all performed automatically. In this way, the mask need only have the pattern for a single group of devices, which is repeated in an array across the wafer.

5.2.5 Nonoptical Lithography

The resolution of photolithography is limited by the wavelength of light; thus the definition of patterns below 0.25 μm becomes difficult using purely optical means. One way around this limitation is through the use of particles with shorter wavelengths than photons—electrons, ions, and x-rays. Electron-beam lithography, referred to as e-beam, is the most widely applied of the three; it is used for direct writing of structures into polymethylmethacrylate (PMMA), a positive resist. Direct writing implies that a mask is not used, but that the electron beam is deflected in the plane of the wafer to write a pattern. This inherently serial process severely limits throughput so that an e-beam lithography step is usually reserved only for those layers that require the smallest critical dimensions. E-beam is also commonly used for the fabrication of normal optical masks, as described earlier, which are then used in conventional optical systems.

By increasing the mass of the incident particles further, the wavelength decreases yet again. Focused ion beam (FIB) systems [12] use Ga or In atoms to expose a resist. The resolution is also enhanced with respect to e-beam due to the reduced scattering of heavy ions. FIB is actually a more flexible technique since, depending on the energy and conditions, it can be used locally not only for exposure but also to implant ions below the surface or mill holes into it. X-rays, with a wavelength of 1 nm, would allow the return to a parallel exposure concept because the wafer would be flood exposed through a mask rather than directly written as with e-beam or FIB. A synchrotron source for x-rays is usually required and fabricating masks is a challenge; Au or W on Si or SiC have been used successfully.

5.3 ETCHING

Etch technology is required to remove material from the wafer surface, and most processes contain several different etch steps. Our sample process of Figure 5.1 used two: one to generate the waveguide ridge and a second to open a contact hole in the nitride. Etch technology may be divided into two categories, dry and wet, each with its own set of advantages and limitations. The former uses a plasma to etch the surface and is typically a balance between chemical and physical etch components. The latter is always chemical and involves immersing the wafers into a liquid bath.

Wet chemical etches are often crystallographic, implying that the etch stops on a particular crystal plane. They are also generally isotropic, meaning the etch rate is roughly the same vertically and horizontally, as shown in Figure 5.5. In general, dry etching allows greater control over the etch shape; proper selection of dry etch parameters can yield anisotropic and vertical profiles, often desired for many applications, but profile control often involves a tricky manipulation of many etch parameters. The extensive research and development work involved in the establishment of etch processes is warranted by the broad spectrum of application in optical and optoelectronic devices [13].

5.3.1 Dry Etching Basics and Machinery

Dry etching of semiconductors or dielectrics is a process that can be well controlled with respect to etch depth and profile shape. It is based on a plasma that is formed by an ac discharge in a reactive or neutral gas at low pressures; depending on the etch machinery and parameters, etching will have physical and chemical components. The physical aspect of dry etching consists of ions from the plasma that bombard the wafer surface and remove material by sputtering it away; this can employ a

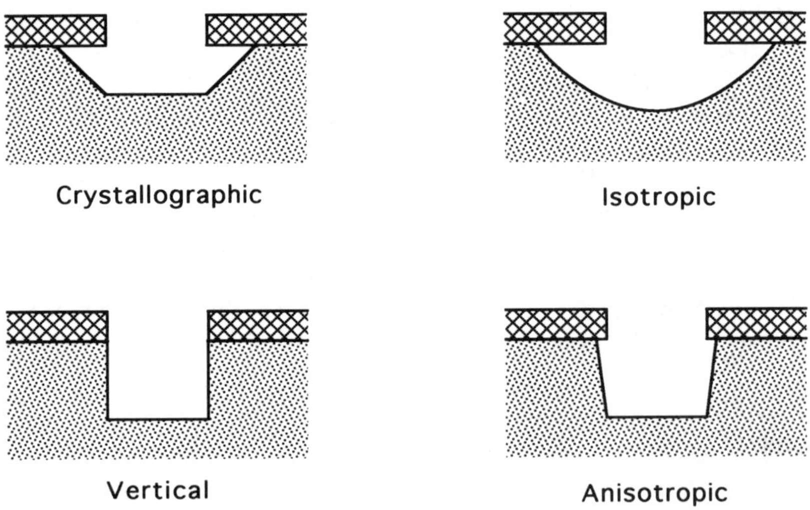

Figure 5.5 A selection of etch profiles. A crystallographic etch stops on a crystal plane and a vertical profile is usually generated by a predominantly physical etch. Isotropic profiles are due to equal horizontal and vertical etch components, whereas anisotropy is usually generated by a physical etch mechanism.

neutral species such as Ar. The chemical aspect is less drastic: reactive species from the plasma form compounds with atoms from the surface, and these either leave the surface of their own accord or are helped along by a bit of bombardment from the plasma. The process is frequently a combination of both mechanisms.

In Table 5.1 we have a list of four broad dry etching categories, which we will discuss further. The techniques are distinguished not only by the configurations of the machinery, but also by the pressures and the dominant etch component. Plasma etching, for example, often uses the popular barrel configuration of Figure 5.6: the wafers sit on a susceptor that floats electrically and the RF power is applied to one

Table 5.1
A Summary of Dry Etch Techniques, Outlining Some of Their Distinguishing Characteristics

Category	*Reactor*	*Etch Type*	*Pressures*	*Profiles*
Plasma	Barrel/parallel plate	Chemical	> 10 Pa	Isotropic
RIE	Parallel plate	Chemical/physical	10 – 100 Pa	Variable
CAIBE/RIBE	Beam	Chemical/physical	0.01 Pa	Anisotropic
IBM/FIB	Beam	Physical	< 0.01 Pa	Anisotropic

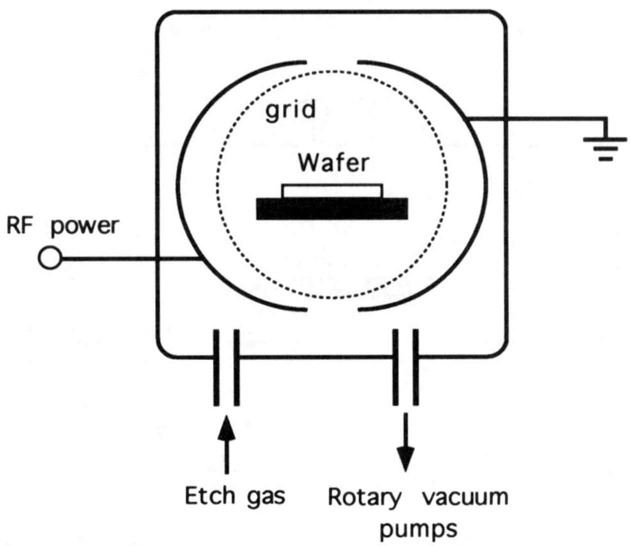

Figure 5.6 Schematic view of a barrel plasma etcher. The wafers are at a floating potential, and the plasma is struck by a discharge between the two semicylindrical electrodes. Etching is chemical and isotropic.

of the electrodes in the chamber. The etch chamber is essentially a large cylinder, thus the label barrel. Pressures are relatively high (10 Pa or more), and etching is purely chemical. The plasma serves to generate a reactive species, but the configuration does not give rise to any substrate bombardment; hence the name "plasma" etching, even though some of the more physical dry etches also make use of a plasma. Barrel ashers, which are barrel etchers using O_2, are popular for photoresist stripping (ashing) and descum.

A parallel-plate configuration can also be used for plasma etching, the sample either sitting on an insulated holder or on the grounded electrode. By placing the wafer on the powered electrode of such an arrangement, as shown in Figure 5.7, we move to reactive ion etching (RIE). In this case, the physical etch component is enhanced when compared to a plasma etcher, as most of the electric field that accelerates the ions in the plasma is at the cathode on which the wafers are placed. This field results in ion directionality at the wafer surface, meaning that the etchant species in the plasma has significant kinetic energy and can contribute to physical removal of the substrate material, particularly at the pressures employed, 10–100 Pa. Etch species are typically reactive gasses, mostly Cl- or F-based. The balance between chemical and physical components is a function of several etch parameters, including pressure, RF power, flow rate of the gasses entering the chamber, and RF frequency.[3]

By removing the reactive species from the plasma altogether and ionizing a neutral gas, such as Ar, we can move to ion beam milling (IBM), which is a completely

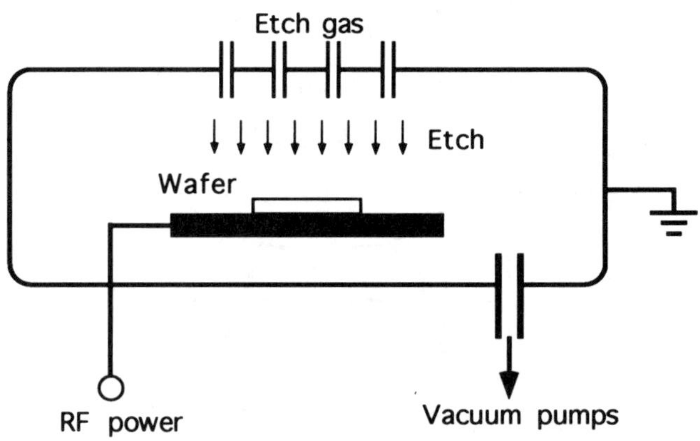

Figure 5.7 In reactive ion etching (RIE), the wafer sits in the powered cathode and is thus subject to ion bombardment from the reactive species in the plasma. The etch profile can be adjusted by varying numerous etch parameters.

[3]See, for example: Sze, S. M.. (ed.), *VLSI Technology*, New York: McGraw Hill, 1983, Chapter 8.5.

physical etch. As is illustrated in Figure 5.8, the plasma, generated by a dc discharge, has ions extracted from it by a biased grid. These ions, accelerated by a bias of 500V or more, are launched through the low-pressure ambient, 0.01 Pa or less, and impale themselves on the wafer surface. The transfer of the kinetic energy from the beam ions to atoms on the surface results in the latters' physical removal, hence etching. This etch is a purely physical milling process, implying that etch rates are slow and that selectivity between photoresist and surface to be etched is poor to nonexistent, but that the profiles are anisotropic and vertical due to the collimated nature of the beam.

If one uses a reactive gas species, such as Cl, with an IBM configuration, a combination of strong physical and chemical etch components can be achieved. This is known as reactive ion beam etching and allows for greater control of the etch profiles than pure IBM.

There are, as the reader has no doubt come to expect, many variations on these themes, some quite popular in research and industry. Electron cyclotron resonance (ECR) plasmas generate the plasma by the use of the cyclotron resonance of the electrons in the plasma [14]. This permits a separate control of ionization density at low pressures and temperatures. It is also attractive for optical and electronic applications [15] due to the low voltage drops and hence reduced substrate damage of the

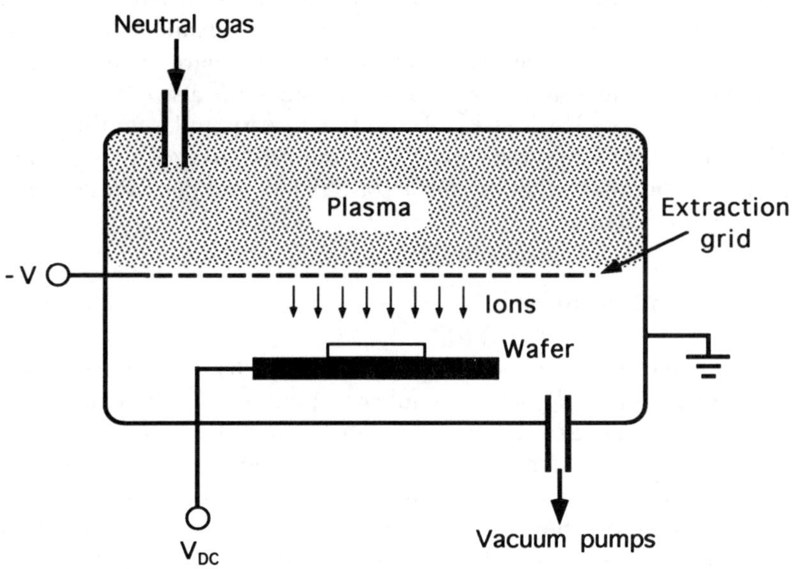

Figure 5.8 Ion beam milling (IBM) uses a neutral plasma whose ions are accelerated by a high voltage and thus etches the wafer surface by a purely physical sputtering mechanism. If a reactive species, such as Cl, is used in this configuration, it is known as reactive ion beam etching (RIBE).

etch. Combining a neutral ion beam, usually Ar, with Cl fed into the plasma near the wafer surface, gives rise to chemically assisted ion beam etching (CAIBE), which has been used to generate very smooth, vertical sidewalls, independent of material composition for the GaAs/AlGaAs system. CAIBE is very attractive for the dry etching of laser facets and has been used successfully in that area [16]. A further variation, magnetic field enhanced reactive ion etching (MERIE), often erroneously referred to as magnetron etching, uses permanent magnets in the etch chamber in order to confine the plasma and enhance ionization without increasing the bias voltages. This is a lucrative alternative for a low-damage etch process [17].

5.3.2 Etch Gasses

The mechanics of dry etching, as we have seen, are a strong function of many machine and process parameters; the different configurations we discussed earlier play as much a role as the gas pressures, applied powers, flow rates, and bias voltages. With the exception of purely physical etches, such as IBM, there is still chemistry involved in dry etching and the reactive gasses employed are a function of the substrate to be etched.

The palette of potential etch gasses is large [18]. For Si, the etchants are Cl- or F-based, with CF_4 and CCl_4 commonly used. GaAs is etched by Cl-containing gasses, such as $SiCl_4$, CCl_4, or CCl_2F_2; the latter has the advantage of etching GaAs selectively over AlGaAs, but the disadvantage of being environmentally unfriendly and is thus no longer in use. Fluorine-based plasmas are used for the etching of dielectrics, such as SiO_2 and Si_3N_4; CF_4, SF_6, and CHF_3 are popular for this application. Finally, for wafer cleaning and stripping of photoresist, oxygen plasmas are ideal, as oxidation of contamination on the surface is efficient and does not give rise to any measurable substrate damage.

5.3.3 Problems and Limitations

Despite the attractiveness and widespread application of dry etching in semiconductor processing, the techniques do have a number of limitations. The issue of substrate damage is an ornery one, as it is often difficult to quantify, and it is the subject of ongoing research.[4] Whereas chemical etches are considered benign, etches with a physical component bombard the surface with a heavy ion, propelling some of the substrate atoms into space but also giving rise to amorphization by displacement of atoms that remain in the crystal. This effect is compounded by implantation of the etchant species and the generation of traps, often by electrons from the plasma. All

[4]For example: Zappe, H. P., "Electronic Study of Plasma-Induced Damage in GaAs Heterostructures," *Microelectronic Engineering*, Vol. 20, 1993, pp. 227–239.

three of these phenomena can have undesirable electrical and optical effects, such as enhanced leakage currents or increased optical losses. For optical applications, degradation due to etch processes must be understood and minimized—often a difficult task because the electrical and optical regimes respond differently to many types of damage.

The redeposition of material during etching is often unavoidable. Redeposited layers can take the form of polymers, fluoro-, or chlorocarbons, either deposited directly from the plasma or formed as a product of etched material and gas-phase components. Etched substrate material can also redeposit directly, which often gives rise to oddly-shaped semiconductor profiles. The crux of the redeposition issue is that most redeposited material cannot be removed, so surface morphology and etch profiles are often drastically affected. Particularly sensitive to this problem are contact holes and recesses subject to further etch processing.

A final issue is one of selectivity: we would like a high etch rate of the semiconductor while not etching the mask at all. Photoresist, which defines the etched regions on a substrate, is usually not attacked by wet chemical etchants. For dry etching, the physical etch component will erode the mask in addition to the wafer, so that selectivity between substrate and mask is poor and etch depths are sometimes limited. In addition, it is often useful to have an etch stop, a layer in the substrate through which the etchant does not proceed (AlGaAs acts as an etch stop in CCl_2F_2, as mentioned earlier). A high selectivity between the layers to be etched and the etch stop is thus required. This action is more difficult to obtain in dry rather than wet etching, since many dry etches happily etch at the same rate through all layers. An etch stop is often useful when an exact etch depth is crucial, and a wet etch may be advantageous for that type of requirement.

5.3.4 Wet Etching

The poorer control of profile and etch depth, the virtually unavoidable mask undercut, and the generally strong dependence on substrate composition and orientation may reflect disfavorably on wet etching for structuring semiconductors. However, one can take advantage of the improved selectivity, the etching along certain crystalline planes, and the lack of substrate damage when applying wet etches.

In general, chemical etching of GaAs is more complex than Si [19]. The etch process in III-Vs is generally an oxidation/dissolution one, meaning an oxide of the surface is first formed and then dissolved. Numerous etchants are popular; NH_4OH:H_2O_2:H_2O gives an undercut profile that is a function of AlGaAs aluminum content. A class of diffusion-limited etches, including H_2SO_4:H_2O_2:H_2O and HCl:H_2O_2:H_2O, have been shown to produce smooth, isotropic etch sidewalls that are useful for the wet-etch fabrication of low-loss waveguides [20]. GaAs and AlGaAs can both be etched with H_3PO_4:H_2O_2:H_2O, the two materials exhibiting the same, albeit highly temperature-dependent, rate; this is useful for uniform etching through various layers of differing composition. Citric acid with hydrogen peroxide, $C_6H_8O_7$:H_2O_2 etches GaAs selectively over $Al_xGa_{1-x}As$ for $x \geq 0.45$ [21]; this is a useful etch, for example, to selectively remove doped surface GaAs cap layers. Finally, the HF-based AB etch removes AlGaAs faster than GaAs and is often employed for staining cleaved layers in preparation for examination by SEM.

When using crystallographic etchants in GaAs, the crystalline orientation of the substrate becomes important. As we see in Figure 5.9, for a (100) surface, etching a waveguide in the $[\bar{1}10]$ direction yields a different profile than etching in the [110] direction, even though these directions are of the same <110> family. This effect arises because chemical etching is slowest on the (111a) planes, the Ga-rich planes, and these are oriented differently if we change directions by 90°.

The etching of silicon by wet chemical means has a long history [22]. The impetus for much of that work was the fabrication of new kinds of transistors and their electrical isolation using anisotropically etched trenches. Anisotropic Si etches, such as KOH:H_2O, stop on the (111) surface, so that V-grooves rather than vertical trenches result when a normal (100) substrate is used, as shown in Figure 5.9b. The

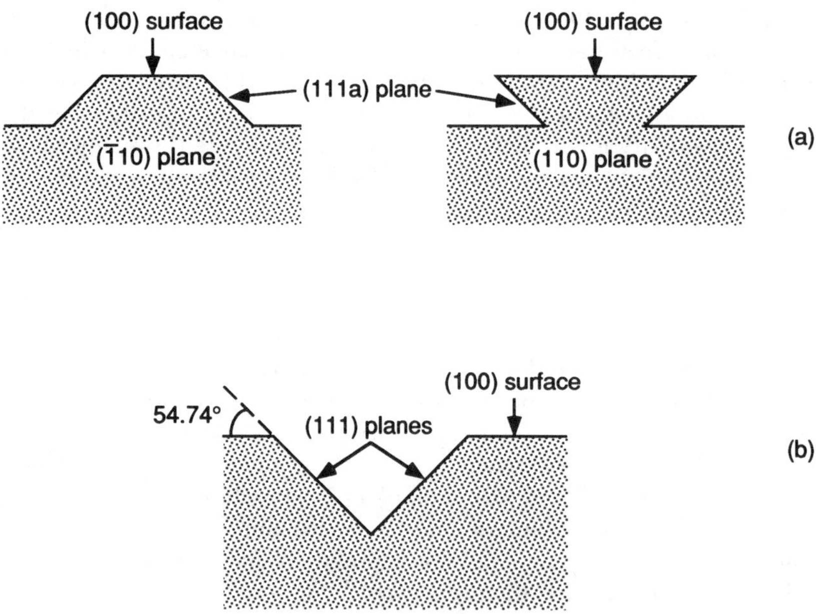

Figure 5.9 (a) Side profile of chemically etched GaAs with a (100) surface: the $[\bar{1}10]$ direction yields a sloped profile while the [110] direction generates an undercut, since etching stops on the Ga-rich (111a) planes. (b) A V-groove etched by anisotropic means into (100) Si. The angle of the groove is constant (54.74°) and defined by the (111) planes on which etching stops.

crystal geometry, namely the angle between the (100) and (111) planes (see Figure 2.4), defines the 54.74° angle, which is characteristic of all Si V-grooves. A vertical wet etch in silicon can only be achieved if a (111) surface is employed, which is undesirable for most electrical applications. $HF:HNO_3:CH_3OOH$ in the ratio 8:75:17, on the other hand, is an isotropic etch for silicon.

Dielectrics and metals are often stripped with wet etchants, more in the Si than in the GaAs world. SiO_2 is removed by HF or buffered (with $NH_4F:H_2O$) HF (BHF); the buffering maintains a constant oxide etch rate with time and etchant depletion. Since Si is not attacked by HF at all, this is a lucrative means to strip SiO_2 selectively from a silicon surface. The removal of Si_3N_4 by wet means is a bit trickier: BHF etches nitride slowly (less than 1 nm/min) and hot H_3PO_4 is often a more useful etch. In Si technology, the Al used for metalization can be patterned by HCl or $H_3PO_4:HNO_3:H_2O$. Since all of these chemicals etch GaAs, dielectrics and metals are not commonly removed by wet means for this material system; nitride is usually dry etched and metal is typically patterned by lift-off.

5.4 DIELECTRICS

In our sample process, a Si_3N_4 dielectric layer was used for insulating the semiconductor surface in preparation for metalization. Both the dielectric and the metal are thin films that can be deposited by a number of techniques. Chemical vapor deposition (CVD), of which MOCVD is the most refined example, is a high-temperature technique that yields layers with the best stoichiometry and structure. Plasma deposition, often termed plasma-enhanced CVD or PECVD, employs similar machinery as dry etching and is essentially a low-temperature process. Finally, sputtering or e-beam evaporation are popular means to deposit metals or dielectrics, as we will see in Section 5.5.

5.4.1 Plasma Deposition

The same types of barrel or parallel-plate reactors we examined for dry etching are used for dielectric layer deposition by PECVD. The substrates typically float electrically and the temperatures are usually below 300°C, despite a high effective ion temperature, an important consideration for GaAs. This advantage is offset by a relatively low throughput, high density of particle contamination, and films that exhibit limited step coverage.

The dielectric film most useful for many III-V applications is Si_3N_4, which is deposited by PECVD using SiH_4 and NH_3 as precursors, both diluted by an inert carrier gas such as Ar or He. These films always have a high hydrogen content, up to 25 atomic percent, and typically have a high stress and difficult stoichiometry control; the latter affects both the film refractive index and etch rate.

Silicon dioxide can also be deposited by using SiH_4 but with an oxygenic ambient, generated either by adding N_2O, CO_2 or plain O_2. Oxygen will react more readily than nitrogen with SiH_4, so that SiO_2 formation dominates over Si_3N_4, but the control is tricky. A mixed film, silicon oxynitride (SiO_xN_y), is thus often used, since by proper choice of precursors and deposition conditions the films can be made stress-free and of repeatable stoichiometry.

Plasma deposition is useful for a wider palette of materials, including SiC, Ta_2O_5, and TiO_2. The latter two have numerous applications in integrated optical structures such as waveguiding films.

5.4.2 Chemical Vapor Deposition

Deposition of dielectrics, metals, and semiconductors from the vapor phase, chemical vapor deposition (CVD), is a widely applied process in microelectronics, predominantly in silicon technology [23]. CVD usually employs a temperature range of 700–950°C, making it of limited to no use for III-V materials, since these degrade rapidly at high temperatures. The gas-phase precursors are often the same as those for PECVD: SiH_4, O_2, and NH_3 for silicon-dioxide, nitride, and oxynitride. Using N_2O as an oxygen source permits oxide deposition at relatively low temperatures, on the order of 450°C, somewhat overcoming the high-temperature disadvantage.

The large quartz, hot-wall furnaces normally employed for CVD, as shown in the sketch of Figure 5.10, permit the batch processing of many substrates, so wafer throughput is high. Operation at low pressures of 10 to 100 Pa also implies low particulate contamination; this pressure regime sometimes earns the designation low-pressure CVD or LPCVD. The gas precursors flow through the long furnace tubes

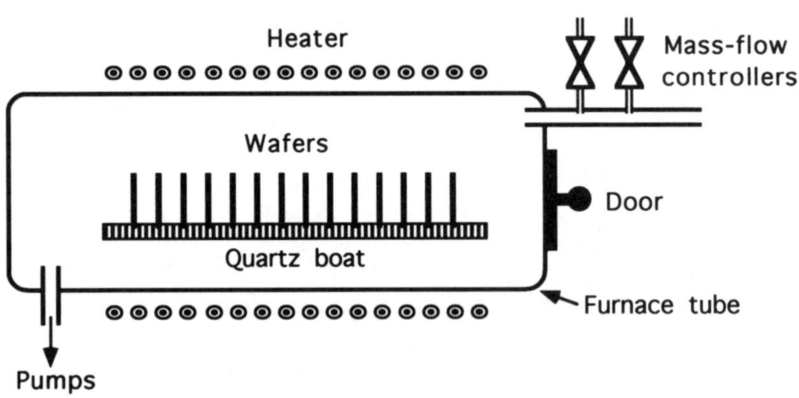

Figure 5.10 Schematic cross-section of a typical low-pressure chemical vapor deposition system. The wafers stand upright in the quartz boat, the gas flow patterns are made uniform by a system of baffles in the reactor (not shown).

and deposit layers with good uniformity, well-controlled composition, and conformal step coverage. The quality of CVD-deposited silicon oxynitride films is high and suitable for optical applications [24], often using silicon as a substrate material.

Silicon technology benefits greatly from numerous CVD processes, including the deposition of polysilicon, a material important for electrical or micromechanical devices but less interesting for optical applications. Si also has a great advantage over other popular semiconducting materials in that it easily grows a native oxide: oxidation of Si is a well-controlled and understood process, taking place in either wet or dry oxygenic ambients at temperatures of 850–1150°C. There is unfortunately no equivalent for the III-V materials, so insulating dielectric layers must usually be deposited by plasma means.

5.5 METALS

One of the final steps in the fabrication of an optical device that requires an electrical contact is the deposition of metalization layers. Our modulator example used a patterned metal pad to contact the surface of the pn junction and a second metalization on the wafer backside. Metals are usually deposited on the semiconductor or dielectric surface by evaporation or sputtering, where both methods effectively generate a stream of metal atoms that coats the wafers mounted in a vacuum chamber.

Evaporation requires a heating of the metallic source either by direct thermal means, using an inductive or resistive heater, or by an electron beam as is illustrated in Figure 5.11. The wafers are placed at some distance from the metal sources on a planetary drive, which rotates the substrates to optimize the coverage of nonplanar surface structures. Resistive heating melts a solid piece of metal on a coil, resulting in high-temperature atoms impinging on the substrate but yielding very conformal profiles; contamination from the coil itself can be a problem, however. The e-beam source uses the electron beam to melt the metal, or a metallic compound, in a crucible; the metallic atoms are again evaporated onto the semiconductor substrate. The electron beam generates x-rays, which can lead to damage in the semiconductor substrate.

Sputtering is a technique that uses the physical energy from neutral atoms in a plasma, usually Ar, to evaporate atoms from a metal target, which then deposit on the wafer surface. This method does not require high temperatures, thus reducing the amount of substrate or photoresist heating. However, it is typically nonconformal, meaning that the sidewalls of steep surface structures are poorly covered.

The metals employed for semiconductor metalization vary somewhat according to application and the nature of the surface to be contacted [25]. Aluminum is the dominant material for the silicon world, although refractory metal silicides ($MoSi_2$, WSi_2) are also useful for VLSI applications. It is relatively easy to ensure that contact to a Si surface is ohmic, meaning that the resistance is low and, more importantly, identical for both polarities of an applied bias; problems only arise if p-type surfaces

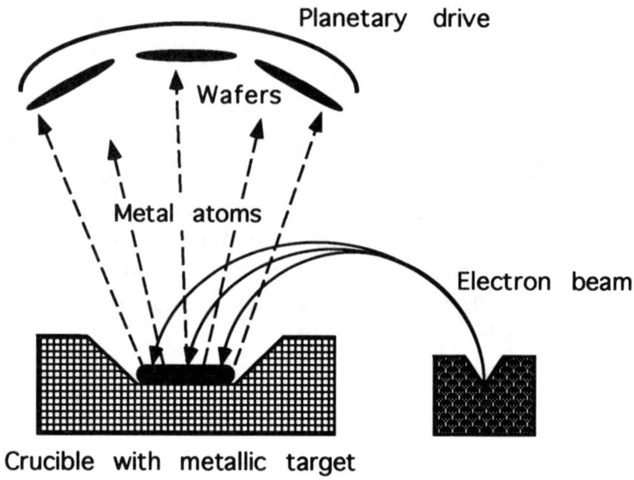

Figure 5.11 An electron beam heater for metal evaporation. The beam is directed onto a metal target by a magnetic field, from which the metal atoms evaporate and deposit on the wafer surface. The wafers are mounted on a planetary drive that rotates the wafers.

are too lightly doped. GaAs is a trickier system, since Schottky barriers tend to form when the doping is under 10^{19} cm^{-3}; as a consequence multiple metal layers are usually required to *auto dope* the surface. For n-type GaAs, a sandwich of GeAuNi is usually employed, the Ge for doping the surface, Ni for adhesion, and Au for low sheet resistance. P-type GaAs surfaces are well contacted by TiPtAu, AgZn, or AuZn employing similar contact mechanisms.

After deposition, low contact resistance is only achieved, for most metals, after a high-temperature anneal. This anneal also promotes adhesion and auto doping in III-Vs. Anneal temperatures vary in the range 380–450°C and use either a neutral (Ar or N_2) or forming-gas (N_2:10% H_2) ambient to reduce the contact resistance and ensure ohmic behavior.

5.6 PROCESSING VARIA

We have discussed a few central aspects of semiconductor processing, albeit concentrating on the most important ones for semiconductor integrated optical applications. Many, particularly electronic, applications employ additional fabrication techniques with which a certain familiarity is useful. In addition, III-V technologies often require some post-processing, much of which is done outside the cleanroom.

5.6.1 III-V Post-Processing

Many semiconductor processes end with some sort of slicing or dicing to separate the individual devices or chips into units suitable for testing or packaging. Much electronic testing can be done at the wafer level, however, so dicing the wafer is often not required for measurement. For optical structures, particularly lasers, exceptionally precise cleaving of the individual devices is necessary because the cleave forms the laser mirror for most types of semiconductor Fabry-Perot lasers; exceptions are those with a dry-etched facet.

Cleaving is nothing more than a careful and controlled breaking of the III-V substrate at precisely defined points; the resultant cavities are typically 200 to 500 μm long, and these lengths must be reproducibly attained. A rule of thumb states that a semiconductor bar can be easily cleaved to a length longer than roughly three times the substrate thickness, so that the completely processed wafers are typically thinned to 100 μm or less. Thinning is a dirty, chemo/mechanical process: the wafers are ground down in a SiC slurry using an apparatus that ensures a uniform and reproducible substrate final thickness. An added fringe benefit for the thinner substrate is that the thermal contact to a device on the surface is improved, especially since GaAs has a relatively (compared to Si) low thermal conductivity.

The complex metal systems employed for III-Vs usually require an additional surface layer to produce contact pads robust enough to withstand bonding. A surface layer of Au is often deposited by electro-plating, to a thickness of 5 μm or more. Plated gold also permits the fabrication of air bridges, metalization structures that can surmount large crevasses or ridges in surface morphology and reduces the sheet resistance of the metal lines.

5.6.2 Ion Implantation

The field of ion implantation is a very large one, but it is less of a process cornerstone for integrated optics than it is for electronics. Implantation essentially involves shooting atoms into a semiconductor at a precisely controlled energy (10 keV to MeV), thus depositing a precisely defined atomic dose (measured in atoms/cm^2) at a precisely controlled depth. Implantation is essential for silicon microelectronics, as it is the preferred technique for generating the complex doping profiles required for many electronic devices. For optoelectronic applications, implantation of H^+ or O^+ is often employed for device isolation or definition of optically active areas.

The implanted atoms are distributed in a Gaussian profile in the substrate, the depth and width depending on the type of ion and its energy. Typical depths can vary from under a nanometer to several microns. The implanted area is defined by a photoresist mask, which must be thick enough to absorb all implanted species in the covered areas; these depths and thicknesses are usually determined by extensive numerical modeling.[5]

5.6.3 Rapid Thermal Annealing

Many of the high-temperature process steps, which are found more in Si technology than in GaAs, are being supplanted by rapid thermal annealing (RTA) or rapid thermal processing (RTP) [26]. These methods involve high temperatures (up to 1200°C) but very rapid temperature ramps (100°C/s) and short times (seconds to several minutes). Although many process steps, in the current move to single-wafer processing, are being replaced by their RTP equivalent, these techniques have been particularly widely applied in the area of annealing—contact, damage removal, and activation anneals are now routinely performed by RTA. Despite the sensitivity of GaAs and related compounds to high temperatures, the choice of proper ambients and conditions allow high-temperature RTA to remain a viable processing technique for III-Vs.

REFERENCES

[1] Brillat-Savarin, *Physiologie de Goût*, Paris: Flammarion, 1982.

[2] Tong, M., D. G. Ballegeer, A. Ketterson, E. J. Roan, K. Y. Cheng, and I. Adesida, "A Comparative Study of Wet and Dry Selective Etching Processes for GaAs/AlGaAs/InGaAs Pseudomorphic MODFETs," *Journal of Electronic Materials*, Vol. 21, 1992, pp. 9–15.

[3] Zucker, J. E., K. L. Jones, B. I. Miller, M. G. Young, U. Koren, B. Tell, and K. Brown-Goebler, "Interferometric Quantum Well Modulators with Gain," *IEEE Journal of Lightwave Technology*, Vol. 10, 1992, pp. 924–932.

[4] Pearton, S. J., F. Ren, S. N. G. Chu, C. R. Abernathy, W. S. Hobson, and R. G. Elliman, "Defects and Ion Redistribution in Implant-Isolated GaAs-Based Device Structures," *Journal of Applied Physics*, Vol. 74, 1993, pp. 6,580–6,586.

[5] Ou, S. S., J. J. Yang, and M. Jansen, "5 W GaAs/GaAlAs Laser Diodes with a Reactive Ion Etched Facet," *Applied Physics Letters*, Vol. 57, 1990, pp. 1,861–1,863.

[6] Kim, J-H., A. Larsson, and L. P. Lee, "Pseudomorphic $In_yGa_{1-y}As/GaAs/Al_xGa_{1-x}As$ Single Quantum Well Surface-Emitting Lasers with Integrated 45° Beam Deflectors," *Applied Physics Letters*, Vol. 58, 1991, pp. 7–9.

[7] Boyd, J. T., R. W. Wu, D. E. Zelmon, A. Naumaan, and H. A. Timlin, "Guided Wave Optical Structures Utilizing Silicon," *Optical Engineering*, Vol. 24, 1985, pp. 230–234.

[8] Tolliver, T. L. (ed.), *Handbook of Contamination Control in Microelectronics*, Park Ridge, NJ: Noyes, 1988.

[9] Colclaser, R. A., *Microelectronics: Processing and Device Design*, New York: John Wiley & Sons, 1980; Hamilton, D. J., and W. G. Howard, *Basic Integrated Circuit Engineering*, New York: McGraw Hill, 1975.

[10] Arnot, H. E. G., H. P. Zappe, J. E. Epler, B. Graf, R. Widmer, and H. W. Lehmann, "Extremely Smooth Sidewalls for GaAs/AlGaAs Ridge Waveguides," *Electronics Letters*, Vol. 29, 1993, pp. 1,131–1,133.

[5] A description of the classic TRIM simulation is found in: Ziegler, J. F., J.P. Biersack, and U. Littmark, *The Stopping and Range of Ions in Solids*, New York: Pergamon, 1985.

[11] Hatzakis, M., B. J. Canavello, and J. M. Shaw, "Single-Step Optical Lift-Off Process," *IBM Journal of Research and Development*, Vol. 24, 1980, pp. 452–460.
[12] Gamo, K., "Focused Ion Beam Technology," *Semiconductor Science and Technology*, Vol. 8, 1993, pp. 1,118–1,123.
[13] Deimel, P. P., "Micromachining Processes and Structures in Micro-Optics and Optoelectronics," *Journal of Micromechanics and Microengineering*, Vol. 1, 1991, pp. 199–222.
[14] Asmussen, J., "Electron Cyclotron Resonance Microwave Discharges for Etching and Thin-Film Deposition," *Journal of Vacuum Science and Technology*, Vol. A7, 1989, pp. 883–893.
[15] Pearton, S. J. and A. Katz, "Dry Etch, Integrated Processing for Micro- and Opto-Electronics," *Microelectronic Engineering*, Vol. 25, 1994, pp. 277–286.
[16] Grande, W. J., J. E. Johnson, and C. L. Tang, "GaAs/AlGaAs Photonic Integrated Circuits Fabricated Using Chemically Assisted Ion Beam Etching," *Applied Physics Letters*, Vol. 57, 1990, pp. 2,537–2,539.
[17] Lehmann, H. W., "Plasma-Assisted Etching," *Thin Film Processes*, San Diego: Academic Press, 1991, pp. 673–748.
[18] Williams, R. E., *Gallium Arsenide Processing Techniques*, Dedham, MA: Artech House, 1984, Chapter 9.
[19] Adachi, S. and K. Oe, "Chemical Etching Characteristics of (001) GaAs," *Journal of the Electrochemical Society*, Vol. 130, 1983, pp. 2,427–2,435.
[20] Seto, M., R. J. Deri, A. Yi-Yan, E. Colas, W. J. Tomlinson, R. Bhat, and A. Shahar, "Fabrication of Submillimeter-Radius Optical Waveguide Bends with Anisotropic Wet Chemical Etchants," *IEEE Journal of Lightwave Technology*, Vol. 8, 1990, pp. 264–270.
[21] de Salvo, G. C., W. F. Tseng, and J. Comas, "Etch Rates and Sensitivities of Citric Acid/Hydrogen Peroxide on GaAs, $Al_{0.3}GaAs$, $In_{0.2}GaAs$, $In_{0.53}GaAs$, $In_{0.52}AlAs$ and InP," *Journal of the Electrochemical Society*, Vol. 139, 1992, pp. 831–835.
[22] Bean, K. E., "Anisotropic Etching of Silicon," *IEEE Transactions on Electron Devices*, Vol. ED-25, 1978, pp. 1,185–1,193.
[23] Kern, W. and G. L. Schnable, "Low-Pressure Chemical Vapor Deposition for Very Large-Scale Integration Processing—A Review," *IEEE Transactions on Electron Devices*, Vol. ED-26, 1979, pp. 647–657.
[24] Gleine, W., and J. Müller, "Low-Pressure Chemical Vapor Deposition Silicon-Oxynitride Films for Integrated Optics," *Applied Optics*, Vol. 31, 1992, pp. 2,036–2,040.
[25] Rhoderick, E. H., and R. H. Williams, *Metal-Semiconductor Contacts*, 2nd Ed., Oxford, England: Clarendon, 1988.
[26] Fair, R. B., *Rapid Thermal Processing*, San Diego: Academic Press, 1993.

Chapter 6

Reflections at Boundaries

Billiard players know all about the effect. They use English to make their cue ball bounce off the first ball it hits, or a cushion, at the angle they need to complete the shot. They even take into account the residual rotation after a first impact. They do not do this by analysis. They accomplish it by alot of practice, superb manual skill, and a type of cerebration which does not seem to be catalogued by the psychologists.

Vannevar Bush, 1965 [1]

Semiconductor optical structures clearly take advantage of the optical characteristics of the material in which the fields are transmitted. Perhaps even more important than the material characteristics, however, are the changes in optical properties at the boundaries between two materials; the reflection, transmission, and phase shift of waves at a dielectric boundary form the basis for many semiconductor optical devices.

In this short chapter, we examine what happens to a propagating wave when it is incident on a boundary between two materials[1] because the phenomena involved are central to much of semiconductor integrated optical design. We begin with a discussion at the level of electric and magnetic fields, but we will be able to derive more phenomenological relationships that facilitate thinking about and calculating what happens at a dielectric interface. Luckily, everyday normal cerebration will be suitable for an analysis of the effects at boundaries.

6.1 ANGLES OF REFLECTION AND REFRACTION

Long before the dawn of billiards, it was observed that a light beam incident at a particular angle on a reflecting surface will be reflected at the same angle. Employing the notation of Figure 6.1, this effect is expressed mathematically simply as

[1]A more extensive discussion may be found in: Hecht, E., *Optics*, 2nd Ed., Reading, MA: Addison Wesley, 1987, Chapter 4.

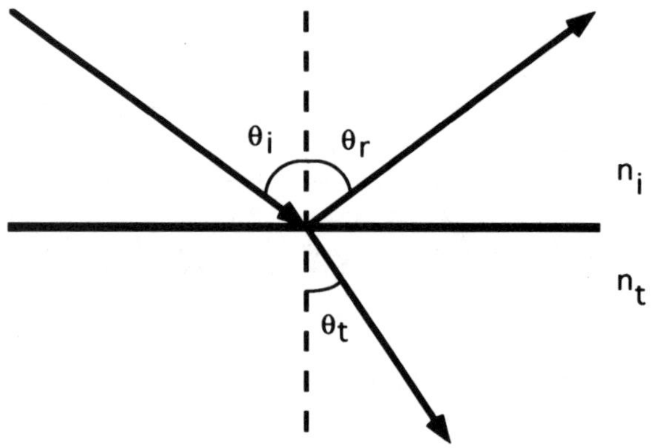

Figure 6.1 Definition of the angles of incidence, θ_i, reflection, θ_r, and transmission, θ_t, for an electromagnetic wave incident from a material with refractive index, n_i, upon a boundary with a second material of index, n_t.

$$\theta_i = \theta_r \tag{6.1}$$

for incidence angle θ_i [rad] and reflection angle θ_r [rad]. This equivalence of incidence and reflection angle is independent of the refractive index values, n_i [dimensionless] and n_t [dimensionless] of the materials for the incident and transmitted waves, respectively. In this chapter, the index values, n, will always be purely real.

If the surface is not perfectly reflecting, then a portion of the wave is transmitted from the region of refractive index, n_i, to that of refractive index, n_t. The relation of the angle of the transmitted component, θ_t, to the incident angle, θ_i, is given by Snell's law, which states

$$n_i \sin\theta_i = n_t \sin\theta_t \tag{6.2}$$

Thus the angle of the transmitted beam *is* a function of the relative refractive indices.

Snell's Law can be derived from consideration of Fermat's Principle. This states that a light beam between two points will follow the path of least time and if two paths require the same time, the beam will follow both simultaneously.

The astute reader will notice that Equation 6.2 has an interesting implication for $n_i > n_t$, which is when a light beam moves from a material of high refractive index to a region with lower index. $n_i > n_t$ implies that $\theta_t > \theta_i$ so that θ_t will approach 90°, for some smaller value of θ_i. $\theta_t = 90°$ means that none of the beam is transmitted into the material of lower refractive index and Snell's law simplifies to

$$\sin\theta_i = \frac{n_t}{n_i} \tag{6.3}$$

This relation allows us to define the critical angle, θ_c, as

$$\theta_c = \sin^{-1}\left(\frac{n_t}{n_i}\right) \tag{6.4}$$

For angles of incidence greater than this critical angle, $\theta_i > \theta_c$, the incident beam experiences total internal reflection—no light is transmitted across the boundary. This phenomenon is basically what makes optical waveguides possible and will be subject to our further attention in the next chapter.

Example

Consider an interface between air ($n_i = 1$) and $Al_{0.3}Ga_{0.7}As$ ($n_t = 3.35$). If we wish to calculate the maximum angle of transmission into the AlGaAs, θ_g, we can use Snell's law (Equation 6.2) and set $\theta_i = 90°$. We then find $\theta_g(\max) = \sin^{-1}(1/3.45) \approx 17°$. This is, of course, exactly the critical angle for total internal reflection as given by Equation 6.4. The implication is that, even for a cone of incidence of 180°, only a narrow cone of light is generated in the $Al_{0.3}Ga_{0.7}As$.

6.2 BOUNDARY CONDITIONS OF ELECTROMAGNETIC FIELDS

The simple picture of the light beam incident on a partially reflecting boundary is only moderately useful when we consider an electromagnetic wave impinging on a surface. We need to consider the electromagnetic picture [2] more carefully, as that is the one most important for a dielectric waveguide based on semiconductors.

The reflection and transmission of an arbitrary TEM wave at a boundary is determined through the application of boundary conditions for electric and magnetic fields. For an electric field, the *tangential* (parallel to the interface) component of $\mathcal{E}$ must be continuous across the boundary and the *normal* component (perpendicular to the interface) of the flux density, $D = \epsilon_m \mathcal{E}$, must be continuous. The latter is only true in the absence of any surface charges, which is usually the case for our considerations. For an interface between two materials 1 and 2, we thus have

$$\mathcal{E}_{1T} = \mathcal{E}_{2T} \tag{6.5}$$

and

$$D_{1N} = \epsilon_1 \mathcal{E}_{1N} = \epsilon_2 \mathcal{E}_{2N} = D_{2N} \tag{6.6}$$

where the subscripts {1,2} correspond to either side of the boundary and the {T,N} correspond to the tangential and normal components, respectively.

For magnetic fields, the tangential components of H must be continuous across the boundary, as must the normal components of the flux density, $B = \mu_m H$. This implies

$$H_{1T} = H_{2T} \tag{6.7}$$

and

$$B_{1N} = \mu_{m1} H_{1N} = \mu_{m2} H_{2N} = B_{2N} \tag{6.8}$$

Application of these boundary conditions to a wave incident from one material onto another will then permit calculation of how much will be transmitted and reflected.

6.3 REFLECTION AND TRANSMISSION COEFFICIENTS

By reducing an arbitrary TEM wave into its electric and magnetic field components, we can calculate its behavior when it comes to a boundary by applying the above boundary conditions. We will now proceed by solving two general configurations since any arbitrary wave may always be reduced to these two cases. By then determining the reflection and transmission coefficients of the electromagnetic wave, we can think about the boundary phenomenologically so that detailed consideration of the fields will no longer be necessary.

6.3.1 TE Configuration

The transverse electric (TE) configuration is illustrated in Figure 6.2. A TEM wave with a propagation constant, k_i, in a material with refractive index, n_i, is incident at an angle, θ_i, on a region with refractive index, n_t. A portion of the wave, with propagation constant k_r is reflected at angle θ_r and the remainder, with propagation constant k_t is transmitted at an angle θ_t into the material with n_t. This TEM wave has $\mathcal{E}$ transverse (or tangential) to the plane of incidence (the interface between the two materials), hence its label as transverse electric. Our assignment, should we decide to accept it, is to find the relative amounts of the incident wave reflected and transmitted.

Because the electric field is always parallel to the interface, we can apply Equation 6.5 to write

$$\mathcal{E}_i + \mathcal{E}_r = \mathcal{E}_t \tag{6.9}$$

The tangential components of the magnetic field are given by $\cos(\theta)H$ yielding

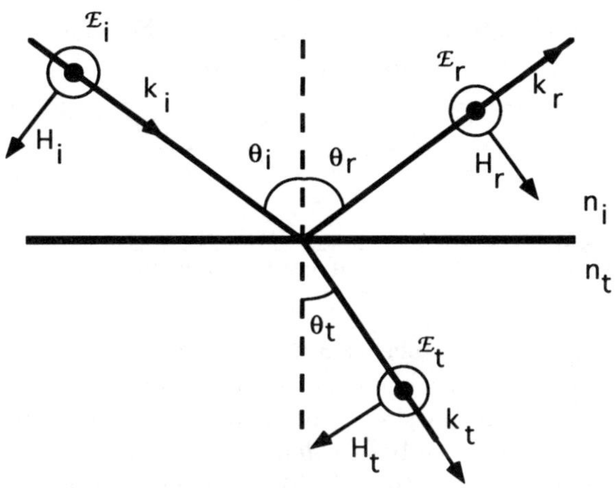

Figure 6.2 Orientation of magnetic and electric fields for the case of TE incidence onto a boundary. TE implies that the electric field is always tangential to the plane of incidence.

$$H_{iT} + H_{rT} = H_{tT} \tag{6.10}$$

$$H_i \cos\theta_i - H_r \cos\theta_r = H_t \cos\theta_t \tag{6.11}$$

The incident and transmitted parts of the tangential magnetic field component point in the $-\mathbf{x}$ direction, whereas the reflected part points in the $+\mathbf{x}$ direction; this explains the minus sign for the second term. (The right-hand rule is indispensable for these exercises.)

Because the electric and magnetic fields are related by the impedance (Equation 1.38)

$$\mathcal{E} = \frac{1}{\sqrt{\epsilon_m \mu_m}} B = \sqrt{\frac{\mu_m}{\epsilon_m}} H \tag{6.12}$$

we can then rewrite Equation 6.11 as

$$\sqrt{\frac{\epsilon_i}{\mu_i}} \mathcal{E}_i \cos\theta_i - \sqrt{\frac{\epsilon_r}{\mu_r}} \mathcal{E}_r \cos\theta_r = \sqrt{\frac{\epsilon_t}{\mu_t}} \mathcal{E}_t \cos\theta_t \tag{6.13}$$

We remember that $\theta_i = \theta_r$, $\epsilon_i = \epsilon_r$, and $\mu_i = \mu_r = \mu_t = \mu_m$ (the latter for nonmagnetic materials, such as semiconductors), and $n = \sqrt{\epsilon}$ for $\epsilon_m = \epsilon \cdot \epsilon_0$ (Equation 3.6), which allows us to simplify Equation 6.13 to

$$n_i \cos\theta_i (\mathcal{E}_i - \mathcal{E}_r) = n_t \cos\theta_t \mathcal{E}_t \tag{6.14}$$

By massaging Equations 6.9 and 6.14, we can define the reflection and transmission coefficients, r_{TE} and t_{TE} [dimensionless], as follows:

$$r_{TE} \equiv \frac{\mathcal{E}_r}{\mathcal{E}_i} = \frac{n_i \cos\theta_i - n_t \cos\theta_t}{n_i \cos\theta_i + n_t \cos\theta_t} \tag{6.15}$$

$$t_{TE} \equiv \frac{\mathcal{E}_t}{\mathcal{E}_i} = \frac{2n_i \cos\theta_i}{n_i \cos\theta_i + n_t \cos\theta_t} \tag{6.16}$$

Note that r_{TE} and t_{TE} define the relative amounts of the fields, not the powers, which are reflected or transmitted. We shall return to that distinction presently. In Figure 6.3, the reflection and transmission coefficients for the TE case are plotted as a function of incident angle θ_i; we have calculated the case for a wave incident from air ($n_i = 1$) onto a piece of GaAs ($n_t = 3.45$). We see from this plot that t_{TE} is always positive and r_{TE} is always negative, from which the phase shifts of waves undergoing reflection and transmission are determined. A transmitted TE wave always has a phase shift of $\Delta\phi = 0$ and a reflected TE wave always has a shift of $\Delta\phi = \pi$.

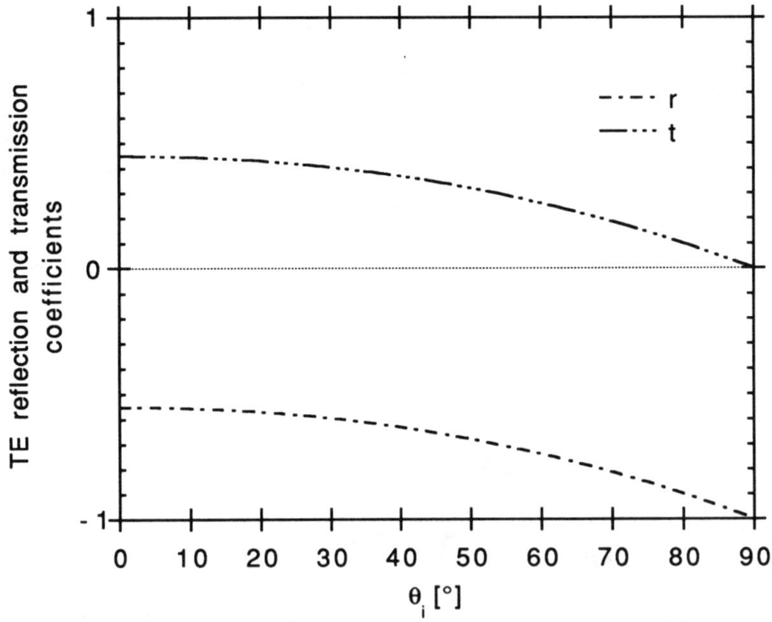

Figure 6.3 r_{TE} and t_{TE} as a function of incidence angle θ_i for a ray incident from air (n = 1) onto GaAs (n = 3.45). Since t_{TE} is always positive, the transmitted wave has zero phase shift; negative r_{TE} implies a phase shift of π for the reflected wave.

6.3.2 TM Configuration

We may now consider the orthogonal configuration, the transverse magnetic (TM) case, for which the magnetic field is always transverse to the plane of incidence as shown in Figure 6.4. We proceed in a like manner as previously; the tangential magnetic fields are continuous

$$H_{iT} + H_{rT} = H_{tT} \tag{6.17}$$

which, using the relationship between $\mathscr{E}$ and H (Equation 6.12), becomes

$$\mathrm{n}_i\mathscr{E}_i + \mathrm{n}_r\mathscr{E}_r = \mathrm{n}_t\mathscr{E}_t \tag{6.18}$$

Since the tangential components of $\mathscr{E}$ are also continuous, this implies

$$\mathscr{E}_i \cos\theta_i - \mathscr{E}_r \cos\theta_r = \mathscr{E}_t \cos\theta_t \tag{6.19}$$

With similar manipulation as before, we can then determine the TM reflection and transmission coefficients to be

$$r_{\mathrm{TM}} \equiv \frac{\mathscr{E}_r}{\mathscr{E}_i} = \frac{\mathrm{n}_t \cos\theta_i - \mathrm{n}_i \cos\theta_t}{\mathrm{n}_t \cos\theta_i + \mathrm{n}_i \cos\theta_t} \tag{6.20}$$

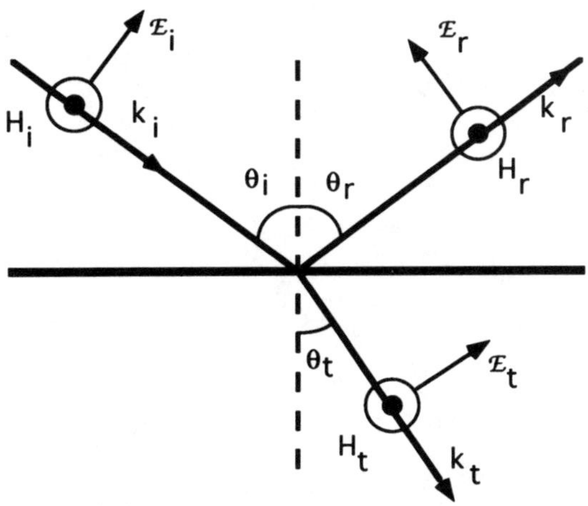

Figure 6.4 Orientation of magnetic and electric fields for TM incidence onto a boundary. TM implies that the magnetic field is always tangential to the plane of incidence.

$$t_{TM} \equiv \frac{\mathcal{E}_t}{\mathcal{E}_i} = \frac{2n_i \cos\theta_i}{n_t \cos\theta_i + n_i \cos\theta_t} \tag{6.21}$$

Figure 6.5 shows the behavior of the TM reflection and transmission coefficients, again for the case of incidence from air onto GaAs. t_{TM} has qualitatively similar behavior to t_{TE}, namely always positive, implying a phase shift of zero for all angles. r_{TM}, however, has a zero crossing at $\theta_i = \theta_{pe} \approx 75°$, and a concomitant change in sign. This has two implications: first, the phase change of a reflected TM wave is zero for $\theta_i < \theta_{pe}$ and becomes $\Delta\phi = \pi$ for $\theta_i > \theta_{pe}$. Second, the angle where $r_{TM} = 0$, namely θ_{pe}, is known as the *polarization angle* or *Brewster angle* and corresponds to an angle of incidence for which a TM wave is transmitted without attenuation (i.e., nothing is reflected). For external incidence, the polarization angle is given by

$$\theta_{pe} = \sin^{-1}\sqrt{\frac{\left(\frac{n_t}{n_i}\right)^2 - 1}{\left(\frac{n_t}{n_i}\right)^2 - \left(\frac{n_i}{n_t}\right)^2}} \tag{6.22}$$

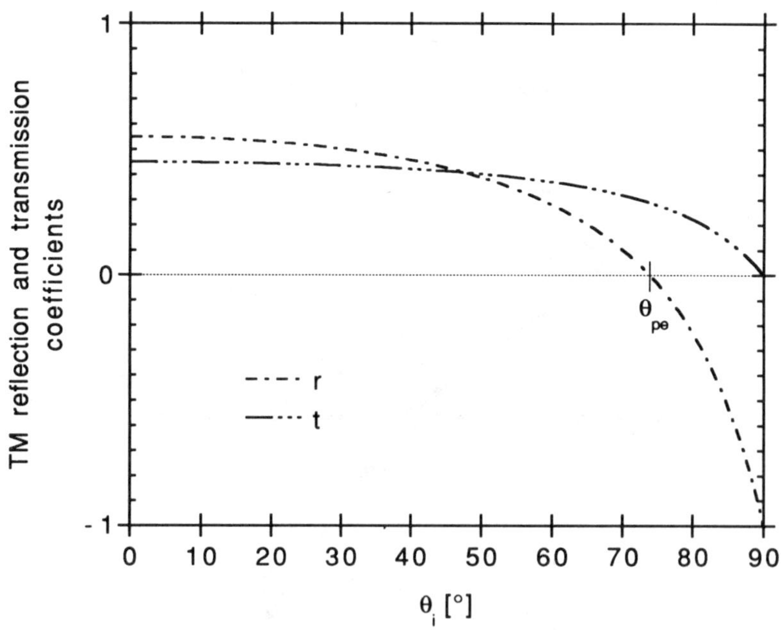

Figure 6.5 r_{TM} and t_{TM} as a function of incidence angle θ_i for a ray incident from air (n = 1) onto GaAs (n = 3.45). The external polarization, or Brewster, angle, θ_{pe}, is at roughly $\theta_i = 75°$.

where we use the subscript "pe" from "polarization, external" to distinguish it from the polarization angle found for internal (i.e., $n_i > n_t$) reflection, which has a different value as we shall discuss below. The Brewster angle is useful for polarizing an incident wave or selecting a particular polarization in a beam; the latter is applied to the Brewster windows in lasers.

The set of equations 6.15, 6.16 and 6.20, 6.21 are known collectively as the Fresnel equations and define the so-called Fresnel reflection and transmission coefficients of a surface.

Example

A glass plate (n = 1.5) is inserted into the beam of a laser at 41° from the beam normal. We note from Equation 6.22 that this angle corresponds to the external polarization angle, implying that the TM mode is completely transmitted but the TE mode has a portion reflected. If this glass plate were inserted inside a laser cavity, the TE mode would be attenuated and only the TM mode would lase. This arrangement is the Brewster window found in many gas lasers, used to filter all but a single mode from the output spectrum.

6.3.3 Power Transmission and Reflection

Since we are usually interested in a measurable quantity, we need to consider the magnitude of power reflected or transmitted at a boundary. Recall that the power density is given by S [W/cm^2] (Equation 1.43)

$$S = \sqrt{\frac{\epsilon_m}{\mu_m}}\,\mathcal{E}^2 \tag{6.23}$$

and the total power by $S' = AS$ [W] (Equation 1.41). The area, A [m^2], subtended by the beam is a function of θ and is given by $A_{(i,r,t)} \propto \cos(\theta_{(i,r,t)})$, as can be seen from the construction of Figure 6.6. We can then write the reflectance and transmittance, R and T [dimensionless], which represent the power reflected and transmitted, respectively, as

$$R = \frac{A_r S_r}{A_i S_i} = \frac{\cos\theta_r \sqrt{\frac{\epsilon_r}{\mu_r}}\,\mathcal{E}_r^2}{\cos\theta_i \sqrt{\frac{\epsilon_i}{\mu_i}}\,\mathcal{E}_i^2} = \frac{\mathcal{E}_r^2}{\mathcal{E}_i^2} = r^2 \tag{6.24}$$

and

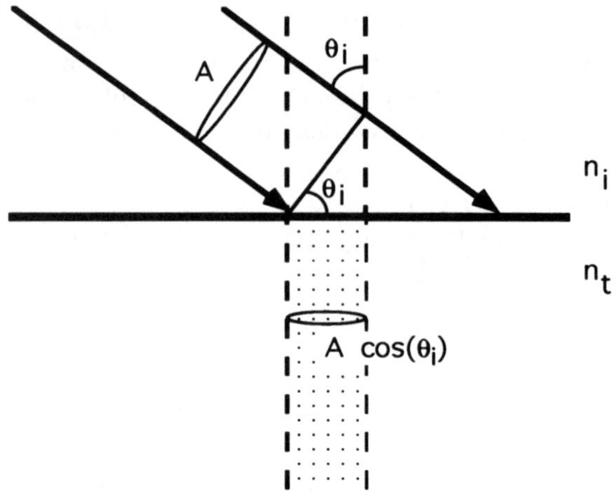

Figure 6.6 Schematic plot of an a wave incident at an angle θ_i showing the cos(θ_i) dependence of the area A subtended by the beam on the surface.

$$T = \frac{A_t S_t}{A_i S_i} = \frac{\cos\theta_t \sqrt{\frac{\epsilon_t}{\mu_t}}\,\mathcal{E}_t^2}{\cos\theta_i \sqrt{\frac{\epsilon_i}{\mu_i}}\,\mathcal{E}_i^2} = \frac{(\mathrm{n}_t \cos\theta_t)\mathcal{E}_r^2}{(\mathrm{n}_i \cos\theta_i)\mathcal{E}_i^2} = \frac{\mathrm{n}_t \cos\theta_t}{\mathrm{n}_i \cos\theta_i} t^2 \tag{6.25}$$

We see that, whereas the reflectance is just given by the square of the reflection coefficient, the transmittance has an additional cos(θ) term. Moreover, the TE and TM components may be written separately, as

$$R_{\mathrm{TE,TM}} = r_{\mathrm{TE,TM}}^2 \tag{6.26}$$

and

$$T_{\mathrm{TE,TM}} = \frac{\mathrm{n}_t \cos\theta_t}{\mathrm{n}_i \cos\theta_i} t_{\mathrm{TE,TM}}^2 \tag{6.27}$$

We can easily see that

$$R + T = 1 \tag{6.28}$$

which also holds for each component individually, namely

$$R_{\mathrm{TE,TM}} + T_{\mathrm{TE,TM}} = 1 \tag{6.29}$$

The TM and TE reflectance are presented graphically in Figure 6.7 for the same materials as before. If we actually did a measurement of transmitted and reflected intensities, comparing input with reflected and transmitted powers, the reflectance and transmittance would result, not the reflection and transmission coefficients.

6.4 TWO SPECIAL CASES

There are two types of wave incidence on a surface that appear frequently in practical situations, and thus deserve to be treated separately. The first is normal incidence, in which the distinction between TE and TM disappears, and the second is total internal reflection, which is the condition most relevant for guided wave devices.

6.4.1 Normal Incidence

We are often interested in the special case for which the wave is normally incident onto the boundary, such as for the end-fire coupling of a laser beam onto a cleaved

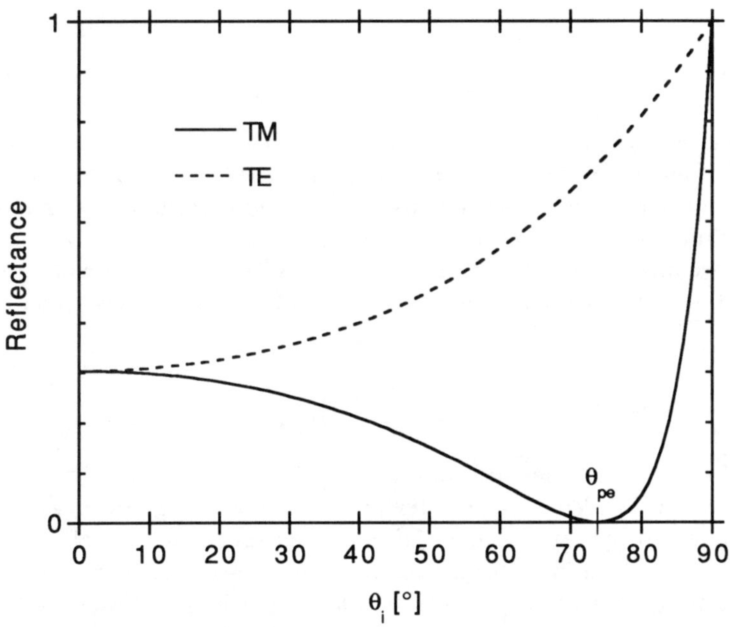

Figure 6.7 The variation of TE and TM reflectance as a function of incidence angle for a wave incident from air onto GaAs. The polarization angle, θ_{pe}, is manifested as a point of zero reflection at $\theta_i = 75°$.

facet or the normal emission from a laser facet. In that case, $\theta_i = \theta_r = \theta_t = 0$ and the transmittance and reflectance take on considerably simpler forms, namely

$$R_{(\theta=0)} = \left(\frac{n_i - n_t}{n_i + n_t}\right)^2 \tag{6.30}$$

and

$$T_{(\theta=0)} = \frac{4n_i n_t}{(n_i + n_t)^2} \tag{6.31}$$

Notice that the distinction between TE and TM disappears for normal incidence, which makes sense because in that case *both* the magnetic and electric fields are always transverse to the surface. We can then just use ratios of the refractive indices to determine reflection and transmission. We also note that R and T are the same for external ($n_i < n_t$) and internal ($n_i > n_t$) reflection; thus, the same fraction is reflected for a wave incident from either side of the boundary. Figure 6.8 shows R as a function of n_t for $n_i = 1$.

Example

Given a boundary between air (n = 1) and GaAs (n = 3.45), we can use Equation 6.30 to determine the reflectance for normal incidence to be 0.3. This value is often quoted for uncoated GaAs cleaved laser mirrors, for example, assuming normal incidence of the laser mode on the interface. Conversely, 30% of a laser beam focused from air normally onto a GaAs facet is also reflected. The maximum expected reflectance for a semiconductor/air boundary would be for Ge (n = 4), with $R = 0.36$.

6.4.2 Total Internal Reflection

The second case that particularly interests us is that of internal reflection ($n_i > n_t$) when the angle of incidence exceeds the critical angle, $\theta_i \geq \theta_c$. In that case, no wave is transmitted into the second material, implying total internal reflection, and the reflection coefficients and transmission coefficients are in general complex. For a wave trying to leave a piece of GaAs into air, for example, Equation 6.4 yields a critical angle of $\theta_c = 17°$, implying that only light with a very steep incidence can escape the crystal.

This behavior becomes clearer if we express the reflection coefficient for TE and TM waves incident onto a boundary as a function of only θ_i as follows:

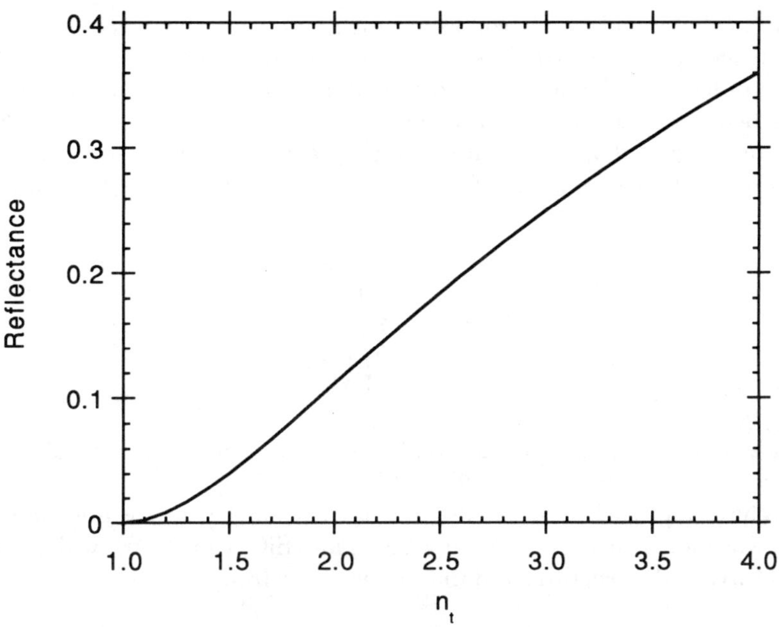

Figure 6.8 Reflectance for normal incidence onto a boundary between air (n_i = 1) and a material with n_t varying from 1 to 4. GaAs, for example, would have n_t = 3.45.

$$r_{\mathrm{TE}} = \frac{\cos\theta_i - \sqrt{\left(\frac{n_t}{n_i}\right)^2 - \sin^2\theta_i}}{\cos\theta_i + \sqrt{\left(\frac{n_t}{n_i}\right)^2 - \sin^2\theta_i}} \tag{6.32}$$

and

$$r_{\mathrm{TM}} = \frac{\left(\frac{n_t}{n_i}\right)^2 \cos\theta_i - \sqrt{\left(\frac{n_t}{n_i}\right)^2 - \sin^2\theta_i}}{\left(\frac{n_t}{n_i}\right)^2 \cos\theta_i + \sqrt{\left(\frac{n_t}{n_i}\right)^2 - \sin^2\theta_i}} \tag{6.33}$$

These relations are easily derived from the expressions for r (Equations 6.15 and 6.20) and the application of a few trigonometric identities plus Snell's law. We can see immediately, since the critical angle is given by $\sin\theta_c = n_t/n_i$, that for $\theta_i > \theta_c$, the argument of the square root becomes negative and hence the expressions for r_{TE} and r_{TM} become complex. Nevertheless, it is easy to verify that $rr^* = 1$ for $\theta_i > \theta_c$.

The reflection behavior, R_{TE} and R_{TM}, of a wave moving from GaAs onto an air interface behaves as a function of incident angle as shown in Figure 6.9. Reflection is unity for angles above the critical angle, $\theta_c = 17°$, very close to the internal polarization, or Brewster, angle of $\theta_{pi} = 16°$; the latter may be found from

$$\theta_{pi} = \cos^{-1}\sqrt{\frac{\left(\frac{n_t}{n_i}\right)^2 - 1}{\left(\frac{n_t}{n_i}\right)^4 - 1}} \tag{6.34}$$

A comparison with Equation 6.22 reveals that the internal and external polarization angles are simply related as $\theta_{pe} + \theta_{pi} = 90°$.

We have repeated these calculations for two further interesting cases, namely that of a wave incident from Si_3N_4 (n = 2.05) into SiO_2 (n = 1.46), which is a typical Si-based waveguide structure, and that of incidence from $Al_{0.3}Ga_{0.7}As$ (n = 3.35) onto

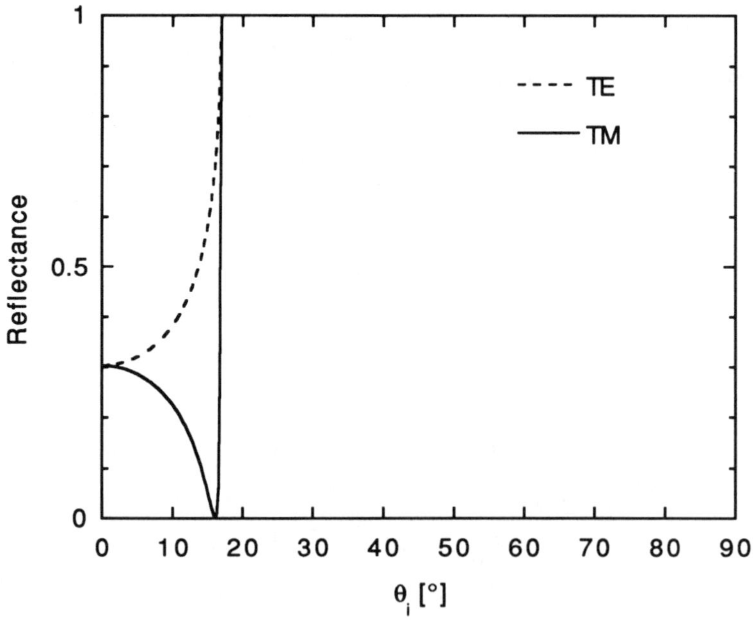

Figure 6.9 Reflectance for both TE and TM incidence for internal reflection of a wave traveling from inside GaAs onto an air interface. Note that $\theta_{pi} = 15°$ and $\theta_c = 16°$.

$Al_{0.8}Ga_{0.2}As$ (n = 3.12), a realistic waveguide sandwich for many III-V based systems. The nitride/oxide waveguide, whose reflectance is shown in Figure 6.10, has an internal polarization angle of 35° and a critical angle of 45°. The $Al_{0.3}Ga_{0.7}As$/$Al_{0.8}Ga_{0.2}As$ waveguide, with the reflectance behavior of Figure 6.11, is similar to that for GaAs and air, albeit with very low reflectance for steep incidence, $\theta_{pi} = 43°$ and a critical angle of 69°. The refractive index values are found from Figure 4.4 evaluated at $\lambda = 850$ nm, and the small index step, 0.2, is typical for structures in this material system.

The critical and internal polarization angles for a GaAs/$Al_xGa_{1-x}As$ waveguide are plotted in Figure 6.12 as a function of aluminum mole fraction, x, in the AlGaAs; the calculations assume incidence from a GaAs layer, x = 0. We see that although the Brewster angle changes little with composition, the small index differences between GaAs and AlGaAs layers yield large critical angles. Because a waveguide relies on total internal reflection to propagate a mode, the large values for θ_c imply that only waves with grazing incidence on the boundary have a chance of being completely reflected, and thus transferred down the waveguide. Modes with steeper angles of incidence are partially transmitted past the boundary and eventually disappear; this will be the subject of the next chapter.

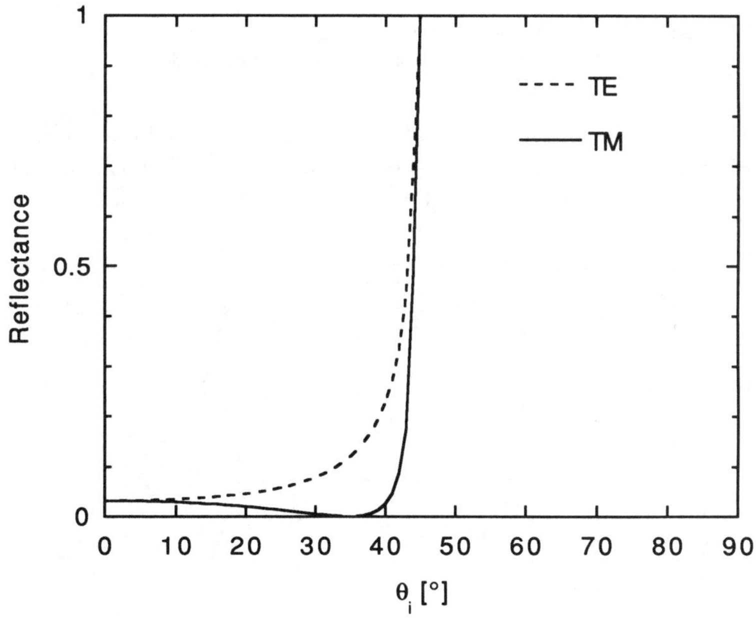

Figure 6.10 Reflectance for a wave incident from Si_3N_4 (n = 2.05) into SiO_2 (n = 1.46); critical angle is 45°.

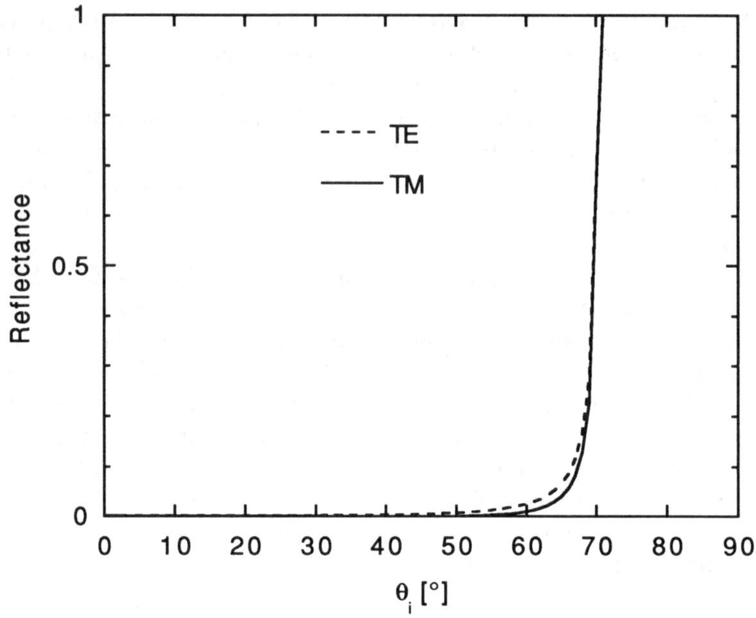

Figure 6.11 Reflectance for a wave incident from $Al_{0.3}Ga_{0.7}As$ (n = 3.35) onto $Al_{0.8}Ga_{0.2}As$ (n = 3.12); critical angle is 69° and reflectance is very low for steep incidence.

Because the reflection and transmission coefficients for total internal reflection are complex, the phase change for the reflected wave will behave in a complicated way. Taking $\Delta\phi$ to be the phase shift of $r = |r|\exp(j\phi)$, we obtain [3]

$$\Delta\phi_{TE} = 2\tan^{-1}\sqrt{\frac{\sin^2\theta_i - \sin^2\theta_c}{\cos^2\theta_i}} = 2\tan^{-1}\sqrt{\frac{\sin^2\theta_i - \left(\frac{n_t}{n_i}\right)^2}{\cos^2\theta_i}} \tag{6.35}$$

and

$$\Delta\phi_{TM} = 2\tan^{-1}\sqrt{\frac{\sin^2\theta_i - \sin^2\theta_c}{\cos^2\theta_i \sin^4\theta_c}} = 2\tan^{-1}\sqrt{\frac{\left(\frac{n_i}{n_t}\right)^2\sin^2\theta_i - 1}{\left(\frac{n_t}{n_i}\right)^2\cos^2\theta_i}} \tag{6.36}$$

which are both shown graphically in Figure 6.13 for $\theta > \theta_c$. Below θ_c, the behavior is essentially the inverse of the external reflection, such that TE, and TM below the

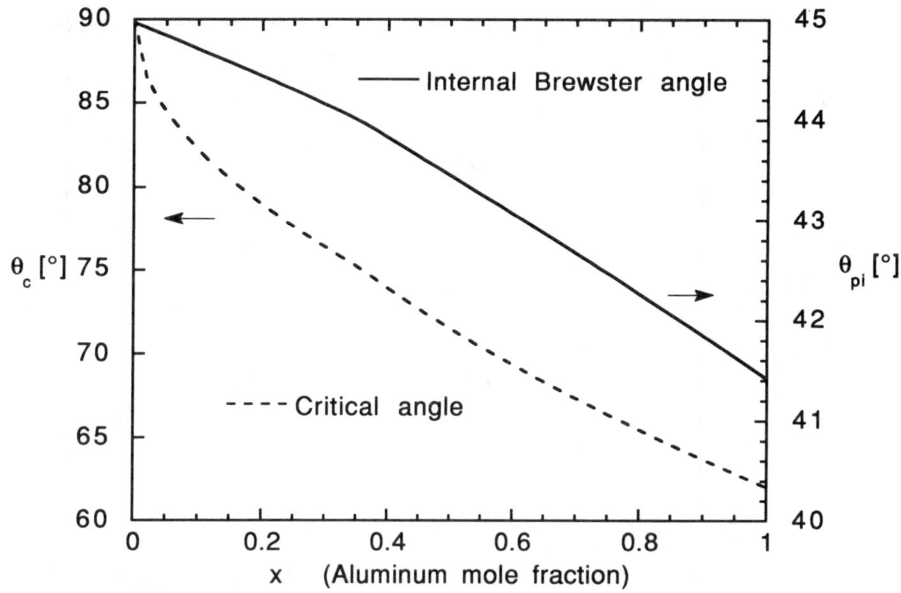

Figure 6.12 Critical and internal polarization angles for waves incident from a GaAs waveguide core (x = 0) onto an interface with $Al_xGa_{1-x}As$ of varying aluminum mole fraction x. Only rather shallow angles of incidence are subject to total internal reflection.

polarization angle, have a phase shift of π. For angles of incidence above the critical angle, the phase shift varies continuously between 0 and π as seen in the figure.

Finally, we must be more specific when we state that no *wave* is transmitted into the material with lower index—strictly, no *power* is transmitted, whereas an evanescent wave does extend beyond the boundary. An evanescent wave decays exponentially with distance y from the interface, as

$$\mathscr{E} = \mathscr{E}_0 e^{-\alpha' \cdot y} \tag{6.37}$$

where α' [m^{-1}], the extinction coefficient, is given by (we will derive this in the next chapter)

$$\alpha' = \frac{2\pi}{\lambda_0}\sqrt{n_i^2 \sin^2\theta_i - n_t^2} \tag{6.38}$$

The 1/e penetration depth beyond the boundary is then simply $1/\alpha'$. Typical values for $1/\alpha'$ are a fraction of the wavelength λ for angles more than several degrees larger than the critical angle. We stress that the exponentially decaying field does not result

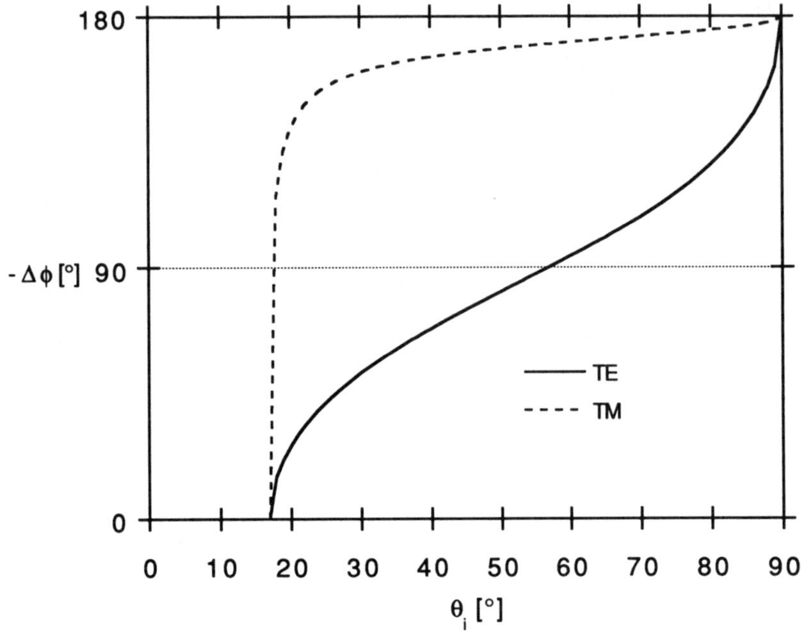

Figure 6.13 The phase shift upon total internal reflection ($\theta_i > \theta_c = 17°$) for TE and TM waves incident from GaAs onto an air interface.

in any power flow through the interface. However, because the field does extend beyond the boundary, the refractive index conditions beyond the boundary do have an effect on the propagation constant of the reflected wave. We will examine this in depth in Chapter 7.

Example

Let's examine a wave with $\lambda = 850$ nm propagating by total internal reflection along a boundary between $Al_{0.3}Ga_{0.7}As$ ($n_i = 3.35$) and $Al_{0.8}Ga_{0.2}As$ ($n_t = 3.12$). We saw in Figure 6.11 that the critical angle is 69°. A wave propagating with an incidence angle of 72° can be seen to have a penetration depth into the $Al_{0.8}Ga_{0.2}As$ of $1/\alpha' = 276$ nm, whereas one with incidence angle 85° penetrates 120 nm into the $Al_{0.8}Ga_{0.2}As$. Thus, as we might expect, the more shallow the incidence, the less the evanescent field penetrates beyond the boundary.

REFERENCES

[1] Bush, V., "When Bat Meets Ball," in *Science Is Not Enough*, New York: William Morrow, 1967.
[2] Kraus, J. D. and K. R. Carver, *Electromagnetics*, 2nd Ed., New York: McGraw Hill, 1973, Chapters 3 and 6.
[3] Born, M. and E. Wolf, *Principles of Optics*, 6th Ed., Oxford, England: Pergamon, 1980, Chapter 1.5.

Chapter 7

Guided Waves and Slab Waveguides

Wir bewegen uns auf Schienen, der Tunnel muβ also irgendwohin führen. Nichts beweist, daβ am Tunnel etwas nicht in Ordnung ist, auβer natürlich, daβ er nicht aufhört.[1]

Friedrich Dürrenmatt, 1952 [1]

A waveguide is a structure that allows us to guide an optical wave by making use of multiple total internal reflections from two or more surfaces. In its simplest configuration, light is guided between two parallel plates, a structure known as a slab waveguide since the light is confined to a slab of material between the plates. If we confine the wave in a further dimension by adding more surfaces, then the wave is guided in two dimensions and a channel waveguide is formed. This represents a true optical *wire*, exemplified by optical fibers and semiconductor strip waveguides. The function of a waveguide is to lead the guided light somewhere, preferably to a useful device while maintaining power density along the wire.

The key to understanding how waveguides work is knowing what happens to a wave at an interface, a subject we studied in the last chapter. The forms of the electric and magnetic fields, which we reviewed in Chapter 1, in a waveguide are also an important quantity. We will now apply the results of both of those chapters to describe the propagation of light in waveguides by means of two distinct models: the ray-optic model and the electromagnetic model. The ray-optic representation allows us think of light rays reflecting from the waveguide boundaries and provides a heuristic, if mathematical, approach to understanding why waveguides have distinct modes and how these behave. The electromagnetic model requires that we solve the wave equation and provides expressions for the electric and magnetic fields in the waveguide, although perhaps this advantage comes at the expense of being slightly less intuitive. The ray-optic model is useful when we want to determine the number and types of modes present in a particular structure, as well as the propagation

[1]"We are moving on tracks, so the tunnel must lead someplace. Nothing indicates that something is wrong with the tunnel, except, of course, that it doesn't end."

constants and effective indices. If the field and intensity distributions need to be known, which are often important for determining the interaction of modes with structures outside the waveguide or for calculating the shape of the field for coupling to another optical structure, the electromagnetic model must be used.

When applying the theory we are about to develop to real waveguide structures, the engineer will typically wish to determine several parameters: the number of allowed modes, the dimensions and materials that permit only a single mode, the effective propagation constants and effective refractive index in the waveguide, the electric and magnetic field distributions of the propagating modes, and the penetration of these fields into material outside of the waveguide. In the following sections, we will see how to determine all of these for one-dimensional slab waveguides. In Chapter 8, we will turn to two-dimensional channel waveguides and discuss their behavior and fabrication.

7.1 PERFECT MIRROR WAVEGUIDES

Although an optical waveguide is a structure for confining and guiding light, there is an important distinction between propagation of light in a waveguide and propagation in free space. Whereas in free space, or in a bulk material, we could characterize our wave by a propagation constant, k, which could take any value; for a guided wave in a waveguide, k becomes discrete—only waves with certain values of propagation constant can actually propagate. The allowed values of k define the waveguide modes, which are likewise discrete.

In order to appreciate the origin of this phenomenon, we will use the ray-optic model for propagation in a waveguide. This model allows us to represent light beams as geometric rays in the direction of propagation, and we recognize in this the famous zig-zag picture of a light beam zigging and zagging down a waveguide by reflecting off the walls. This model says nothing about electric or magnetic fields, but provides a useful representation for the propagation of optical fields.

7.1.1 Discrete Waveguide Modes

We begin by simplifying our waveguide structure by assuming that the boundaries are perfectly reflecting mirrors; i.e., $R = 1$ for all incidence angles. This kind of structure is not unlike that of microwave waveguides[2] allowing no field penetration into the walls. In later sections, we will analyze the more realistic dielectric waveguide, where perfect reflection is not assumed.

---

[2]See for example: Ramo, S., J. R. Whinnery, and T. Van Duzer, *Fields and Waves in Communication Electronics*, New York: John Wiley & Sons, 1965, Chapters 7 and 8.

In Figure 7.1(a), we have two perfectly reflecting parallel mirrors separated by a material with refractive index n = 1 and thickness t [m]. This is a slab waveguide because it is only bound in the **y** direction and is infinite in the **x** and **z** directions. If we permit a plane wave to propagate through this structure and trace this propagation using the ray-optics representation, we see that the wave will be reflected at some angle, θ, each time it bounces off the mirror because the incidence angle equals the reflection angle. This is seen in the cross section of Figure 7.1(b). However, even though we might illuminate the edge of the waveguide and excite waves at arbitrary angles, only certain values of θ are allowed. We can see this by realizing that the part of the wave that undergoes two consecutive reflections must be in phase with the part that had not been reflected; if this were not the case, destructive interference between reflected and nonreflected parts would cause the wave to eventually disappear.

This condition should become clear in Figure 7.2. The original TE wave with two parts, A and B, whose crests are shown by parallel lines, is reflected once at

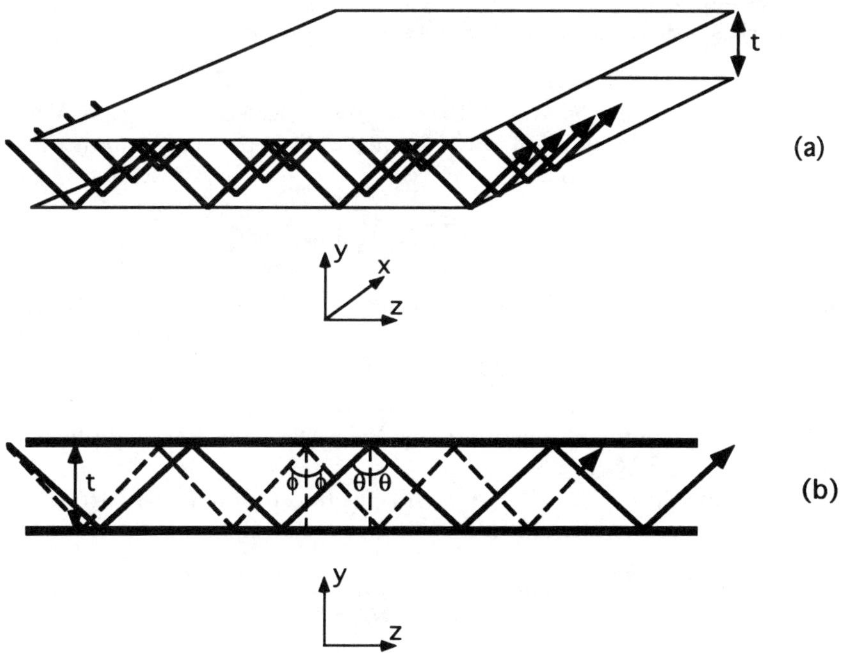

Figure 7.1 The parallel-mirror slab waveguide: (a) The two perfectly reflecting mirrors are separated by a distance, t. (b) A cross section of the waveguide showing the ray-optic representation of two waves propagating in the **z** direction. Incidence and reflection angles are the same, but only certain discrete values of θ are allowed.

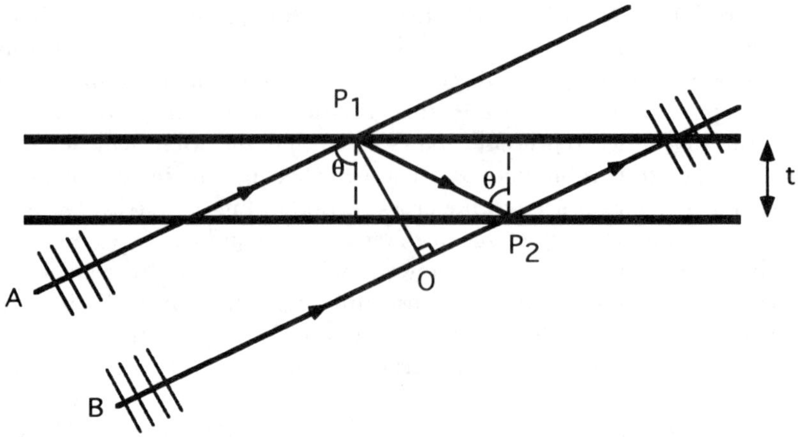

Figure 7.2 Geometric construction showing the phase-matching condition of a wave propagating in the slab waveguide. The portion of a TE wave from A undergoes reflection at P_1 and P_2 and must be in phase with the portion from B after the second reflection.

point P_1 and once at point P_2. After reflection, the reflected crests from A must be in phase with those of the original wave from B. Thus, the path length difference between parts A and B, ΔL [m], is given by

$$\Delta L = \overline{P_1P_2} - \overline{P_2O} \tag{7.1}$$

and a bit of trigonometry (fiendishly left to the reader) shows that this is given by

$$\Delta L = 2t\cos\theta \tag{7.2}$$

The number of wavelengths in this interval is then $\Delta L/\lambda$ and the total phase shift is $2\pi\Delta L/\lambda$ for wavelength λ.

If we now take into account that each reflection results in a phase shift of π, the condition that the waves A and B are in phase after P_2 becomes

$$\frac{4\pi t}{\lambda}\cos\theta - 2\pi = 2\mathrm{m}\pi \tag{7.3}$$

where m is an integer, m = 0,1,2, ... We thus have a condition on θ, namely

$$\theta_{m'} = \cos^{-1}\left(\frac{\lambda \mathrm{m}'}{2t}\right) \tag{7.4}$$

where m′ = m + 1, m′ = 1,2,3, . . . The $\theta_{m'}$ are the discrete allowed values of reflection angle, and each of these corresponds to an allowed waveguide mode. Thus we have a key result: propagation in a waveguide takes place in the form of discrete modes, each with its own reflection and propagation conditions. The integers m′ are known as the mode number; alternatively, we speak of the mth order mode.

Because the argument of the $\cos^{-1}$ in Equation 7.4 must be less than unity, this places limitations on the maximum permitted value of m′. In particular, we are restricted to

$$m' \leq \frac{2t}{\lambda} \tag{7.5}$$

remembering of course that m′ must be an integer greater than or equal to unity. The number of allowed modes is a function of the relative sizes of wavelength and waveguide width. If $2t/\lambda$ is large, many modes will be supported; as the waveguide shrinks (or the wavelength lengthens), the number allowed will decrease.

At the limit, where $\lambda > 2t$, *no modes* will be allowed because $2t/\lambda < 1$ but m′ > 1. This condition is known as cutoff and implies that for wavelengths longer than a certain cutoff wavelength, $\lambda_c = 2t$, no waves will propagate. Thus for certain conditions, the waveguide will not transmit any light. We recall that this parallel mirror structure is a very idealized model, and that for real semiconductor waveguides, the cutoff condition will be more complicated.

Example

If we analyze a mirror waveguide with a width of $t = 2\ \mu$m at a wavelength of 850 nm, we find from Equation 7.5 that the maximum value of m′ is 4. Thus, four distinct modes are supported in this guide, with reflection angles, found from Equation 7.4, of $\theta_1 = 78°$, $\theta_2 = 65°$, $\theta_3 = 51°$, and $\theta_4 = 33°$. We see that the incidence angle increases as we move to higher order modes.

7.1.2 Propagation Constants and Effective Index

In addition to calculating the reflection angles of the allowed modes in a waveguide, we are quite eager to determine the propagation constant. The optical wave is moving down the waveguide in the **z** direction, but doing a lot of zigging and zagging on the way. The amount of zigging and zagging is a function of the mode, each one different, so we expect that each mode also has its own propagation constant.

The triangles sketched in Figure 7.3(a) illustrate this. If there were no waveguide, the wave would be propagating happily along with propagation constant k_0, where we recall that $k_0 = 2\pi/\lambda_0$, still assuming that the refractive index of the medium

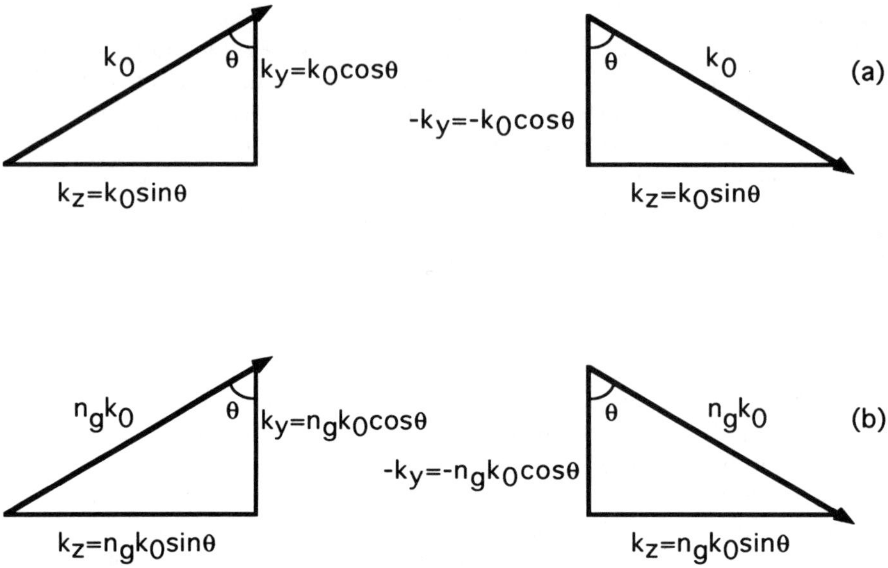

Figure 7.3 The relationships between propagation constants: $k_0 = 2\pi/\lambda_0$, representing the actual propagation constant of the wave bouncing down the waveguide, k_z is the effective propagation constant in the **z** direction, and k_y is the component directed normal to the reflecting planes: (a) in free space, or with n = 1; (b) in a waveguide with refractive index n_g.

is n = 1. Since the waveguide causes the wave to zig and zag in the **y** direction, there is a component of the propagation constant in that direction, k_y, and only a portion directed down the waveguide, k_z. Since k_y is given by

$$k_y = k_0 \cos\theta_{m'} \tag{7.6}$$

our friend Pythagoros and Equation 7.4 allow us to write

$$k_z = k_0\sqrt{1 - \frac{\lambda^2 m'^2}{4t^2}} = k_0 \sin\theta_{m'} \tag{7.7}$$

We see two things immediately: the propagation constant down the waveguide is also discrete, and its value decreases as the order of the mode (the value of m′) increases.

This relationship becomes even more useful if we dispense with the assumption that n = 1 and include the dielectric constant of the waveguide material, $n = n_g$. In a material with arbitrary n_g, the group velocity, v, wavelength, λ, and propagation constant, k, become

$$v_n = \frac{c}{n_g}$$
$$\lambda_n = \frac{\lambda_0}{n_g} \tag{7.8}$$
$$k_n = n_g k_0$$

respectively. Since we employ the wavelength a great deal, however, we usually write $\lambda \equiv \lambda_n$, and use λ_0 only when we explicitly mean the vacuum wavelength. Therefore, in a material of refractive index n_g, k_0 becomes $k_n = n_g k_0 = 2\pi n_g/\lambda_0$, so that the propagation constant in the **z** direction, from Equation 7.7, may be written now as

$$k_z = n_g k_0 \sqrt{1 - \frac{\lambda^2 m'^2}{4t^2}} = n_g k_0 \sin\theta_{m'} \tag{7.9}$$

By defining

$$N = n_g \sin\theta_m' \tag{7.10}$$

where N [dimensionless] is referred to as the mode effective index, we can write the **z**-directed propagation constant as

$$\beta \equiv k_z = N k_0 \tag{7.11}$$

where the variable β [m^{-1}] is often used in lieu of k_z, and we shall do so below. The mode in the waveguide thus sees an effective refractive index of N, which means that all the zigging and zagging can be ignored and the wave is imagined to propagate straight down a waveguide with index N. The effective mode propagation constant is then given by β. We remember, of course, that every mode has a different N and β, implying that each mode moves with a different speed through the waveguide.

7.2 DIELECTRIC WAVEGUIDES: THE RAY OPTIC MODEL

Perfectly reflecting mirrors represent a useful idealization, albeit one that obscures many properties. We will now treat a more realistic optical waveguide structure, as shown in Figure 7.4. Rather than relying on reflecting mirrors, the waves will now be guided by total internal reflection from the interfaces between the materials of varying refractive index. The waveguiding layer, or core, has an index, n_g, the lower cladding, or substrate, n_s, and the upper cladding, referred to as the superstrate or cap, n_c. In order to guide a wave in the core, we require that $n_g > n_s$, n_c. For many semiconductor waveguide structures, the upper and lower claddings will be identical

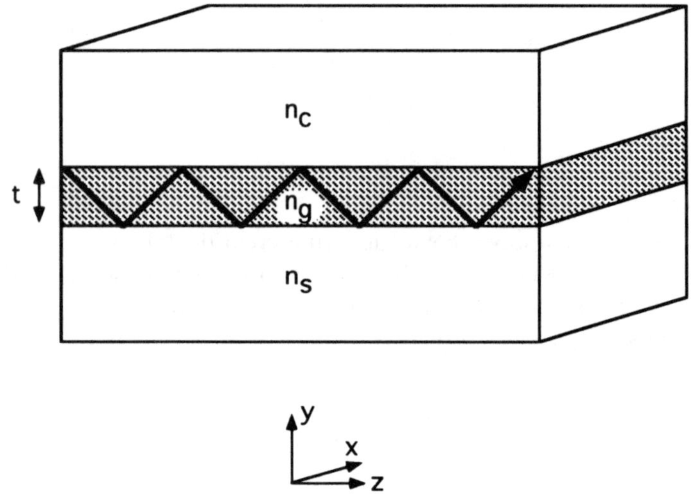

Figure 7.4 A dielectric waveguide, with a waveguide core of $n = n_g$, an upper cladding, or cap, with $n = n_c$, and a lower cladding, or substrate, with $n = n_s$. The symmetric waveguide has $n_c = n_s$.

so that $n_c = n_s$ for a so-called symmetric waveguide. In the case where the upper cladding is air, which also occurs frequently, we would have $n_g > n_s > n_c$. We usually find, then, that

$$n_g > n_s \geq n_c \tag{7.12}$$

As we will see in greater detail in the next chapter, this dielectric sandwich can be fabricated in a number of ways. For III-V based waveguides, for example, the dielectric slabs usually correspond to MOCVD- or MBE-grown layers. A typical structure might include a lower cladding of $Al_{0.8}Ga_{0.2}As$, one to several microns thick, a waveguide core of $Al_{0.3}Ga_{0.7}As$, typically several hundred nanometers thick, and a final $Al_{0.8}Ga_{0.2}As$ upper cladding, again about a micron thick. Such a structure represents one useful waveguide for the 800-nm wavelength range, but a great number of variations exist on this design. In a silicon-based waveguide system, we might employ a Si_3N_4 core on a SiO_2 lower cladding.

7.2.1 Total Internal Reflection

We can begin an analysis of the dielectric slab using the tools of Chapter 6 to obtain a new picture of the maximum number of modes that can be supported in a waveguide structure with nonperfectly reflecting walls; the need for total internal reflection now also restricts the reflection angle of the modes. In Equation 7.4, the maximum mode number was limited because of the argument that the $\cos^{-1}$ must be less than unity. In the dielectric sandwich of Figure 7.4, total internal reflection is required for propagation of a mode, and we remember from the last chapter that this took place for angles greater than the critical angle, θ_c. For the two interfaces we have now, the critical angles are

$$\theta_{c(\text{substrate})} = \sin^{-1}\left(\frac{n_s}{n_g}\right) \tag{7.13}$$

and

$$\theta_{c(\text{cap})} = \sin^{-1}\left(\frac{n_c}{n_g}\right) \tag{7.14}$$

where $\theta_{c(\text{substrate})} < \theta_{c(\text{cap})}$ for $n_s > n_c$. Thus, as the order of the mode increases and θ decreases, the angle of incidence for the lower interface will eventually drop below the critical angle. The wave will begin to leak into the substrate, and as the order of the mode increases further, the wave will finally leak into the cap as well.

A true dielectric waveguide, then, will require the use of both phase-matching and total-internal-reflection arguments to determine the allowed modes. We will now undertake this analysis for a number of cases.

7.2.2 Symmetric Waveguides

In Section 7.1, we determined that only discrete modes would propagate in the parallel mirror waveguide due to arguments about phase matching between successively reflected wavefronts in the guide. For the dielectric waveguide, this is not quite so simple due to the more complex dependence of phase shift on incident angle. Recall that in Equation 7.3, we had a phase shift of π for each reflection, hence the factor 2π in the left-hand side. But in Chapter 6, Equations 6.35 and 6.36 as summarized in Figure 6.13 gave us expressions for the phase shift upon total internal reflection when the walls are not perfectly reflecting. Thus we have to write the phase matching condition (Equation 7.3) as

$$\frac{4\pi t}{\lambda}\cos\theta - \phi_u - \phi_l = 2m\pi \tag{7.15}$$

where ϕ_u and ϕ_l [rad] are the phase shifts at the upper and lower interfaces, respectively, and $m = 0,1,2, \ldots$.

Since in general $n_s \neq n_c$, ϕ_u is not necessarily the same as ϕ_l. We can postpone that more complex calculation to a later section and treat the symmetric waveguide first, so that $n_s = n_c$ and $\phi_l = \phi_u$. In that case, Equation 7.15, using Equation 6.35 and assuming a TE incident wave, becomes

$$\frac{4\pi t}{\lambda}\cos\theta - 4\tan^{-1}\sqrt{\frac{\sin^2\theta - \sin^2\theta_c}{\cos^2\theta}} = 2m\pi \tag{7.16}$$

where θ is the incident angle and θ_c the critical angle, $\theta_c = \sin^{-1}(n_s/n_g)$; we have taken the cladding on both sides of the core to have index n_s. This relation may be usefully be massaged into the form

$$\tan\left(\frac{\pi t}{\lambda}\cos\theta - \frac{m\pi}{2}\right) = \frac{\sqrt{\sin^2\theta - \left(\frac{n_s}{n_g}\right)^2}}{\cos\theta} \tag{7.17}$$

yielding an eigenvalue equation in which the only variable is θ, but which must be subdued by numerical techniques. The case for a TM incident wave is similar; using Equation 6.36 for the phase shift ϕ, its eigenvalue equation is seen to be

$$\tan\left(\frac{\pi t}{\lambda}\cos\theta - \frac{m\pi}{2}\right) = \frac{\sqrt{\left(\frac{n_g}{n_s}\right)^2 \sin^2\theta - 1}}{\left(\frac{n_s}{n_g}\right)\cos\theta} \tag{7.18}$$

again only a function of θ. It is important to note in both these expressions is that λ is the wavelength in the material, so that $\lambda = \lambda_0/n_g$. Equations 7.17 or 7.18 are tackled numerically to yield the value of θ corresponding to a particular m^{th} order mode. Once we know θ, the effective index and propagation constant can be immediately determined, as we will see below.

We can analyze Equation 7.17 somewhat without actually solving it. The left-hand side and right-hand side are equal only for discrete values of θ, each for a particular value of m. The left-hand side is periodic with

$$\cos\theta = m\frac{\lambda}{2t} \tag{7.19}$$

with $m = 0,1,2 \ldots m_{max}$ and the right-hand side decreases monotonically as θ decreases with a minimum value for θ of $\theta = \theta_c$. We remember that the lowest order mode (smallest m) has the largest θ and θ decreases as m increases. Realizing this, we can see that the maximum value for m is given by

$$m_{max} = \text{int}\left[\frac{2t\cos\theta_c}{\lambda}\right] = \text{int}\left[\frac{2t}{\lambda}\sqrt{1 - \frac{n_s^2}{n_g^2}}\right] \tag{7.20}$$

where int means the greatest integer value of the expression. The total number of allowed modes is then given by $m_{max} + 1$, which includes the mode m = 0. Note that, unlike Equation 7.4, Equations 7.17 and 7.18 allow a solution for m = 0 which implies that the lowest order mode is always allowed; in other words, there is *no cutoff* for the lowest order mode in a symmetric dielectric waveguide.

To illustrate how the mode angles behave, we have solved Equation 7.17 and 7.18 and plotted the reflection angle, θ, as a function of cladding refractive index, n_s, for $n_g = 3.35$, waveguide thickness $t = 500$ nm, and wavelength $\lambda = 850$ nm in Figure 7.5. Both the TE and TM polarizations for the m = 0 and m = 1 modes are shown. The difference between polarizations is relatively small, but it is clear that

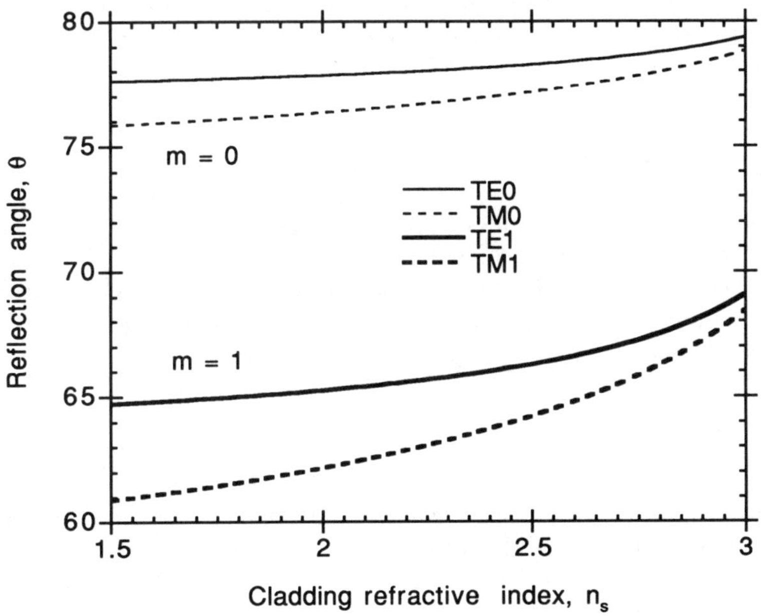

Figure 7.5 Reflection angle θ derived from solutions of Equations 7.17 and 7.18 as a function of cladding refractive index, n_s, for $n_g = 3.35$, $t = 500$ nm, and $\lambda = 850$ nm. This is the symmetric waveguide, implying that the core is surrounded on both sides with material of $n = n_s$. The 0th order TE0 and TM0 plus 1st order TE1 and TM1 modes are shown.

the higher order modes have considerably smaller incidence angles. In addition, as the cladding index increases, the incidence angle rises, as we would expect since we approach grazing incidence for closely matched core and cladding indices.

Example

Assume we have a 1,000-nm thick AlGaAs waveguide with an $Al_{0.3}Ga_{0.7}As$ core (n_g = 3.35) and an $Al_{0.8}Ga_{0.2}As$ cladding (n_s = 3.12) propagating light at λ_0 = 850 nm. The effective wavelength in the waveguide is given by $\lambda = \lambda_0/n_g = 254$ nm. From Equation 7.20, we find that $m_{max} = 2$, implying that the total number of allowed modes is 3 including the 0th order mode.

If we now want to make a monomode waveguide from these materials, we find from Equation 7.22 that the maximum allowed thickness of the waveguide core would be 329 nm. If, on the other hand, we want a waveguide thickness of 1,000 nm but are willing to choose other materials in order to guarantee monomode operation, Equation 7.23 tells us that the maximum allowed index step is $\Delta n = 0.03$. If we stick with the $Al_{0.3}Ga_{0.7}As$ core, that would require a cladding of about $Al_{0.35}Ga_{0.65}As$. We see from this example a bit of the interplay of the various parameters with which we can design a waveguide. One typically has both material types as well as layer thicknesses to juggle about, giving useful flexibility in the design.

7.2.3 Monomode Conditions

For many applications, waveguides must be designed to allow only the single 0th order mode; this is important for interferometry or structures in which the dispersion due to the variable velocity of different modes is important. For TE modes, we can see from Equation 7.17 that a dielectric waveguide is monomode (only the m = 0 mode allowed) when

$$\cos\theta_c < \frac{\lambda}{2t} \tag{7.21}$$

This inequality can be rewritten as a condition on thickness or refractive index difference between the waveguide and cladding. The former yields

$$t < \frac{1}{2}\lambda\left(1 - \frac{n_s^2}{n_g^2}\right)^{-\frac{1}{2}} \tag{7.22}$$

where t is the maximum thickness for which the guide is still monomode; for larger t, higher order modes will be allowed. Alternatively, we can define

$$\Delta \mathrm{n} \equiv \mathrm{n}_g - \mathrm{n}_s < \mathrm{n}_g\left(1 - \sqrt{1 - \frac{\lambda^2}{4t^2}}\right) \tag{7.23}$$

as the maximum tolerable index step if only a single mode is to be allowed. If the index difference between core and cladding is higher, again, more than one mode will be guided.

These equations imply that, for a given set of waveguide materials, thinner waveguides will support less modes, as we expect. Alternatively, for a given waveguide thickness, a smaller index step Δn will reduce the number of modes supported. A larger index difference between core and cladding will always increase the number of modes allowed, which will also become important later in designing ridge waveguide width.

7.2.4 Propagation Constants and Effective Index, Again

We can define the effective propagation constant of a wave zigging and zagging through a dielectric waveguide as we did for the perfect mirror guide. As before (Equations 7.9–7.11), we have

$$\beta \equiv k_z = \mathrm{n}_g k_0 \sin\theta = \mathrm{N} k_0 \tag{7.24}$$

since

$$\mathrm{N} = \mathrm{n}_g \sin\theta \tag{7.25}$$

as before. The values for θ must now be obtained by solving Equations 7.17 or 7.18 numerically. However, we can place bounds on β—since we know that $\theta \geq \theta_c$, we can obtain the condition on the minimum value of β

$$\beta \geq \mathrm{n}_g k_0 \sin\theta_c = k_0 \mathrm{n}_s \tag{7.26}$$

giving a lower bound on the effective propagation constant for confined modes. Note that the refractive index of the cladding, n_s, is the relevant factor.

We can also see, by consulting Figure 7.3b, that the maximum value of β is $\mathrm{n}_g k_0$, which would occur for $\theta = 90°$. Thus we have a range of allowed, discrete effective propagation constant values, namely

$$k_0 \mathrm{n}_g \geq \beta \geq k_0 \mathrm{n}_s \tag{7.27}$$

which directly permits us to define the range of effective waveguide indices, N, as

$$n_g \geq N \geq n_s \tag{7.28}$$

Thus, not surprisingly, the refractive indices of both the core and cladding define the propagation behavior of the transmitted modes.

We can see this clearly in Figure 7.6, which plots the effective index of the waveguide that we solved in Figure 7.5 as a function of the cladding index n_s by simple application of Equation 7.25. The 0th order mode changes little with cladding index, slowly approaching $n_g = 3.35$ as the cladding index rises.

7.2.5 Asymmetric Waveguides

Earlier, we considered the symmetric dielectric waveguide, which has the same cladding material on both sides of the core, implying $n_c = n_s$. The asymmetric waveguide, where the upper and lower claddings are different, can be solved in precisely the same manner, except that Equation 7.15 will have two different terms for phase shift upon reflection [2]. The eigenvalue equation for TE polarization then becomes

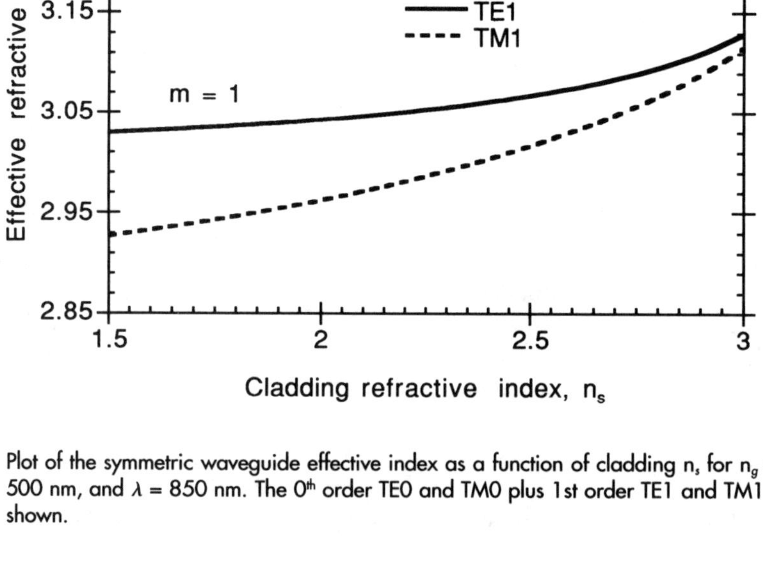

Figure 7.6 Plot of the symmetric waveguide effective index as a function of cladding n_s for $n_g = 3.35$, t = 500 nm, and λ = 850 nm. The 0th order TE0 and TM0 plus 1st order TE1 and TM1 modes are shown.

$$\left(\frac{2\pi t}{\lambda}\right)\cos\theta - m\pi = \tan^{-1}\sqrt{\frac{\sin^2\theta - \left(\frac{n_c}{n_g}\right)^2}{\cos^2\theta}} + \tan^{-1}\sqrt{\frac{\sin^2\theta - \left(\frac{n_s}{n_g}\right)^2}{\cos^2\theta}} \quad (7.29a)$$

which can be transformed into

$$\tan\left(\frac{2\pi t}{\lambda}\cos\theta - m\pi\right) = \frac{\cos\theta\left[\sqrt{\sin^2\theta - \left(\frac{n_c}{n_g}\right)^2} + \sqrt{\sin^2\theta - \left(\frac{n_s}{n_g}\right)^2}\right]}{\cos^2\theta - \sqrt{\left(\sin^2\theta - \left(\frac{n_c}{n_g}\right)^2\right)\left(\sin^2\theta - \left(\frac{n_s}{n_g}\right)^2\right)}} \quad (7.29b)$$

For TM polarization, the same considerations lead to its own eigenvalue equation

$$\left(\frac{2\pi t}{\lambda}\right)\cos\theta - m\pi = \tan^{-1}\sqrt{\frac{\left(\frac{n_g}{n_c}\right)^2 \sin^2\theta - 1}{\left(\frac{n_c}{n_g}\right)^2 \cos^2\theta}} + \tan^{-1}\sqrt{\frac{\left(\frac{n_g}{n_s}\right)^2 \sin^2\theta - 1}{\left(\frac{n_s}{n_g}\right)^2 \cos^2\theta}} \quad (7.30a)$$

or equivalently

$$\tan\left(\frac{2\pi t}{\lambda}\cos\theta - m\pi\right) = \frac{\cos\theta\left[\frac{n_g}{n_c}\sqrt{\left(\frac{n_g}{n_c}\right)^2 \sin^2\theta - 1} + \frac{n_g}{n_s}\sqrt{\left(\frac{n_g}{n_s}\right)^2 \sin^2\theta - 1}\right]}{\cos^2\theta - \frac{n_g^2}{n_c n_s}\sqrt{\left[\left(\frac{n_g}{n_c}\right)^2 \sin^2\theta - 1\right]\left[\left(\frac{n_g}{n_s}\right)^2 \sin^2\theta - 1\right]}} \quad (7.30b)$$

Numerical solution of Equations 7.29 and 7.30 yields the variation of effective index as a function of changing cap refractive index, n_c, while keeping the substrate constant at $n_s = 3.12$ and with a guiding layer $n_g = 3.35$, as shown in Figure 7.7. We see that the effective index, N, rises from near its minimum value of n_s toward its maximum n_g, the guide index, as the cap increases in refractive index. As the cap index approaches the value for the core, the mode will be increasingly poorly guided. Despite the same large variation of upper cladding index as in Figures 7.5 and 7.6, the change in effective index is much smaller because the fixed lower cladding limits the possible reflection angles.

A relation similar to Equation 7.28 can be written for the asymmetric waveguide, namely

$$n_g \geq N \geq n_x \quad (7.31)$$

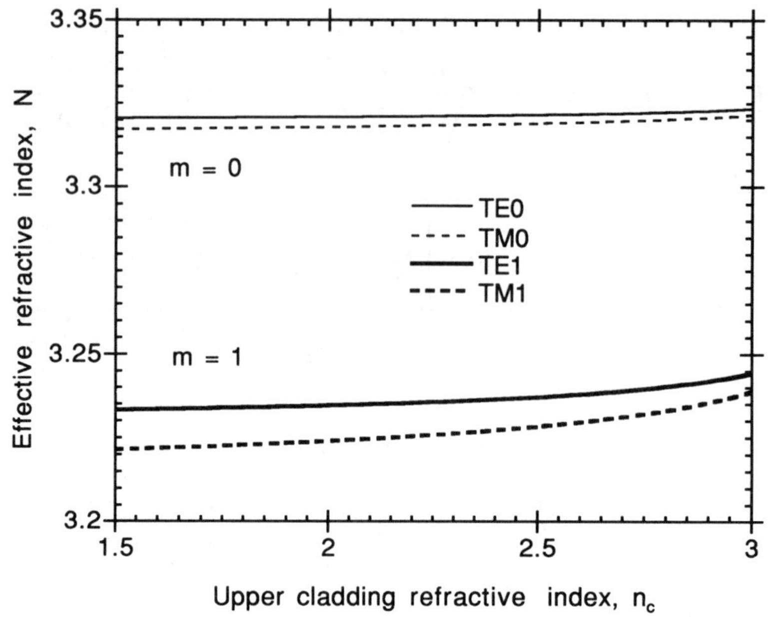

Figure 7.7 Variation of effective index of an asymmetric waveguide as a function of cap refractive index, n_c, with $n_s = 3.12$ and $n_g = 3.35$, $t = 800$ nm, and $\lambda = 850$ nm. The 0th order TE0 and TM0 plus 1st order TE1 and TM1 modes are shown.

where n_x is the *larger* of n_c, n_s. The effect of two different cladding layers on the propagation of a waveguide mode may be clarified by slowly increasing the angle of incidence, θ, from 0. Assuming that the upper cladding is of lower index than the lower, $n_c < n_s$, as is the case for most real structures, we see in Figure 7.8(a) that for small angles, the mode leaks out through the surface and through the substrate, so we cannot speak of a guided mode at all. When the angle between the guide and cap reaches $\theta_{c(\text{cap})}$, the wave is internally reflected from the top surface, as seen in Figure 7.8(b), but still leaks out through the substrate. This is because, since $n_c < n_s$, $\theta_{c(\text{cap})} < \theta_{c(\text{substrate})}$. This case is known as a substrate radiation mode. Finally, as we continue to increase θ until $\theta > \theta_{c(\text{substrate})}$, we have internal reflection at the lower interface as well, and the mode is properly confined and can happily propagate along without leaking out anywhere. This is the desired condition for most waveguide structures.

7.3 ELECTROMAGNETIC MODEL

The ray-optic model has been a useful tool to visualize the existence of and calculate the modes in both mirror and dielectric waveguides. An electromagnetic calculation

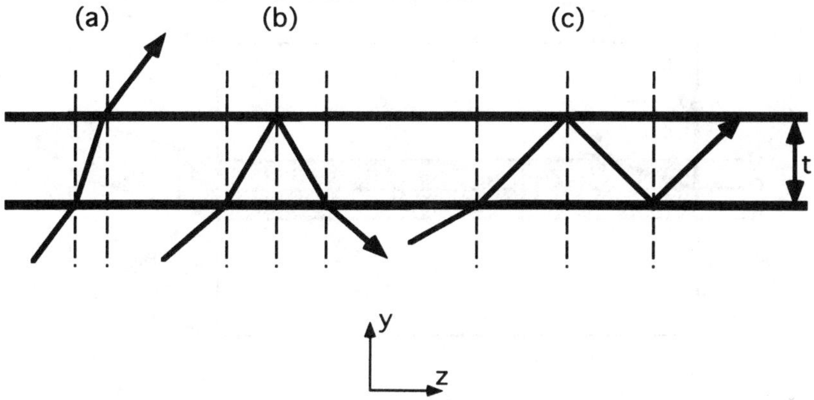

Figure 7.8 Cross section of a slab waveguide showing the three possible types of modes for the asymmetric waveguide: (a) nonguided mode, (b) substrate mode, and (c) guided mode.

based on solution of the wave equation from Chapter 1 will allow us to calculate the distribution of the fields. Knowledge of the field distribution is important, for example, when calculating the overlap integral for coupling out of a waveguide or determining the confinement factor for a laser. We will approach this task in two parts: first we will derive another simple eigenvalue equation to determine what kinds of solutions we expect, and then we'll solve the wave equation explicitly to determine the fields.

7.3.1 Eigenvalue Equation for a Slab Waveguide

Recall that we had the wave, or Helmholtz, equation in Chapter 1 (Equation 1.15),

$$\nabla^2 \mathcal{E} = \epsilon_m \mu_m \frac{\partial^2 \mathcal{E}}{\partial t^2} = \frac{1}{v^2} \frac{\partial^2 \mathcal{E}}{\partial t^2} \tag{7.32}$$

where velocity, v, is given by c/n for a material with refractive index n. Since this equation has solutions of the form

$$\mathcal{E}(r, t) = \mathcal{E}_0(r) e^{j\omega t} \tag{7.33}$$

we can write an eigenvalue equation, by remembering that $k_0 = \omega/c$,

$$\nabla^2 \mathcal{E} + (\mathrm{n} k_0)^2 \mathcal{E} = 0 \tag{7.34}$$

Consider the dielectric waveguide structure of Figure 7.9, which is the same as the one that we studied using the ray-optic model. Letting the wave propagate only in the **z** direction yields solutions to 7.34 of the form

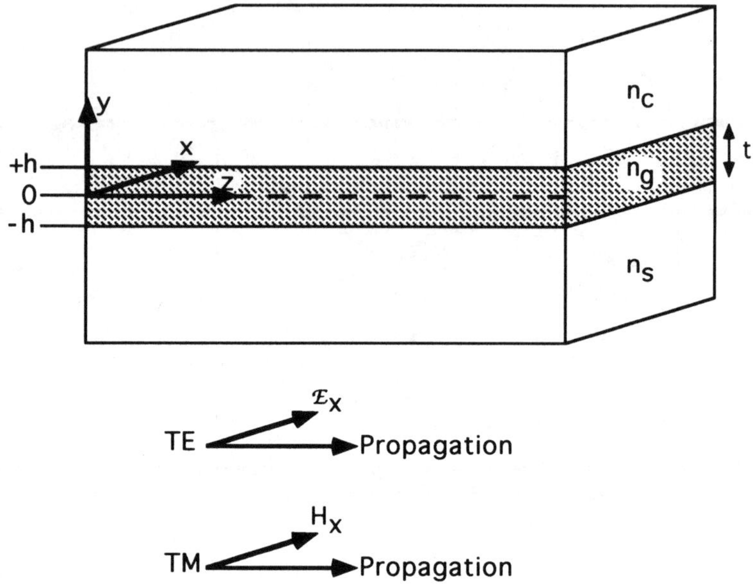

Figure 7.9 The dielectric waveguide, with a waveguide core $n = n_g$, a cap $n = n_c$, and a substrate $n = n_s$, which provide the boundary conditions for solution of the wave equation. The polarizations for TE and TM are also shown.

$$\mathscr{E}(r, t) = \mathscr{E}_0(x, y)e^{-j\beta z}e^{j\omega t} \tag{7.35}$$

remembering that $\beta \equiv k_z$ in the waveguide layer; notice the similarity to Equation 3.48. Since the electric field extends to infinity in **x**, it is uniform in that direction, and the only variation of $\mathscr{E}$ is in the **y** direction, allowing the differential to reduce to

$$\nabla^2\mathscr{E} \rightarrow \frac{\partial^2\mathscr{E}}{\partial y^2} - \beta^2 \tag{7.36}$$

The eigenvalue equation then becomes

$$\frac{\partial^2\mathscr{E}}{\partial y^2} + (k_0^2 n_i^2 - \beta^2)\mathscr{E} = 0 \tag{7.37a}$$

or equivalently, by defining $k_{yi}^2 = k_0^2 \cdot n_i^2 - \beta^2$,

$$\frac{\partial^2\mathscr{E}}{\partial y^2} + k_{yi}^2\mathscr{E} = 0 \tag{7.37b}$$

In this case, the subscript i stands for i = c,g,s for the cap, guide, and substrate portions of the waveguide, respectively; we have had to add to our notational pudding since we will determine both **y** and **z** directed propagation constants in all three regions.

Equation 7.37 is then the wave equation for an **x**-directed wave propagating in the **z** direction and guided by reflecting dielectric boundaries in the **y** direction. Before solving it explicitly, we can look at the types of solutions that are expected.

7.3.2 Propagating and Decaying Solutions

Solutions for Equation 7.37 can take two forms, depending on the sign of the second term. For $(k_0^2 n_i^2 - \beta^2) > 0$, $\mathcal{E}$ must have a solution of the form $\mathcal{E}_0 \exp(\pm jk_{yi}y)$, corresponding to a sum of sinusoids. On the other hand, for $(k_0^2 n_i^2 - \beta^2) < 0$, $\mathcal{E}$ must have a solution of the form $\mathcal{E}_0 \exp(\pm k_{yi}y)$ that corresponds to an exponential decay in the **y** direction. Thus the relation of the mode propagation constant, β, to the material characteristic propagation constant $(k_0 n_i)$ defines whether the mode propagates (as a sinusoid) or decays (exponentially) and disappears. We can now address the specific cases that arise in a waveguide structure. If, for the waveguide of Figure 7.9, we have $n_g > n_s > n_c$, which is often an intended design, there are four possibilities for the propagation constant β:

1. $\beta > k_0 n_g$: The solutions in all three areas (cap, guide, and substrate) are of the form $\mathcal{E}_0 \exp(\pm k_{yi}y)$, where we take the + solution for $y < 0$ and the – one for $y > 0$ in order to avoid an unbounded increase of the field away from the boundaries, which would be unphysical. These exponentially decaying solutions imply that no wave propagates in the waveguide, and any injected mode will decay to zero, so it rarely represents an interesting case. It corresponds qualitatively to Figure 7.8(a), derived from our ray optic model.
2. $k_0 n_g > \beta > k_0 n_s$: In this case, the guide region has a solution $\mathcal{E}_0 \exp(\pm jk_{yi}y)$, which is a propagating sinusoid. In the cap and substrate, the solutions are still exponentially decaying. This is usually the most interesting case because it corresponds to a mode only in the center of the waveguide and a decrease of field on the other sides of the boundary; it is the condition of Figure 7.8(c).
3. $k_0 n_s > \beta > k_0 n_c$: Here propagating solutions of the form $\mathcal{E}_0 \exp(\pm jk_{yi}y)$ exist in the guide and in the substrate, whereas the decaying case of $\mathcal{E}_0 \exp(\pm k_{yi}y)$ is still found in the cap. This situation is known as a *substrate radiation mode* and implies loss of mode energy into the substrate, as we would have in Figure 7.8(b).
4. $k_0 n_c > \beta$: Finally, this case provides propagating solutions in all three layers—cap, guide, and substrate. Such a situation of oscillatory behavior in all layers implies that there is no guiding and is thus also not usually interesting; it again corresponds to Figure 7.8(a).

7.3.3 Propagation and Decay Constants

From the wave equation (Equation 7.37) and the above discussion, we can easily write down handy expressions for the propagation and decay constants, k_c, k_g, and k_s in the **y** direction. The second item in the list, which we'll call Case 2, is usually the intended situation for a waveguide designer, and if we assume that the situation $k_0 n_g > \beta > k_0 n_s$ as satisfied, we obtain

$$k_{yc} = \sqrt{\beta^2 - k_0^2 n_c^2} \tag{7.38}$$

$$k_{yg} = \sqrt{k_0^2 n_g^2 - \beta^2} \tag{7.39}$$

$$k_{ys} = \sqrt{\beta^2 - k_0^2 n_s^2} \tag{7.40}$$

The first and last represent exponentially decaying fields, and the middle represents a propagating one. Remembering that $k_0 = \omega/c$ and $\beta = Nk_0$, we can rewrite the **y**-directed propagation constant in the waveguide, k_{yg}, as

$$k_{yg} = \sqrt{k_0^2 n_g^2 - \beta^2} = k_0 \sqrt{n_g^2 - \frac{\beta^2}{k_0^2}} = k_0 \sqrt{n_g^2 - N^2} \tag{7.41}$$

and likewise the decay constants in the claddings

$$k_{yc} = k_0 \sqrt{N^2 - n_c^2} \tag{7.42}$$

$$k_{ys} = k_0 \sqrt{N^2 - n_s^2} \tag{7.43}$$

As before, when we used the ray-optic model and did not concern ourselves with electric or magnetic fields, the values for N lie in the range $n_s < N < n_g$ for the usual situation where $n_s > n_c$.

Recalling that $N = n_g \cdot \sin\theta$, we can write the decay constant into the substrate (decay rather than propagation because we have an $\exp(-k_{ys})$ behavior) as

$$k_{ys} = k_0 \sqrt{n_g^2 \sin^2\theta - n_s^2} \tag{7.44}$$

Because the angle of incidence, θ, becomes smaller as the mode number increases, it is clear that higher order modes have progressively smaller values for k_{ys}. That means greater penetration of the mode into the substrate and thus poorer confinement. The same argument holds for the decay of the mode into the cap.

Example

We can look at the penetration of a mode into the $Al_{0.8}Ga_{0.2}As$ cladding ($n_c = 3.12$) of a 500-nm thick waveguide with a $Al_{0.3}Ga_{0.7}As$ core ($n_g = 3.35$) at a wavelength of $\lambda = 850$ nm. By solving Equation 7.17, we determine an effective index of N = 3.299 for the m = 0 mode and N = 3.166 for the m = 1 mode. Because $k_{ys} = k_0\sqrt{N^2 - n_s^2}$, we find that $k_{ys} = 7.92\ \mu m^{-1}$ for the 0th order and $k_{ys} = 3.97\ \mu m^{-1}$ for the 1st order mode. The 1/e penetration depths are then $1/k_{ys}$ and have values of 126 nm and 252 nm, respectively. It is clear from these results that higher order modes are progressively more poorly confined. We also see that the decay constants, k_{ys} and k_{yc}, are identical to the extinction coefficient, α', from the end of Chapter 6.

7.4 DIELECTRIC WAVEGUIDES: ELECTRIC AND MAGNETIC FIELDS

Solving the wave equation explicitly now allows us to determine the fields in the waveguide, and we can thereby verify the results we derived earlier using those more heuristic techniques [3]. We will treat the case for guiding in the core layer (Case 2 in Section 7.3.2) because that is usually the desired situation. In Figure 7.9, we had sketched our dielectric slab waveguide of thickness t; for mathematical convenience, let's take the extent of the core to be $-h \le y \le +h$ so that the total thickness is $t = 2h$.

Both the TE ($\mathcal{E}_x$, H_y, H_z) and TM (H_x, $\mathcal{E}_y$, $\mathcal{E}_z$) field components are indicated in the figure. TE implies that the electric field is in the plane of the waveguide; TM means that the magnetic field is oriented in that direction. Let us direct our attention to the TE case.

7.4.1 Field Solutions

For TE, we only have the electric field component $\mathcal{E}_x(y, z, t)$, so the solutions of the wave equation (Equation 7.32) are of the form

$$\mathcal{E}(r, t) = \mathcal{E}_x(y)e^{-jk_z z}e^{j\omega t} \tag{7.45}$$

This implies an **x**-directed field with a sinusoidal time dependence, which propagates sinusoidally in the **z** direction. We now need to solve for its behavior in the **y** direction using the boundary conditions imposed by the waveguide structure. Assuming a propagating solution only in the core, we have three cases for the three regions of Figure 7.9:

$$\text{Cladding} \qquad \mathcal{E}_x(y) = \mathcal{E}_c e^{-k_{yc}(y-h)} \qquad y \ge h \tag{7.46}$$

Core $\quad \mathscr{E}_x(y) = \mathscr{E}_g \cos(k_{yg}y + \phi) \quad h \le y \le -h \quad (7.47)$

Substrate $\quad \mathscr{E}_x(y) = \mathscr{E}_s e^{k_{ys}(y+h)} \quad y \le -h \quad (7.48)$

where the k_{yg}, k_{yc}, and k_{ys} are given by Equations 7.38–7.40. The prefactors $\mathscr{E}_c$, $\mathscr{E}_g$, and $\mathscr{E}_s$ are constants that we will now determine. The phase ϕ allows us to take a single sinusoidal solution as opposed to the general sum of sinusoids.

We now apply the boundary conditions that the electric field, $\mathscr{E}$, and its derivative, $\partial\mathscr{E}/\partial y$, the latter being equivalent to H_z (see Equation 1.35), must be continuous at the boundaries y = $\pm h$. The continuity of electric field yields

$$\mathscr{E}_c = \mathscr{E}_g \cos(k_{yg}h + \phi) \qquad \text{at } y = h \qquad (7.49)$$

and

$$\mathscr{E}_s = \mathscr{E}_g \cos(-k_{yg}h + \phi) \qquad \text{at } y = -h \qquad (7.50)$$

From the continuity of the derivative, we can derive

$$\mathscr{E}_c = \frac{k_{yg}}{k_{yc}}\mathscr{E}_g \sin(k_{yg}h + \phi) \qquad \text{at } y = h \qquad (7.51)$$

and

$$\mathscr{E}_s = \frac{k_{yg}}{k_{ys}}\mathscr{E}_g \sin(k_{yg}h - \phi) \qquad \text{at } y = -h \qquad (7.52)$$

By manipulation of these four boundary conditions, we can then derive the eigenvalue equation from which we can find the propagation constants again. Equating 7.49 and 7.51, we find

$$\tan(k_{yg}h + \phi) = \frac{k_{yc}}{k_{yg}} \qquad (7.53)$$

and by equating 7.50 and 7.52, we are left with

$$\tan(k_{yg}h - \phi) = \frac{k_{ys}}{k_{yg}} \qquad (7.54)$$

Taking $\tan^{-1}$ and adding Equations 7.53 and 7.54 then leaves the eigenvalue equation

$$2k_{yg}h = \tan^{-1}\left(\frac{k_{yc}}{k_{yg}}\right) + \tan^{-1}\left(\frac{k_{ys}}{k_{yg}}\right) + m\pi \tag{7.55}$$

where m = 0,1,2, ... For the special case of the symmetric waveguide, where $k_c = k_s$, as we also considered in an earlier section, this eigenvalue equation simplifies a touch to

$$2k_{yg}h = 2\tan^{-1}\left(\frac{k_{ys}}{k_{yg}}\right) + m\pi \tag{7.56}$$

The analysis for TM modes proceeds in precisely the same manner. We have an additional $1/n^2$ term from the boundary conditions of continuous H_x and $\mathcal{E}_z$, as we can see from Equation 1.32. The TM eigenvalue equation is then given by

$$2k_{yg}h = \tan^{-1}\left(\frac{n_g}{n_c}\right)^2\left(\frac{k_{yc}}{k_{yg}}\right) + \tan^{-1}\left(\frac{n_g}{n_s}\right)^2\left(\frac{k_{ys}}{k_{yg}}\right) + m\pi \tag{7.57}$$

which reduces to

$$2k_{yg}h = 2\tan^{-1}\left(\frac{n_g}{n_s}\right)^2\left(\frac{k_{ys}}{k_{yg}}\right) + m\pi \tag{7.58}$$

for the case of a symmetric waveguide.

By expressing all the k_y terms as a function of β (Equations 7.38–7.40), by writing β in terms of N (Equation 7.24 and 7.25) and by realizing that the wavelength λ is given by λ_0/n_g, we can rewrite the asymmetric TE and TM eigenvalue equations in yet another form, namely

$$2hk_0\sqrt{n_g^2 - N^2} - m\pi = \tan^{-1}\left(\sqrt{\frac{N^2 - n_c^2}{n_g^2 - N^2}}\right) + \tan^{-1}\left(\sqrt{\frac{N^2 - n_s^2}{n_g^2 - N^2}}\right) \tag{7.59}$$

for TE and

$$2hk_0\sqrt{n_g^2 - N^2} - m\pi = \tan^{-1}\left(\left(\frac{n_g}{n_c}\right)^2\sqrt{\frac{N^2 - n_c^2}{n_g^2 - N^2}}\right) + \tan^{-1}\left(\left(\frac{n_g}{n_s}\right)^2\sqrt{\frac{N^2 - n_s^2}{n_g^2 - N^2}}\right) \tag{7.60}$$

for TM.

By remembering how all the k's, N, and θ are related, it is easy to show that Equations 7.29, 7.55, and 7.59 are all exactly equivalent expressions for the TE modes in an asymmetric waveguide and Equations 7.30, 7.57, and 7.60 are all equivalent expressions for the TM modes. Thus the eigenvalue equations for determining the propagation constants of various modes in a waveguide can equivalently be derived by our ray-optic model and the electromagnetic model. Once solved for β (or θ or N), these equations then yield all the necessary k_y values for the field equations Equations 7.46–7.48, and the only remaining unknown is the constant term, $\mathcal{E}_g$.

7.4.2 Power Normalization

This remaining constant $\mathcal{E}_g$ is to some extent arbitrary, since we are usually only interested in relative intensities in the three regions. However, it is customary to normalize the field intensities by defining the power propagating down the waveguide as unity (i.e., 1 W/m, power per waveguide width); this power is calculated by the integral

$$\frac{1}{2}\int_{-\infty}^{\infty} \mathcal{E}_x H_y^* dx = 1[\text{W/m}] \tag{7.61}$$

Because Maxwell's equations relate the two field components by

$$H_y = -\frac{j}{\omega\mu_m}\frac{\partial \mathcal{E}_x}{\partial z} = \frac{k_z}{\omega\mu_m}\mathcal{E}_x \tag{7.62}$$

we can then write the integral as

$$\frac{k_z}{2\omega\mu_m}\int_{-\infty}^{\infty} |\mathcal{E}_x^2| dy = 1 \tag{7.63}$$

Because we know $\mathcal{E}_x$ for all three regions (Equations 7.46–7.48), and because the constants $\mathcal{E}_c$ and $\mathcal{E}_s$ are given by Equations 7.49 and 7.50, and finally because all the k's can be expressed as a function of β, Equation 7.63 can be evaluated easily. For example, using the field distributions of Equations 7.46–7.48, we can write $\mathcal{E}_g$ as

$$\mathcal{E}_g = \sqrt{\frac{2c\mu_0}{\text{N}\left[\frac{1}{2k_{ys}} + \frac{1}{2k_{yc}} + \frac{1}{2k_{yg}}\sin(2k_{yg}h)\right]}} \tag{7.64}$$

for speed of light, c, and free-space permeability, μ_0. We note, however, that this normalization is arbitrary, based on our initial condition of Equation 7.61.

7.4.3 Symmetric Waveguide: Cutoff Condition

Through approximation, our eigenvalue equation for the symmetric waveguide ($n_c = n_s$), Equation 7.56, can yield an analytic solution if we postulate a mode close to cutoff. At cutoff, the modes will begin to have real (oscillatory) solutions in the substrate and cap, which implies that

$$\beta \rightarrow k_0 n_c = k_0 n_s \tag{7.65}$$

so that $k_{ys} = k_{yc} \rightarrow 0$. Thus, the argument of the $\tan^{-1}$ in Equation 7.56 approaches zero and we are left with the simple relation

$$2k_{yg}h = m\pi \tag{7.66}$$

Because the propagation constant in the core may then be written as

$$k_{yg} = \sqrt{k_0^2 n_g^2 - k_z^2} = \sqrt{k_0^2 n_g^2 - k_0^2 n_s^2} = \frac{2\pi}{\lambda_0}\sqrt{n_g^2 - n_s^2} \tag{7.67}$$

we can write an expression for the *minimum* refractive index step between core and cap required for an mth order mode to be guided, namely

$$\Delta n = n_g - n_s > \frac{m^2 \lambda_0^2}{16h^2(n_g + n_s)} \tag{7.68}$$

This equation becomes a little more easy to use if we assume that $n_g \approx n_s$ so that we have the approximate relation

$$\Delta n > \frac{m^2 \lambda_0^2}{32h^2 n_g} \tag{7.69}$$

recalling that h is half the core width, $h = t/2$.

Equation 7.69 is a complementary relationship to the one of Equation 7.20, which, from our ray-optic model, gave us the *maximum* allowable index step if the symmetric waveguide was to be monomode. From Equation 7.69, we may calculate the minimum step necessary for mode m to be guided. This relationship also confirms for us that the symmetric waveguide has no cutoff: we see from the equation that, for m = 0, any $\Delta n > 0$ will permit a guided mode, at any wavelength. Thus, the lowest order mode is never cut off, even if confinement is poor.

Example

If we take $n_g = 3.39$ for a 300-nm-thick $Al_{0.2}Ga_{0.8}As$ waveguide core at $\lambda_0 = 850$ nm, Equation 7.69 lets us calculate quickly that an index step of $\Delta n > 0.30$ is required for the m = 1 (first order) mode to be guided. This is a large index jump and implies that the 300-nm-wide waveguide needs an $Al_xGa_{1-x}As$ cladding with $x > 0.9$ to support more than a single mode.

7.4.4 Asymmetric Waveguide: Cutoff Condition

The asymmetric waveguide cutoff condition may also be approximated by an analytic expression, although this requires a bit of mathematical trickery [4]. We just quote the result here, assuming that $n_g \approx n_c$ as before,

$$\Delta n > \frac{(2m + 1)^2 \lambda_0^2}{32 h^2 n_g} \tag{7.70}$$

again the minimum Δn required for mode m to be guided. Equivalently, we can restrict h as

$$h > \frac{(2m + 1)\lambda_0}{\sqrt{32 \Delta n \cdot n_g}} \tag{7.71}$$

giving the minimum core width for a mode to be guided. We note from these expressions, however, that for m = 0, the minimum allowed values for Δn and h are not zero, so that the lowest order mode is not always guided and a long wavelength cutoff exists for an asymmetric waveguide.

7.4.5 Confinement Factor

Since the electric field of the mode is given by the set of equations 7.46–7.48, which have an exponentially decaying field component outside the waveguide core, it is clear that only a portion of the mode energy is actually in the core region. Knowing what fraction of the total energy in the waveguide propagates in the core is useful, as this tells us how effectively the waveguiding layers confine the mode. To this end, we define the confinement factor Γ as

$$\Gamma = \frac{\int_{-h}^{h} \mathcal{E}_x^2(y)\,dy}{\int_{-\infty}^{\infty} \mathcal{E}_x^2(y)\,dy} \tag{7.72}$$

where the electric fields as a function of **y** are given by Equations 7.46–7.48.

Once one of our eigenvalue equations has been solved for β, the **y** component of the propagation constants, the k_y terms, can be determined and Equation 7.72 integrated. We have evaluated Γ for two waveguide structures, assuming the fields given above, as a function of core thickness in Figure 7.10. A thinner core confines a mode less well than a thicker one, especially as cutoff is approached for the asymmetric waveguide. It is also clear that TE and TM modes have differing confinement factors.

A value of Γ close to 1, implying that the mode is strongly confined, is often desirable, if not always practically achievable. A well-confined mode, for example, is less subject to scattering or absorption from imperfections outside the waveguide and can be more easily modulated by an applied electric field; we will see these applications in the next few chapters.

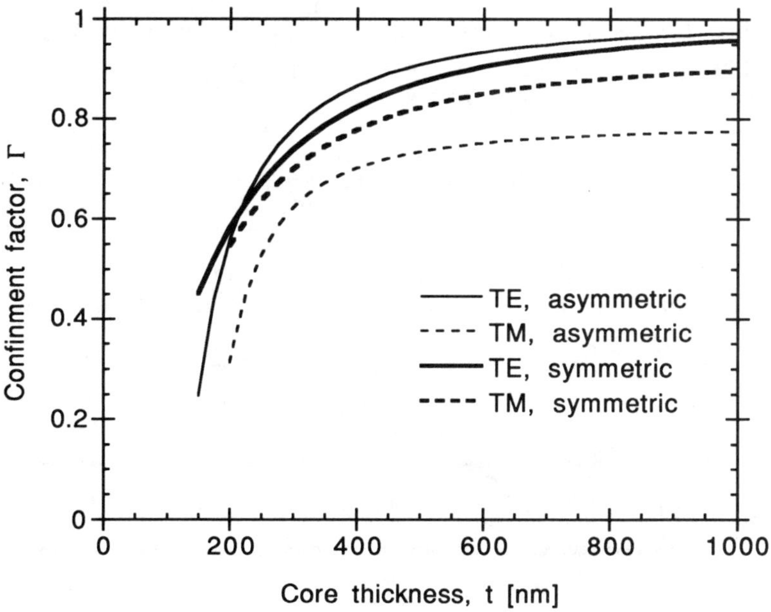

Figure 7.10 Confinement factor for TE and TM modes as a function of core thickness in two different waveguides, both with an $Al_{0.3}Ga_{0.7}As$ core (n_g = 3.35) and an $Al_{0.8}Ga_{0.2}As$ lower cladding (n_s = 3.12). The symmetric guide also has an $Al_{0.8}Ga_{0.2}As$ upper cladding and the asymmetric guide has air (n_c = 1) directly above the core.

7.5 GOOS-HÄNCHEN SHIFT

Our ray-optic and electromagnetic pictures can coalesce somewhat when we consider in detail what happens when a mode propagating down a waveguide bounces off the wall. The former model tells us this occurs by total internal reflection at the boundaries of the core. The latter model said that the mode penetrates beyond the wall and decays exponentially beyond the boundary. In actuality, both of these pictures are correct: the reflection at the interface occurs as sketched in Figure 7.11, where we see that the ray penetrates a distance equal to the inverse of the decay constant into the cladding.

From the figure, it is clear that this reflection is accompanied by a lateral shift of the ray between incidence and reflection, labeled S_{GH} and known as the Goos-Hänchen Shift [5]. Through a bit of trigonometry, we can easily find that the shift is given by

$$S_{GH} = \frac{2}{k_{yc}} \tan\theta = \frac{2}{k_{yc}} \frac{\beta}{k_{yg}} \tag{7.73}$$

Because the effective index relates the **z**-directed propagation constant to the free-space propagation constant, $\beta = Nk_0$, and by applying the definitions of k_{yc} and k_{yg} from Equations 7.41–7.43, we can rewrite this as

$$S_{GH} = \frac{2N}{k_0\sqrt{N^2 - n_c^2}\sqrt{n_g^2 - N^2}} \tag{7.74}$$

which is only a function of $k_0 = 2\pi/\lambda_0$ plus the actual and effective refractive indices. We can see from the example that a considerable lateral shift takes place at each

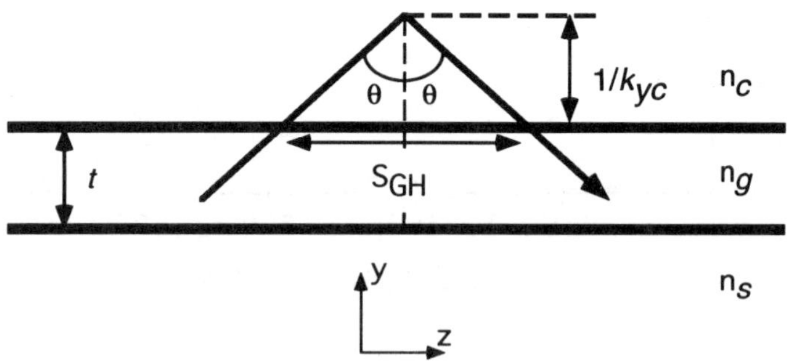

Figure 7.11 Origin of the Goos-Hänchen Shift at the waveguide/cladding interface. The mode penetrates a distance $1/k_{yc}$ into the cladding, giving rise to a lateral shift S_{GH} at each reflection.

reflection, much more than the waveguide thicknesses. It can be shown that higher order modes traverse this distance more quickly than lower order ones, such that their velocity down the waveguide is also higher, leading to dispersion [6].

Example

Take the same waveguide structure we studied in Example 7.3.3. We can calculate the Goos-Hänchen Shift from the effective indices, N = 3.299 for the 0th order and N = 3.166 for the 1st order mode. We find that S_{GH} = 1,430 nm for m = 0 and S_{GH} = 1,455 nm for m = 1. Thus the Goos-Hänchen Shift does not vary as greatly with mode number as the penetration depth does, but the lateral shift has a considerable magnitude.

REFERENCES

[1] Dürrenmatt, F., *Der Tunnel*, Zürich: Diogenes, 1980.

[2] Marcuse, D., *Theory of Dielectric Optical Waveguides*, San Diego: Academic Press, 1974, Chapter 1.

[3] Kogelnik, T., "Theory of Optical Waveguides," in Tamir, T. (ed.), *Guided Wave Optoelectronics*, Berlin: Springer-Verlag, 1988, Chapter 2.3.

[4] Hunsperger, R. G., *Integrated Optics: Theory and Technology*, 3rd Ed., Berlin: Springer-Verlag, 1991 pp. 35–36.

[5] Lotsch, H. K. V., "Reflection and Refraction of a Beam of Light at a Plane Interface," *Journal of the Optical Society of America*, Vol. 58, 1968, pp. 551–561.

[6] Saleh, B. E. A., and M. C. Teich, *Fundamentals of Photonics*, New York: John Wiley & Sons, 1991, p. 257.

Chapter 8

Optical Channel Waveguides

And all of it is held together with light and bones and roots and mud and bubblegum and tar . . .

Lawrence Ferlinghetti, 1959 [1]

As interesting and useful as the slab waveguide is, it is not until the integrated optical engineer can fabricate a channel waveguide that the capabilities of integrated optical technology fully unfold. Whereas the slabs of Chapter 7 guided light in a single dimension, the **y** direction, channel waveguides confine the optical modes in the x direction as well, so that a true optical *wire* is formed. These two-dimensional waveguides form the basis for many types of optical integrated circuits because they permit the flexible routing of light around a semiconductor surface.

In this chapter we will begin by taking a good look at the kinds of channel waveguide structures that are actually fabricated; a slab waveguide is formed during MBE or MOCVD growth of a III-V substrate, for example, but a channel waveguide often requires the full palette of processing capabilities (which we examined in Chapter 5). Once a waveguide has been made, holding together by whatever means necessary, an analysis of the modes can be made, allowing us to design a monomode structure, should that be a design goal. We turn from there to a look at how one can couple light into a waveguide, characterize the losses, and take a glance at bends, couplers, and mode splitter structures.

8.1 TWO-DIMENSIONAL WAVEGUIDE STRUCTURES

For an electrical connection in an integrated circuit, we generally employ a metal wire, a strip of a conductor that confines the electrons in two dimensions and allows them to flow in a third. An optical connection using a waveguide is in principle identical, although the means for confining the optical field in the two dimensions are quite varied. In the previous two chapters, we saw that relatively small refractive index steps between two materials are sufficient to allow total internal reflection for large incidence angles and thus waveguiding. Index steps less than 1% are often enough to generate optical waveguides; optical fibers often rely on a step of 10^{-3}. In this chapter, we will see how one generates these index steps in the **x** direction.

Our discussion of slab waveguides showed that the vertical confinement is typically accomplished by material growth or deposition. For III-Vs, this is typically a set of layers with variable material mole fractions, so that the core region has the highest refractive index. For example, an $Al_{0.3}Ga_{0.7}As$ core, 200–500 nm wide, surrounded on either side by roughly 1-μm-thick $Al_{0.8}Ga_{0.2}As$ cladding has been used in the GaAs/AlGaAs system; an $In_{0.73}Ga_{0.27}As_{0.6}P_{0.4}$ waveguide core with InP cladding of about the same thicknesses can be used for the InGaAsP/InP system. The layer compositions and thicknesses are a function of details in design and application, and one rarely finds two identical structures in the literature.

For the Si world, silicon-based dielectric sandwiches are often used. The higher refractive index of silicon nitride with respect to silicon dioxide, for example, is the basis of an oft-used layer structure; we will discuss further techniques later. Structuring the surface of these III-V and Si slab waveguides, by patterned etching, deposition, or implantation, can form a two-dimensional waveguide.

8.1.1 Etched Structures

The use of epitaxially grown GaAs and AlGaAs layers for vertical guiding has a long history [2], and the development of new and clever techniques to promote lateral guiding is an active research field. In Figure 8.1 we have sketched a set of general etching schemes used for III-V-based waveguide structures. We use a generic layer structure with a waveguide core of refractive index n_g, a lower cladding or substrate of n_s, an upper cladding of n_c, and a cover medium of n_a (the latter usually air with $n_a = 1$).

Figure 8.1(a) shows the ridge structure [3] that results from the etching away of the guiding layers in all regions except for the waveguide. The cover medium generates a large index step at the lateral edges of the waveguide and thus confines the mode effectively in that direction. Etching can be by one of the dry or wet means discussed in Chapter 5.

A variation on this theme, the rib structure of Figure 8.1(b), dispenses with the upper cladding layer and uses n_a for upper confinement [4]. Lateral guiding is accomplished by a thinning of the waveguide core in regions outside the waveguide; the net effect of this process is to raise the effective refractive index, N, in the thicker region with respect to the etched areas, thereby confining the mode. The step in effective index between the center and sides is smaller than for the ridge waveguide, so that mode confinement is poorer. We will see in later sections why this works. An inverted rib [5], as shown in Figure 8.1(c), performs the same function: a recess is etched into the lower cladding before the waveguiding layer is grown or deposited.

It is often desirable, for loss considerations as we shall see, to avoid etching the waveguide core altogether. The strip-loaded waveguides of Figures 8.1(d) and 8.1(e)

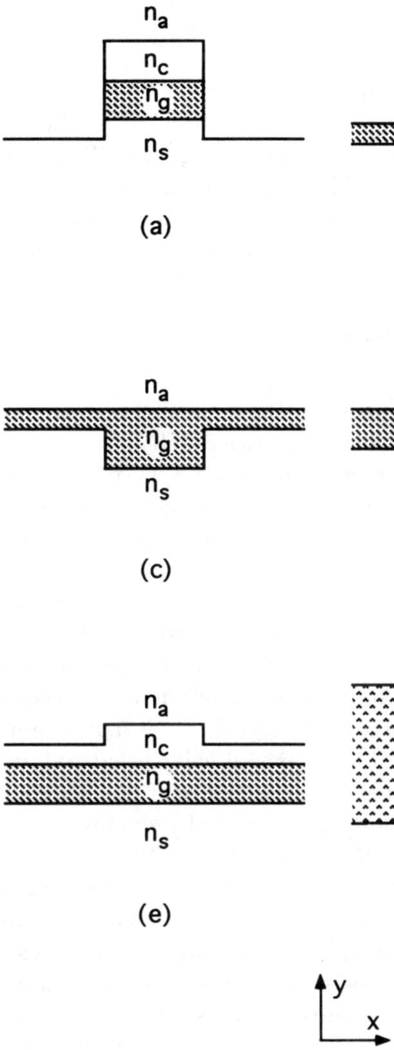

Figure 8.1 Etched III-V waveguide structures: (a) ridge; (b) rib; (c) inverted rib; (d) and (e) strip loaded; and (f) etched and regrown. The waveguide core is shaded in each structure.

achieve lateral guiding by patterning the upper cladding, either by removing the cladding completely [6] or thinning it [7] in all regions except over the waveguide. Analogously to the rib structure, the effective index in the waveguide is raised under the thicker cladding, again since the index of the cover, n_a, is smaller than that of the clad, n_c. Confinement of the mode is also not as strong as it is for the deeply etched ridge.

Finally, a buried structure as shown in Figure 8.1(f) can achieve the strong confinement of a ridge structure but reduce losses by covering the etched sides [8]. After initial layer growth and ridge etch, a cover layer, labeled n_{c2}, is regrown on the surface, burying the ridge. The index of the cover medium can be chosen to optimize the refractive index step between the guide and the side regions. Such a structure has been widely applied to many InP-based optical integrated circuits [9].

8.1.2 Other III-V Waveguide Techniques

Etching waveguide structures is likely the most popular waveguide-fabrication technique. Nevertheless, a host of other clever schemes for the generation of two-dimensional waveguides has also been developed in the III-V system, particularly for the generation of buried waveguides less prone to the enhanced losses that can bedevil etched structures. Some of these tend to be of predominantly academic interest, but we will discuss several to provide a taste of the variety.

Although the III-V compound semiconductors do not oxidize as readily as Si, techniques have been developed to grow thermal oxides in AlGaAs; this employs oxidation steps at 400°C in H_2O ambients. Because the aluminum oxides have a lower refractive index (about 1.6) than the semiconductor, a recessed oxide layer can be used to generate a lateral waveguide [10] as seen in Figure 8.2(a). This structure again has the advantage that the waveguide is buried and its sidewall losses are defined by the smooth oxide/semiconductor interface.

Implantation, or surface diffusion, is conceptually similar, namely through the generation of lateral buried layers with reduced index. Implanting H^+ (protons) as shown in Figure 8.2(b) causes a compensation of dopants in the semiconductor [11]. Compensation reduces the free electron concentration that leads to a rise in the refractive index, a mechanism also employed in some Si-based waveguides as we shall see. Alternatively, implantation or surface diffusion of certain atomic species can lead to a change in composition and thus index, as was demonstrated in the ZnSe/CdSe system [12]. Both of these approaches date from the early days of semiconductor waveguides and are seldom used at present. A more modern refinement of this approach relies on the controlled disordering of superlattice structures in a waveguide substrate through the use of patterned implantation by focused ion beams [13].

One can also take advantage of the slight rise in refractive index that occurs under tensile stress in III-V semiconductors [14]. By the deposition and patterning of certain metals on the surface, for example W as shown in Figure 8.2(c), tensile stress can be induced in the material. The local rise in index, due to the photoelastic effect, is small but sufficient for lateral guiding, with the added advantage that the waveguide core region is unaffected by any processing steps.

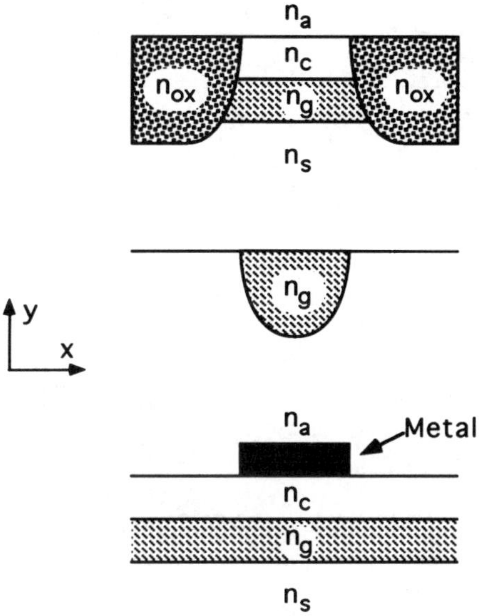

Figure 8.2 Nonetched III-V waveguide structures: (a) oxidized; (b) implanted; and (c) photoelastic.

8.1.3 Silicon-Based Waveguides

The great flexibility in composition and thus refractive index afforded by the controlled growth of III-V materials of varying composition is not generally available to silicon, although most of the etch procedures we outlined above can still be used. Vertical guiding in silicon-integrated optics is usually accomplished by the use of various dielectric layers,[1] as we mentioned above, or through variation of Ge composition in Si/Ge waveguides. Although the usable wavelength range for Si itself is $\lambda \geq 1.2\ \mu m$, the dielectrics are transparent well into the visible wavelength range. For this reason, Si-based waveguide structures have been employed for coupling to optical fibers, an application for which the mode size ($8 \times 8\ \mu m$ is typical) is well suited. III-V waveguides typically have much smaller modes. The low cost of silicon has also prompted the use of passive Si optical elements for optical interconnects [15], a subject to which we will return in Chapter 12.

Because Si_3N_4 has a higher refractive index than SiO_2 (2.05 vs. 1.46 at $\lambda = 800$ nm), the etched structure of Figure 8.3(a) forms a suitable nitride waveguide [16]. The Si_3N_4 layer is patterned by dry or wet etch techniques and can subsequently be covered by a second oxide layer [17] as shown in 8.3(b) to improve the interface

[1]For a review: Hickernell, F. S., "Optical Waveguide on Silicon," *Solid State Technology*, Vol. 31, 1988, pp. 83–88.

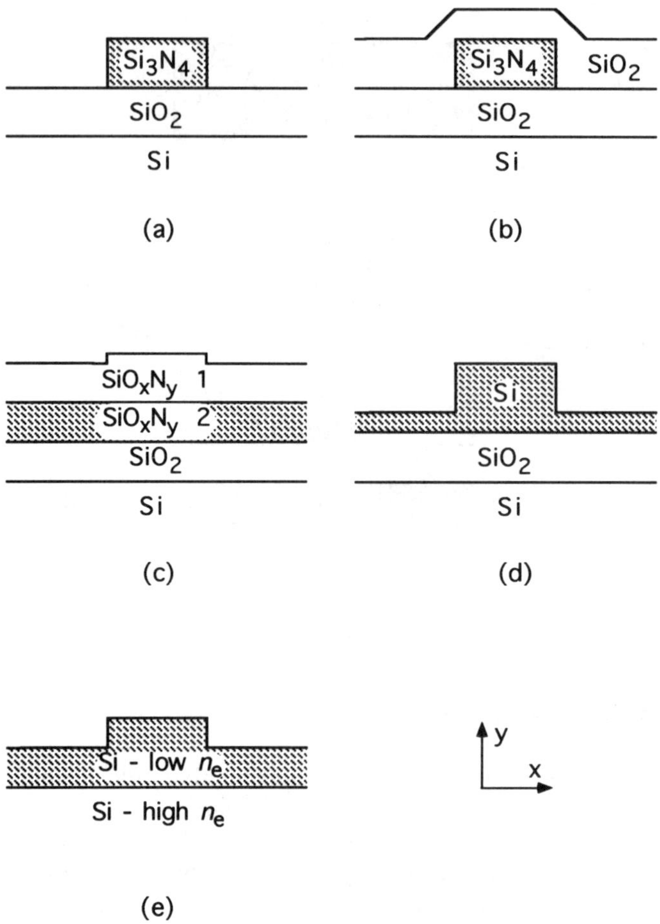

Figure 8.3 Silicon- and dielectric-based waveguides: (a) nitride ridge; (b) buried nitride; (c) oxynitride rib; (d) epitaxial silicon rib; and (e) silicon doping rib.

between the core and cladding. The same ridge or buried ridge configurations that we saw for III-V waveguides are then generated.

The lower SiO_2 cladding must be several microns thick to minimize absorption by the Si substrate, especially when wavelengths less than 1.2 μm are employed. Deposition of such thick nitride and oxide is typically by CVD means, although SiO_2 can also be deposited by flame hydrolysis [18]. This latter technique decomposes $SiCl_4$ in an O_2/H_2 ambient, resulting in an oxide film on the Si surface after consolidation at 1,200°C. This procedure is derived from the manufacture of optical glass fibers and is useful for the deposition of very thick (~100 μm) oxide layers.

Changes in relative oxygen and nitrogen concentrations in silicon oxynitride, SiO_xN_y, also give rise to changes in refractive index. This can be used to fabricate the oxynitride-based waveguide [19] of Figure 8.3(c) where the relative refractive indices can be freely chosen in the range 1.46 to 2.05. The topmost layer is patterned into a rib structure, again relying on the increased effective waveguide index under the thicker portion for lateral guiding. Si_3N_4 films typically have high residual stress; however, by properly choosing the relative concentrations of oxygen and nitrogen in SiO_xN_y, virtually stress-free films can be deposited—a condition important for many applications.

Using silicon itself as a waveguide has the advantage that carrier-induced electro-optical or thermal effects can be employed for the fabrication of active devices; these properties are a given in III-Vs but are more difficult to achieve in Si. Taking advantage of silicon-on-insulator (SOI) technology developed for isolation in electrical applications, we can grow an epitaxial Si layer, with refractive index 3.44, on an oxide cladding as shown in Figure 8.3(d) [20]. Alternatively, we can implant oxygen to generate a buried oxide layer [21]. The lower cladding must again be thick enough to prevent leakage of the optical mode into the lower Si substrate. Lateral patterning then makes use of the simple rib configuration.

Ribs are also used for purely silicon waveguides based on the variations of refractive index with carrier concentration. We can see the origin of this mechanism by harking back to the definitions of susceptibility in Chapter 3: the real part of the susceptibility due to the presence of electrons of density n_e [cm^{-3}] was given by (Equation 3.32)

$$\chi_{eR} = -\frac{q^2 n_e}{\epsilon_0 m \omega^2} \tag{8.1}$$

where we assume that the excitation frequency is much higher than the damping. Recalling that $\omega = 2\pi c/\lambda_0$, we can thus write the refractive index as

$$\mathrm{n}^2 = \mathrm{n}_0^2 - \frac{n_e q^2 \lambda_0^2}{4\pi^2 c \epsilon_0 m^*} \tag{8.2}$$

where n_0 is the index of the material in the absence of charge carriers, q [C] is the charge, ϵ_0 [F/m] is the free space dielectric constant, c [m/s] is the speed of light, and m^* [kg] is the electron effective mass. Assuming that the change in index due to carriers is small compared to the index value itself, n_0, we can approximate

$$\mathrm{n} \cong \mathrm{n}_0 - \frac{n_e q^2 \lambda_0^2}{8\pi^2 c \mathrm{n}_0 \epsilon_0 m^*} \tag{8.3}$$

Thus an increase in the electron concentration leads to a decrease in the material refractive index. For the doping ranges typically employed, < 10^{15} to > 10^{19} [cm^{-3}], the change in refractive index is roughly $\Delta n \sim 10^{-2}$, so that the more lightly doped surface layer has the larger index and thus acts as the waveguide core [22] as seen in Figure 8.3(e). Alternatively, the doping polarity can also be changed [23] resulting in a lightly doped p-type surface over a heavily doped n-type substrate, the former acting as the waveguide layer.

Combining Si with Ge, as we saw in Section 4.5.3, begins to provide for silicon some of the structural flexibility found in III-Vs. Germanium, closely lattice matched to silicon and with an index of 4.0, can be used to form a Si/Ge heterostructure with an index higher than Si alone. As we see in Figure 8.4(a), the resultant slab structure is made into a waveguide by the etching of a rib [24]. Alternatively, Ge can be implanted or diffused into Si [25] to give the buried profile of Figure 8.4(b), albeit with a more diffused boundary than its etched equivalent and thus less tightly confining the mode.

The use of germanium is not confined to Si, but can also be employed in its oxidized form. Diffusing Ge into SiO_2 results in the local GeO_2 region [26] seen in Figure 8.4(c), again with a rather broad laterally confined region. Silicon dioxide can also be altered by diffusion of phosphorous, a dopant that raises the value of refractive index [27].

Although the use of the silicon-based dielectrics, SiO_2 and Si_3N_4, is widespread due to their well-developed technology and varied applications in other areas, numerous other dielectrics have been developed for integrated optics applications, for both

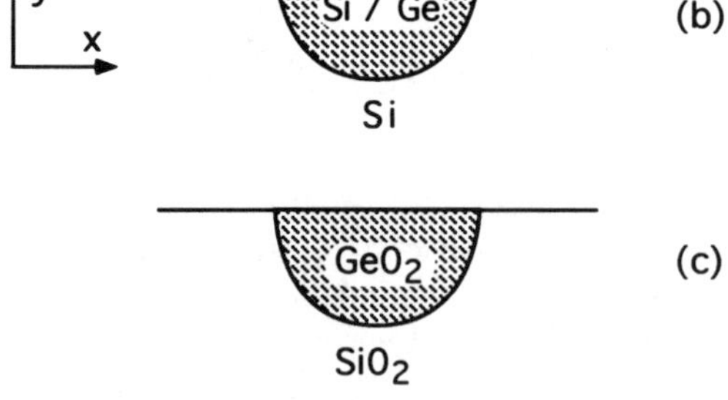

Figure 8.4 Silicon/germanium waveguides: (a) heterostructure rib; (b) diffused silicon; and (c) diffused silicon dioxide.

Si and III-Vs [28]. In particular, TiO_2 and Ta_2O_5 have found widespread use, as has sapphire, Al_2O_3. The latter is also well known to the silicon crowd from its use as an insulating substrate for epitaxial Si growth.

8.2 TWO-DIMENSIONAL WAVEGUIDE ANALYSIS

Now that we have seen how a channel waveguide looks and how it is made, we are still faced with the problem of analyzing it. In particular, one often wants to design a waveguide that supports only a specified number of modes, or determine the propagation constants and effective index. The one-dimensional problem of Chapter 7 is now a two-dimensional one and, in general, solutions can not be written down analytically. For that reason, real two-dimensional waveguide structures are usually analyzed by numerical simulation. This is a field unto itself, and a discussion of state-of-the-art numerical modeling approaches falls outside of the scope of this text.

We can, however, offer two approximate techniques that yield useful results. Both approaches essentially reduce the analysis to coupled one-dimensional problems and yield approximate—but adequate—values for propagation constants and effective index under many conditions. Both methods begin to fail as the modes of interest approach cutoff or the physical waveguide structure becomes excessively complicated.

8.2.1 Effective Index Approximation

We can make an initial simplification if we ignore the fields in the waveguide and attempt to reduce the refractive index distributions in two dimensions into one. As shown in Figure 8.5, the general channel waveguide can be thought of as the combination of two slab waveguides in orthogonal directions. From a cut in the **y** direction we can determine an effective index, n_{eff}, which then becomes the index of the center section in the orthogonal cut, **x** directed. This second one-dimensional problem is then solved to determine the two-dimensional waveguide effective index, N_{wg}.

We proceed as follows: taking the TE mode, with the electric field **x** directed, we solve the three-layer slab waveguide in the **y** direction using the asymmetric TE eigenvalue equation, Equation 7.55, and thereby determine k_{yg}. Since we remember that (Equation 7.41)

$$k_{yg} = \mathrm{k}_0\sqrt{\mathrm{n}_g^2 - \mathrm{N}^2} \tag{8.4}$$

we can write the effective index, n_{eff}, as

$$\mathrm{n}_{\mathrm{eff}}^2 \equiv \mathrm{N}^2 = \mathrm{n}_g^2 - \left(\frac{k_{yg}}{k_0}\right)^2 \tag{8.5}$$

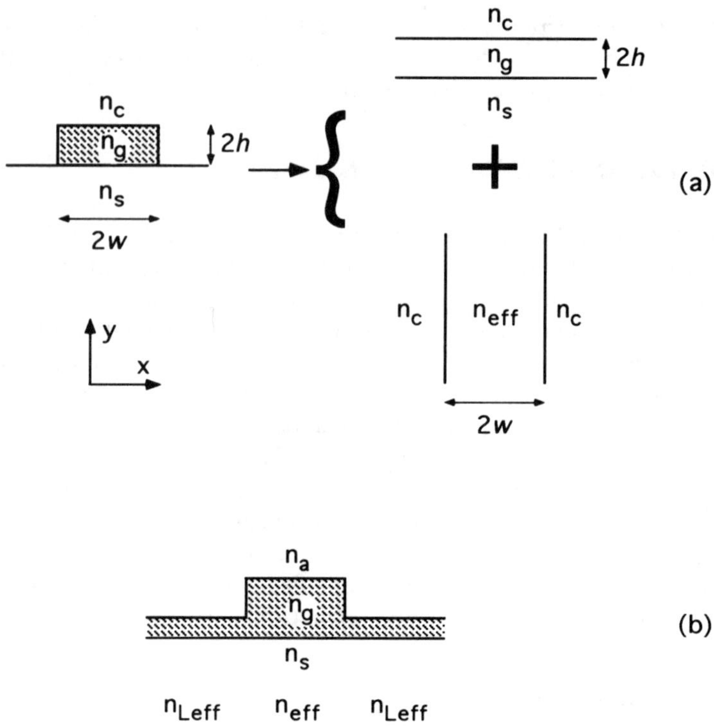

Figure 8.5 (a) Decomposition of the two-dimensional ridge waveguide structure into two one-dimensional slab waveguides that can then be solved by the effective index method; and (b) a generic rib structure.

With this value in hand, we now turn to the orthogonal slab waveguide in the **x** direction and solve the symmetric TM eigenvalue equation (Equation 7.58). TM is used because the fields are oriented as they were for the previous solution, but the slabs are of course rotated by 90°, so **y** becomes **x**. The center slab now has an index of n_{eff}, so that we solve

$$2k_{x\text{eff}}w - m\pi = 2\tan^{-1}\left(\frac{n_{eff}}{n_c}\right)^2\left(\frac{k_{xL}}{k_{x\text{eff}}}\right) \tag{8.6}$$

for $k_{x\text{eff}}$, which is the **x**-directed propagation constant in the center region of index n_{eff}. k_{xL} is the decay constant in the lateral cladding region, n_c, where we call it k_{xL} rather than k_{xc} to agree with notation we will use below. From our propagation constant, $k_{x\text{eff}}$, we can then directly write down the effective two-dimensional waveguide index, N_{wg}, as

$$N_{wg}^2 = n_{eff}^2 - \left(\frac{k_{xeff}}{k_0}\right)^2 \quad (8.7)$$

For a rib structure as shown in Figure 8.5(b), we need to take two steps before solving Equation 8.6. First, we determine the effective index in the center region, n_{eff}, and then we do the same for the side regions to determine n_{Leff}. The second step is necessary because there is no single index at the sides of the waveguide. We then plug n_{eff} and n_{Leff} into the TM eigenvalve equation,

$$2k_{xeff}\, w - m\pi = 2\tan^{-1}\left(\frac{n_{eff}}{n_{Leff}}\right)^2\left(\frac{k_{xL}}{k_{xeff}}\right) \quad (8.8)$$

and determine N_{wg} as we did previously.

The effective-index method is a simple and fast approach to determining the propagation constants and effective indices of a two-dimensional waveguide. Using it, we have solved a ridge waveguide of the type in Figure 8.1(e), where the upper clad has been completely etched away except for the ridge. Using an $Al_{0.3}Ga_{0.7}As$ waveguide core, $n_g = 3.35$, and a $Al_{0.8}Ga_{0.2}As$ upper and lower cladding, $n_s = 3.12$, and $n_a = 1$, the variation of waveguide effective index for the lowest order mode as a function of core thickness and ridge width is shown in Figure 8.6. We see that the strongest variation in effective index with both parameters is for small thickness and width values.

Numerous general solutions for realistic waveguide structures have been calculated [29] and even diffused waveguide profiles can be tackled by this technique [30]. The method is only an approximation, however; for large index steps, complex structures, or poorly confined modes, the effective-index approach may be inaccurate. If the fields in the waveguide need to be known, however, the reader is faced with having to continue with next section.

Example

Let's analyze a waveguide of the type in Figure 8.1(b) in two dimensions for the lowest order TE mode at $\lambda = 850$ nm by the effective index method. If we take a guide index of $n_g = 3.35$ ($Al_{0.3}Ga_{0.7}As$), a lower cladding with $n_s = 3.12$ ($Al_{0.8}Ga_{0.2}As$) and air above the rib ($n_a = 1$), we have to solve the asymmetric TE eigenvalue equation (Equation 7.55) twice—assuming a rib height of 500 nm, the first solution yields the effective index under the rib, $n_{effg} = 3.287$. If the rib is etched 150 nm so that the sides have a remaining height of 350 nm, the second solution tells us that the effective index outside the rib is then $n_{effL} = 3.247$. We then plug those two values into the symmetric TM eigenvalue equation (Equation 7.58), which we solve in the **x** direction using n_{effg} as the guide index and n_{effL} as the cladding index. From that we determine the two-dimensional waveguide effective index, namely $N_{wg} = 3.264$.

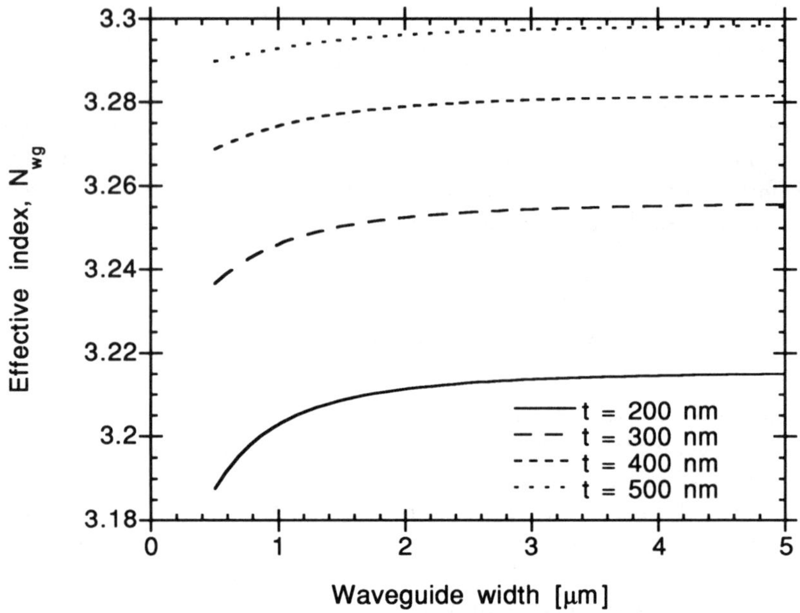

Figure 8.6 Variation of two-dimensional effective index with waveguide rib width and core thickness. The calculation was performed for a waveguide of the type in Figure 8.1(e) with the clad etched down to the core outside the ridge. An $Al_{0.3}Ga_{0.7}As$ waveguide core, $n_g = 3.35$, $Al_{0.8}Ga_{0.2}As$ upper and lower cladding, $n_s = 3.12$ and $n_a = 1$, were assumed.

From N_{wg}, we can directly determine the **x**-directed decay constant outside the guide, $k_{xL} = k_0\sqrt{(N_{wg}^2 - n_{effL}^2)} = 2.49\ \mu m^{-1}$, the **x**-directed propagation constant under the ridge, $k_{xg} = k_0\sqrt{(n_{effg}^2 - N_{wg}^2)} = 2.89\ \mu m^{-1}$, and the **z**-directed propagation constant, $\beta = k_0 N_{wg} = 24.13\ \mu m^{-1}$.

8.2.2 Method of Field Shadows

An elegant approximation for determining the fields in a two-dimensional waveguide was developed by Marcatili [31] and is occasionally referred to as the method of field shadows. If we represent our channel waveguide as shown in Figure 8.7, we see that for well-confined modes far from cutoff, most of the power is transmitted in n_g and that the fields in the regions n_c, n_s, n_L, and n_R are exponentially decaying. The approximation is based on being able to ignore the fields in the cross-hatched areas, the so-called field shadows, which is acceptable if the modes are indeed far from cutoff. In that case, we can again solve two one-dimensional problems and combine the solutions.

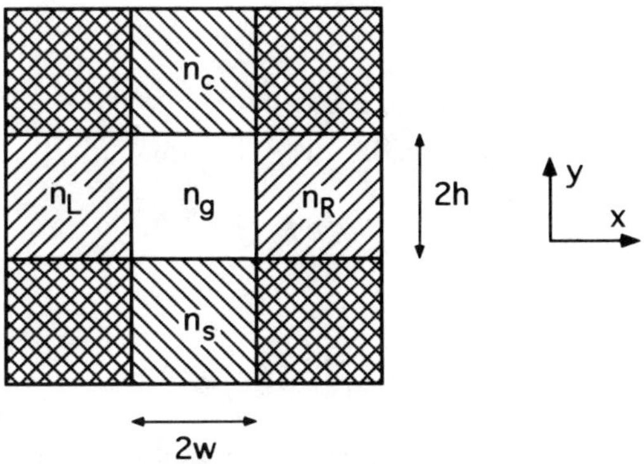

Figure 8.7 The channel waveguide as analyzed by the method of field shadows. Solutions need only be determined in the horizontal and vertical directions, ignoring the fields in the cross-hatched regions.

The two one-dimensional problems require the determination of the "TE-like" fields, $\mathcal{E}_x$, H_y, where the electric field will be denoted $\mathcal{E}^x{}_{p,q}$, and the "TM-like" fields, $\mathcal{E}_y$, H_x with its electric field $\mathcal{E}^y{}_{p,q}$. The notation p and q for the fields refers to the mode number in the **x** and **y** directions, respectively. The labels TE and TM are somewhat vague, since, for a square waveguide cross section, the definitions of transverse and lateral become arbitrary. However, we usually refer to the transverse direction as the one normal to the wafer surface and lateral to the one parallel to the wafer surface. The total field is usually a mixture of both TE and TM, but we will take one or the other to dominate the behavior of the mode.

To illustrate the method, we will now derive the solution for a TM mode, assuming that only $\mathcal{E}_y$ and H_x are nonzero. If we determine solutions for the wave equation, as we did for the slab waveguide in Section 7.4, we have:

$$\mathcal{E}_y = A\cos(k_{xg}x + \phi_1)\cos(k_{yg}x + \phi_2) \qquad \text{in } n_g \qquad (8.9)$$

$$\mathcal{E}_y = B\cos(k_{xg}x + \phi_1)\, e^{-jk_{yc}y} \qquad \text{in } n_c \qquad (8.10)$$

$$\mathcal{E}_y = C\cos(k_{xg}x + \phi_1)\, e^{jk_{ys}y} \qquad \text{in } n_s \qquad (8.11)$$

$$\mathcal{E}_y = D\cos(k_{yg}y + \phi_2)\, e^{jk_{xL}x} \qquad \text{in } n_L \qquad (8.12)$$

$$\mathcal{E}_y = E\cos(k_{yg}y + \phi_2)\, e^{-jk_{xR}x} \qquad \text{in } n_R \qquad (8.13)$$

In the above expressions, A-E are constants and the $\exp(-j\beta z) \cdot \exp(j\omega t)$ dependence has been ignored.

Analogously to Equations 7.38–7.40, the **x**-, **y**-, and **z**-directed propagation constants are related as

$$(k_0 \mathrm{n}_i)^2 = k_{xi}^2 + k_{yi}^2 + k_{zi}^2 \tag{8.14}$$

where the subscript i denotes the region, (g, c, s, L, or R). Looking at the boundary conditions of the structure in Figure 8.7, we see immediately that

$$k_{xg} = k_{xc} = k_{xs} = k_x \tag{8.15}$$

and

$$k_{yg} = k_{yL} = k_{yR} = k_y \tag{8.16}$$

so that we can define a single k_x for **x**-directed propagation in the core, upper cladding, and substrate, and a single k_y for **y**-directed propagation in the core and the lateral claddings, L and R. From Equation 8.14, we can write the mode propagation constant, β, as

$$\beta \equiv k_z = \sqrt{k_o^2 \mathrm{n}_g^2 - k_x^2 - k_y^2} \tag{8.17}$$

which we see is only a function of k_x and k_y.

We need to find k_x and k_y to determine the effective propagation constant and thus the effective index of the waveguide. Because the mode is assumed to be well guided, we employ the same eigenvalue equations as previously (Equations 7.55 and 7.58) for the two orientations. Using the boundaries at $\pm w$, the TE case, we have then

$$2k_x w - \mathrm{p}\pi = \tan^{-1}\left(\frac{k_{xR}}{k_x}\right) + \tan^{-1}\left(\frac{k_{xL}}{k_x}\right) \tag{8.18}$$

for p = 0,1,2, . . . , which simplifies to the symmetric waveguide case if regions R and L are identical. We then solve the orthogonal, TM, case using

$$2k_y h - \mathrm{q}\pi = \tan^{-1}\left(\frac{\mathrm{n}_g}{\mathrm{n}_c}\right)^2\left(\frac{k_{yc}}{k_y}\right) + \tan^{-1}\left(\frac{\mathrm{n}_g}{\mathrm{n}_s}\right)^2\left(\frac{k_{ys}}{k_y}\right) \tag{8.19}$$

where q = 0,1,2, ... These two solutions yield values for k_x and k_y, which then directly, by means of Equation 8.17, permit determination of β. And since $\mathrm{N} = \beta/k_0$, we also have a value for the effective index.

8.2.3 Waveguide Modes

Through the use of approximations or numerical simulation, the distribution of electric and magnetic fields in the two-dimensional waveguide can be calculated. We have plotted three mode intensity distributions in Figure 8.8, calculated by numerical simulation for a simplified waveguide structure: a core with $2w = 1.5\ \mu$m, $2h = 500$ nm, and $n_g = 3.35$ surrounded on all sides with a uniform cladding of $n_s = 3.12$. The lowest order transverse and lateral mode, $\mathcal{E}^y{}_{0,0}$, as shown in 8.8(a) is of course the monomode case. A single lateral, dual-transverse mode ($\mathcal{E}^y{}_{0,1}$) is shown in 8.8(b) and single transverse, triple-lateral mode ($\mathcal{E}^y{}_{2,0}$) is shown in 8.8(c). The modes plotted are only a subset of the ones supported in this structure; the highest order mode is (2,1), implying three lateral and two transverse modes.

8.2.4 Monomode Waveguides

For many integrated optical circuit applications, a monomode two-dimensional waveguide is the necessary basic building block. Through an analysis not unlike the one we undertook in Section 7.4.3 plus the effective-index representation, we can write down an approximate set of conditions for monomode behavior.

We again reduce the two-dimensional analysis to two one-dimensional problems; assuming a symmetric waveguide structure, we realize that the first slab waveguide of Figure 8.9 has, at cutoff, $k_{ys} \to 0$. Since (Equation 7.43) $k_{ys} = k_0\sqrt{(N^2 - n_s^2)}$, cutoff implies that $N \to n_s$ and thus $k_{yg} = k_0\sqrt{(n_g^2 - N^2)} = k_0\sqrt{(n_g^2 - n_s^2)}$. If we allow the m = 0 mode only, the TE eigenvalue equation has limits

$$2\tan^{-1}\left(\frac{k_{ys}}{k_{yg}}\right) < 2k_{yg}h < \pi + 2\tan^{-1}\left(\frac{k_{ys}}{k_{yg}}\right) \tag{8.20}$$

where we realize that both $\tan^{-1}$ terms go to zero since $k_{ys} \to 0$. So we can write

$$0 < 2k_0h\sqrt{n_g^2 - n_s^2} < \pi \tag{8.21}$$

which may be written as a condition on wavelength

$$0 < 4h\sqrt{n_g^2 - n_s^2} < \lambda_0 \tag{8.22}$$

Equation 8.22 is the condition for a single mode in the **y** direction and is essentially identical to Equation 7.68.

Following the same procedure in the **x** direction using the second equivalent slab structure of Figure 8.9, we obtain the condition for only a single mode in that direction, namely

$$0 < 4w\sqrt{n_{\text{eff}}^2 - n_L^2} < \lambda_0 \tag{8.23}$$

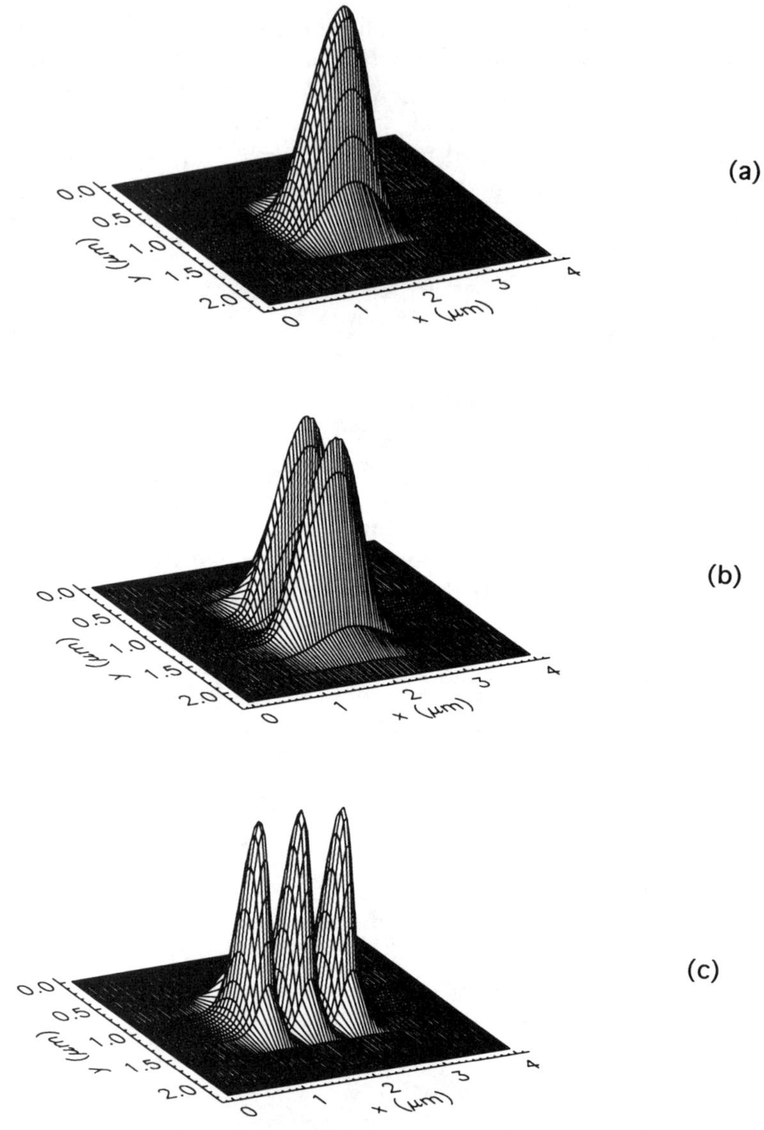

Figure 8.8 Mode intensities in an $Al_{0.3}Ga_{0.7}As$ waveguide core ($2w = 1.5$ μm, $2h = 500$ nm) surrounded on all sides by $Al_{0.8}Ga_{0.2}As$, as determined by numerical simulation: (a) lowest order transverse and lateral mode ($\mathcal{E}^y_{0,0}$); (b) single lateral, dual-transverse mode ($\mathcal{E}^y_{0,1}$); and (c) single transverse, triple-lateral mode ($\mathcal{E}^y_{2,0}$). The vertical axis is intensity in arbitrary units.

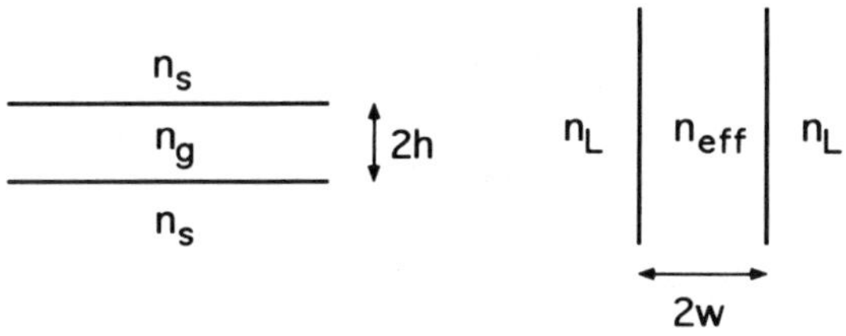

Figure 8.9 Two equivalent one-dimensional slabs for the determination of two-dimensional monomode behavior.

where the effective index, n_{eff}, is given by

$$n_{eff} = \sqrt{n_g^2 - \frac{k_{yg}^2}{k_0^2}} \tag{8.24}$$

To determine k_{yg}, the first one-dimensional slab structure must be solved in any case. However, once n_{eff} is known, these simple relationships permit a quick estimate as to whether or not a two-dimensional waveguide will be monomode.

Example

For an etched waveguide of the form of Figure 8.1(d), it is interesting to examine what physical dimensions yield a monomode waveguide. We will take the same materials as in the previous example ($n_g = 3.35$, $n_s = 3.12$) and add an upper cladding of $n_c = 3.12$, all at $\lambda = 850$ nm. For a core of thickness $2h$, we find from Equation 8.22 that for a single transverse mode $2h < 348$ nm; let's then take a core with $2h = 300$ nm to be safe. Using that thickness, we then solve the symmetric TE eigenvalue equation under the ridge to determine the effective index, $n_{effg} = 3.257$; we repeat the calculation outside the ridge to determine the lateral effective index, $n_{effL} = 3.225$. Those values are directly plugged into Equation 8.23 to yield the condition on waveguide width, namely $2w < 0.933$ μm, so that only a single lateral mode is guided. Thus, a rib width of 0.9 μm would again be on the safe side.

We could just have well solved the eigenvalue equations for k_{yg} and used Equation 8.24 to determine the effective indices; that approach is entirely equivalent.

8.2.5 Lateral Waveguiding Mechanisms

The reason for lateral guiding in an etched ridge structure appears fairly obvious, namely the index step achieved by etching through the waveguide core giving rise to total internal reflection at the lateral boundary, as we see in Figure 8.10(a). For a strip-loaded or rib structure, however, the origin of lateral confinement is not as

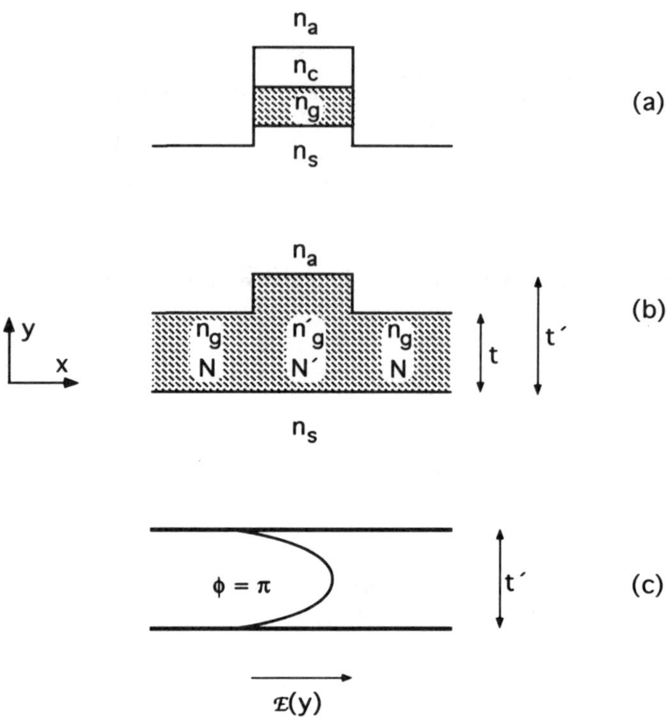

Figure 8.10 (a) An etched ridge waveguide, where lateral guiding is accomplished through the relative index values $n_g > n_c, n_s, n_a$; (b) a rib waveguide for which lateral confinement is based on $N' > N$; and (c) the **y**-directed electric field with a single peak at the center of the waveguide core.

clear—the waveguide is either covered by a different medium or merely thicker in the guiding region.

The effective index model showed us that the index under the rib of Figure 8.10(b), n_g', is different than that outside the guide, n_g—but must $n_g' > n_g$? We see that as t increases, from t to t', k_{yg} decreases. This becomes clear if we take a $\cos(k_{yg}y)$ dependence for **y**-directed electric field in the guide and see that, for a single maximum at the center of the core, as shown in Figure 8.9(c), the total phase of the wave $\phi = \pi$, and thus $k_{yg} = \pi/t$. So we have $k_{yg}' < k_{yg}$, and since $\beta^2 = \sqrt{(n_g^2 k_0^2 - k_{yg}^2)}$, $\beta' > \beta$. Finally, since $N = \beta/k_0$, it is clear that $N' > N$. Thus, the effective index under the ridge is indeed larger than outside the guide, and guiding should occur.

We can employ further arguments to clarify the point. Referring back to our ray-optic waveguide model, we can think of a larger required zig-zag path under the rib and thus a larger effective index. Alternatively, we can solve the eigenvalue equations in both regions and see that a larger t leads to a larger N.

The strip-loaded waveguide can be understood by similar heuristic arguments. Since the index, n_c, of the unetched cap defining the lateral waveguide is larger than the index of the surrounding medium, n_a (often air or a dielectric), penetration of the evanescent tail of the optical mode into the strip is larger than penetration into n_a. As a result, the path a zig-zagging wave needs to cover while moving down the waveguide is, in analogy to the rib waveguide, again larger, and thus the index is increased. While an approximate analysis of the strip-loaded guide is possible by effective-index or field-shadow means [32], numerical simulation is generally applied to these structures.

8.3 INPUT/OUTPUT COUPLING

Once a two-dimensional waveguide structure has been fabricated, the integrated optician is faced with the problem of actually getting light into and out of it. Recalling that the typical lateral sizes of semiconductor-based waveguides are in the micrometer regime, and the layer structures for III-V-based waveguides are often only several hundred nanometers thick, we see that specialized techniques are required. Although conceptually easy, monolithic integration of light source and detector is not entirely trivial from a technological point of view, as we shall see in later chapters. We shall now have a look at several means for coupling external optical fields into a waveguide. Each method has advantages and difficulties associated with it, so the optimum approach is a function of the device structure and intended application.

8.3.1 Butt Coupling

In theory, the simplest approach to coupling light into a waveguide would be to place the source as close as possible to the end of a waveguide. In Figure 8.11, we see a cleaved laser facet in close proximity to a cleaved waveguide facet. The two are separated by a distance, *d*; and typical values for *d* are 1 μm or less. This form of butt coupling is useful for joining two devices of similar structure—either two waveguides [33] or a semiconductor laser with a semiconductor waveguide [34]. Due to the small dimensions involved and the necessary lateral and transverse tolerances required for high-efficiency, successful butt-coupling usually calls for piezo-driven mounts, often with a feedback network to optimize the relative positions of source and waveguide [35].

The efficiency of coupling light from a source into a waveguide by this technique depends on numerous factors. The spacing *d* must be minimized, particularly since semiconductor lasers and waveguides emit an uncollimated beam with a very large (40°) transverse divergence. The lateral and transverse position must be precisely controlled, particularly the latter because the waveguide layers are often only a fraction of a micron thick. Efficiency is highest for similar vertical layer structures,

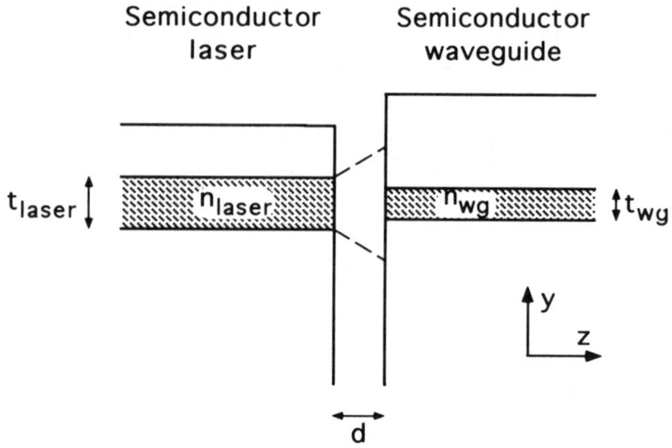

Figure 8.11 Butt coupling between a semiconductor laser and a semiconductor waveguide. The laser emission has a large transverse divergence so that the spacing *d* is typically a micrometer or less. The relative widths of the laser and waveguide core are arbitrary.

since the mode-mismatch is minimized; calculation of the overlap integral between the modes of source and waveguide will allow an estimate of the best coupling achievable while using different layer structures.

The Fresnel reflectance at each surface, in the absence of any antireflection coatings, presents a fundamental constraint, since even for normal incidence only a fraction of the incident power will be transmitted into the waveguide. In addition, power reflected from the waveguide facet back into the laser can critically affect the latter's stability. A final problem is that the two reflecting surfaces, those of the source and the waveguide, can form a Fabry-Perot cavity, so that the transmission oscillates through minima and maxima as d varies; this places a further demand on physical positioning.

Despite the apparently daunting technical challenges, butt coupling is widely employed for the permanent mounting of optical components. The so-called pigtailing technique is well established for the attachment of optical fibers to semiconductor optical circuits [36], where the fabrication of physical alignment structures often facilitates the process.

8.3.2 End-Fire Coupling

A practical technique for the evaluation of waveguide structures in the laboratory involves the focusing of an external source by means of bulk optics onto the cleaved facet of a semiconductor waveguide. In practice, a microscope objective is often used to focus the external laser onto the waveguide facet as shown in Figure 8.12; tapered

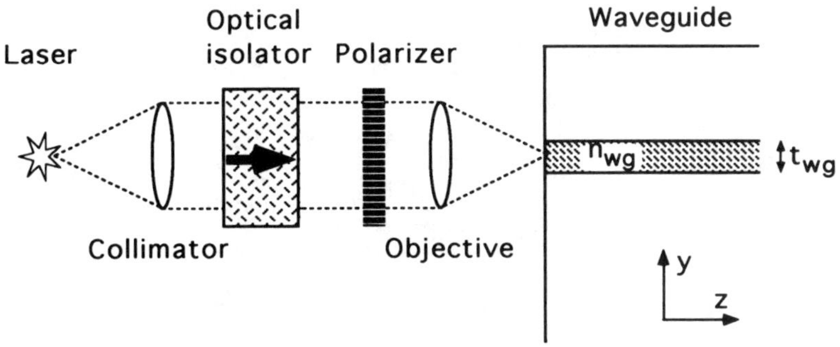

Figure 8.12 End-fire coupling an external laser into a semiconductor waveguide. The bulk-optical approach allows the positioning of an optical isolator and polarizer in the beam to decouple reflections from the laser source.

fibers are also used for this purpose. End-fire coupling retains a good deal of flexibility because nonsemiconductor laser sources can be coupled into waveguides and the polarization, source stability, and feedback insensitivity of these sources can be controlled externally. However, the optics must perch on an optical table to eliminate vibrations.

Similar considerations to those for butt-coupling apply in the determination of end-fire coupling efficiency. The overlap integral between the coupled beam and the waveguide mode must be optimized as the beams are usually Gaussian, which couple well to the lowest order TE or TM modes. Fresnel reflectivity again constrains the maximum efficiency achievable, but optical isolation techniques can reduce both the problem of reflection into the laser source and the annoyance of Fabry-Perot oscillations. Specialized fabrication techniques, such as the use of tapered waveguide structures for the coupling of lasers to fibers, are discussed in Chapter 12.

8.3.3 Grating Coupling

Gratings, periodic structures with a period on the order of optical wavelengths, have numerous applications in integrated optics [37] as functional elements in an optical circuit. Grating couplers may also be used to couple light in and out of waveguides. For the latter application, a grating structure is either etched into the surface of the waveguide itself, as illustrated in Figure 8.13, or formed by the patterned deposition of a surface material, either a metal or dielectric. Typical grating periods, Λ, are several hundred nanometers, in the regime of λ_0/n_g for waveguide index n_g, and the etch depths required are usually small.

Calculations concerning grating structures are in general quite involved; we can, however, briefly illustrate how one is used for input/output coupling. The presence of

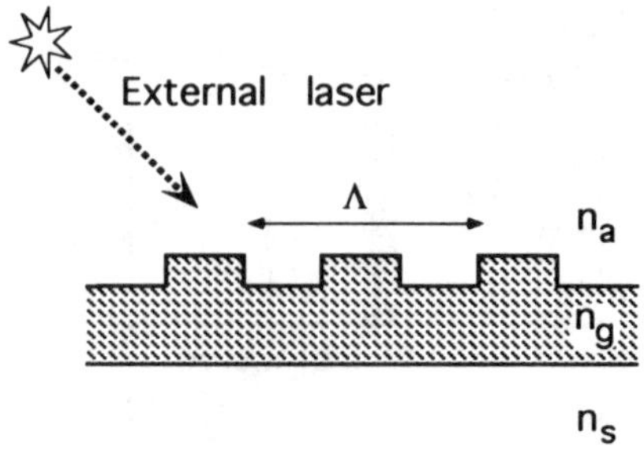

Figure 8.13 Grating coupling an external laser into a waveguide; a grating with period Λ has been etched into the surface, and coupling only takes place for certain values of incidence angle.

a grating on the surface gives rise to a periodic modulation of the waveguide effective index, and thus spatial harmonics in the guide. For an optical mode with $\beta = \beta_0$, corresponding to the propagation constant in the absence of a grating, the propagation constant becomes

$$\beta_q = \beta_0 + \mathrm{q} \cdot \frac{2\pi}{\Lambda} \tag{8.25}$$

for index q = ±1, ±2, ... If $|\beta_q| < \mathrm{n}_a k_0$, then the qth harmonic radiates into the surface region n_a; we have ignored substrate modes, which are also present. The modes radiated upward do so at an angle θ_q where

$$\mathrm{n}_a k_0 \sin\theta_q = \beta_q \tag{8.26}$$

The inverse also holds: for a collimated optical field incident on a grating at an angle θ_q, the corresponding mode will be excited in the waveguide. Thus grating couplers can be used for coupling light into and out of a waveguide.

Grating couplers have the advantage that the coupling structure (i.e., the grating) is permanently fixed onto the waveguide and is thus always aligned to it. However, the incident beam must be well collimated, and the incident angle must be precisely maintained.

8.3.4 Prism Couplers

A similar approach to the grating coupler, one that enjoys popularity in a number of integrated optics technologies, is the use of a prism coupler. A prism is placed in contact with the waveguide surface, and an external light source, again incident at a particular angle and well collimated, excites a waveguide mode through an evanescent component in the prism [38]. Index matching, however, requires the prism material to have a refractive index higher than that of the waveguide in which the mode is to be excited, which presents a problem for most semiconductors. With the possible exception of Ge-based optical components in the infrared wavelength range, no suitable prism materials exist for coupling into III-V- or Si-based waveguides. As a result, this method is rarely used for semiconductor-based integrated optics.

8.4 WAVEGUIDE LOSSES

The waveguide widths and thicknesses can be calculated to reflect the demands of a particular design, and these can be reasonably accurately fabricated. However, imperfections in processing and nonideal material properties also affect the performance of semiconductor waveguides. One of the primary effects is the attenuation of mode intensity as it propagates due to waveguide optical loss. Because these losses must be sufficiently low so that enough of the optical mode initially launched into the waveguide will be available to perform the function for which the device was designed, an understanding of their origin is important.

As we saw in Section 3.5 (Equation 3.51), a wave propagating through a material with initial intensity, $\mathscr{I}_0$, is attenuated with distance, z, as

$$\mathscr{I}(z) = \mathscr{I}_0 e^{-\alpha z} \tag{8.27}$$

where α is the loss factor, or extinction coefficient, with traditional units [cm^{-1}]. Loss is also often quoted in units of dB/cm, where we recall that a dB is defined as the ratio of transmitted to incident intensity (or power),

$$dB = 10\log\left(\frac{\mathscr{I}_{\text{out}}}{\mathscr{I}_{\text{in}}}\right) \tag{8.28}$$

The two expressions for loss are then related as

$$\alpha_{\text{dB}} = \frac{10}{\ln 10}\alpha \tag{8.29}$$

where α_{dB} has units dB/cm and α has units cm^{-1}. The total loss factor is typically the sum of optical losses from a number of sources, which we will now briefly discuss.

8.4.1 Band-Edge Absorption

In Chapter 3, we saw that photons may be absorbed by promoting an electron from the valence to the conduction band. Band-to-band absorption is inherent to a semiconductor and increases rapidly as the energy of the propagating photon approaches the bandgap energy of the material. Thus, band-edge absorption provides a fundamental lower wavelength limit for any semiconductor, as wavelengths shorter than the absorption edge will be strongly absorbed. For GaAs, the absorption edge is about 870 nm at room temperature; for Si, wavelengths shorter than 1.2 μm are absorbed.

Unfortunately, the existence of band tails (Section 3.2.3) implies that considerable absorption takes place even for wavelengths longer than the absorption edge. In GaAs, for example, $\alpha > 1$ cm^{-1} for $\lambda = 950$ nm, which is 100 nm longer than the expected wavelength limit [39]. This is due to the slow decay of band-to-band absorption with increasing wavelength and implies that low-loss waveguides should be operated at wavelengths well removed from the band edge of the material.

8.4.2 Free-Carrier Absorption

Whereas band-edge absorption is an interband transition, the absorption of photons by electrons that remain in the same band is also a significant loss component in many semiconductor waveguides. As we saw in Section 3.2.4, these intraband transitions are due to electrons that are promoted to higher levels in the conduction band, only to thermalize back to lower energies. The resultant photon absorption is usually a broad, featureless spectrum, which increases monotonically with wavelength.

We can analyze free-carrier absorption by again referring back to the polarization effects we derived in Chapter 3. The imaginary component of susceptibility due to free electrons was given by Equation 3.32, ignoring the damping term,

$$\chi_{el} = \frac{q^3 n_e}{\mu \epsilon_0 m^{*2} \omega^3} \tag{8.30}$$

for electron density, n_e; electronic charge, q; vacuum permitivity, ϵ_0; mobility, μ; electron effective mass, m^*; and excitation angular frequency, ω, where the latter term may be related to the wavelength λ as $\omega = 2\pi c/\lambda_0$ for the speed of light, c. By definition, the absorption coefficient, α, is related to the imaginary susceptibility as [40]

$$\alpha = \frac{\omega}{c}\left(\frac{1}{\mathrm{n}_0}\right)\chi_{el} \tag{8.31}$$

where n_0 is the refractive index of the material in the absence of free carriers. We can thus write free-carrier absorption, α_{fc}, as

$$\alpha_{fc} = \frac{q^3 \lambda_0^2 n_e}{4\pi^2 c^3 \mathrm{n}_0 \epsilon_0 \mu m^{*2}} \tag{8.32}$$

As we had already suggested in Section 3.2.4, the wavelength dependence of free-carrier absorption is in general complicated, and in practice the λ^2 dependence predicted by Equation 8.32 is not seen. In GaAs, a λ^3 dependence seems more typical [41]. The type of scattering that provides the necessary change in momentum for the absorbed electron, whether from optical phonons, piezoelectric effects, or charged impurities, will define the wavelength dependence measured [42].

Example

The magnitude of free-carrier absorption as a function of doping can easily be determined from Equation 8.32 by employing the constants $q = 1.6 \cdot 10^{-19}$ C, $\lambda =$ 850 nm, $c = 3 \cdot 10^8$ m/s, $\epsilon_0 = 8.87$ pF/m, $\mu = 8000$ cm^2/Vs, $n_0 = 3.45$, and the effective mass $m^* = (0.07) \cdot 9.1 \cdot 10^{-31}$ kg, for a piece of GaAs. The expression then yields $\alpha_{fc} \sim 3 \cdot 10^{-19} \times n_e$ for dimensions of n_e in cm^{-3}. Thus, a doping of 10^{18} cm^{-3} can yield to free-carrier absorption losses in the range of 1 dB/cm. This calculated value is, however, a function of quantities such as mobility and effective mass, which are difficult to determine with certainty, so that the calculation must be approached with caution.

8.4.3 Scattering

An unevenness or roughness in the waveguide can lead to optical losses due to scattering. In good quality optical material, such scattering centers are usually only found at the boundaries of the waveguide. At an uneven waveguide edge, the optical field will see irregular changes in refractive index and can thus have a portion of the guided mode scattered out of the waveguide.

This issue is a rather tricky one to analyze mathematically [43], due in part to the difficulty in characterizing what is essentially a random structural phenomenon. A simple, closed-form representation for scattering loss, based on specular reflection from irregularities at the waveguide edge, was proposed by Tien [44]. In this model, a waveguide of index n_g with thickness variations Δt, has a scattering loss component given by

$$\alpha_{sc} = \frac{\cos^3\theta}{2t_{\text{eff}}\sin\theta}\left(\frac{4\pi\Delta t n_g}{\lambda_0}\right)^2 \tag{8.33}$$

where θ is the incident angle on the waveguide boundary. t_{eff} is the effective waveguide thickness; $t_{\text{eff}} = (t + k_{ys}^{-1} + k_{yc}^{-1})$ for physical width t and decay constants k_{ys} and k_{yc} into the substrate and cap, respectively. We note from Equation 8.33 that the scattering loss increases as Δt^2 and decreases as λ^2; clearly larger irregularities with respect to wavelength imply more scattering. Higher order modes, with smaller incident angles θ, are also more prone to scattering losses, which is expected since these interact more with the boundary than lower order modes. We also see that well-confined modes, with a smaller t_{eff}, also suffer more from scattering since a greater portion of the mode energy is concentrated at the waveguide edge. By extension, scattering increases as the index step between core and cladding increases, usually as Δn^2, again due to enhanced confinement.

Although the losses for dielectric waveguides are usually dominated by scattering considerations, in general, scattering is a relatively small problem for semiconductor waveguides. MBE and MOCVD layer growth techniques yield vertical boundaries that are flat to within a monolayer of material, so that scattering in the plane of the wafer is usually negligible. For etched ridge waveguides, where the core outside the waveguide is removed, the roughness of the sidewall can lead to scattering centers and thus enhanced losses; optimization of wet and dry waveguide etches is required to minimize this loss component.

Example

We can use Tien's approximation for loss due to scattering from uneven sidewalls as an estimate of the magnitude of this loss component. If we take a roughness of Δt = 1 nm for a 500-nm-thick symmetric waveguide with $n_g = 3.35$ and a cladding of index $n_s = 3.12$, we can perform the calculation for two supported TE modes. The 0th order mode has N = 3.299, implying an incidence angle on the sidewall of 80°; the decay constant into the upper and lower cladding is $k_{yc} = k_{ys} = k_0\sqrt{(N^2 - n_s^2)}$ = 7.92 μm^{-1} so that t_{eff} = 0.752 μm. The scattering approximation (Equation 8.33) then yields a loss of 0.087 cm^{-1} per nanometer of waveguide roughness. Repeating the calculation for the 1st order mode, with N = 3.166, θ = 71°, $k_{yc} = k_{ys}$ = 3.97 μm^{-1}, and thus t_{eff} = 1.0 μm, we find a considerably larger value for scattering loss, namely 0.45 cm^{-1} per nanometer roughness. It appears that indeed higher order modes are more susceptible to scattering loss.

8.4.4 Radiation Losses

For well-designed semiconductor-based waveguides, optical losses due to radiation are minimal. Radiation in this case implies a coupling of energy from the guided mode either into the substrate or into the space above the waveguide. In certain situations, scattering can lead to mode conversion (transfer of energy from a lower to a higher order mode that is beyond cutoff) so that some modal energy can be lost by radiation. Equivalently, some models of scattering treat waveguide roughness as a uniform distribution of radiating points. In general, radiation losses only become relevant for curved waveguides, as we will discuss.

8.4.5 Typical Loss Values

Many of the developments in waveguide-fabrication technology are spurred by the need to reduce the optical losses. As a reference value, glass optical fibers routinely offer loss values below 1 dB/km = 10^{-5} dB/cm, a value that semiconductor-based integrated optical waveguides do not approach. Low-loss waveguides are typically desired to not excessively limit the usable size of optical integrated circuits. A loss of 3 dB/cm implies that half of the optical intensity has been lost after the field has traversed a 1-cm-long chip, which is not an unreasonable size for many applications.

The losses in III-V semiconductor waveguides have been reduced by improvements in growth technology. These improvements have resulted in the availability of low-defect, lightly doped waveguide layers and refinements in etch technology, which have reduced sidewall roughness [45]. Although specialized structures, with dimensions and doping levels not suitable for integrated optical devices, are often employed to demonstrate record low losses, values of 0.1 dB/cm have been achieved [46]. Best results are usually obtained from buried structures and those employed at longer wavelengths.

For the silicon system, SiO_2/Si_3N_4-based waveguides also benefit from enhancement of layer smoothness and improvements in etched sidewall quality. For the nitride waveguide perched on an oxide buffer over a silicon substrate, the thickness of this buffer must be sufficient to prevent absorption by the Si. Thicknesses greater than 5 μm are often required. In addition, oxynitride films suffer from losses due to incorporated hydrogen, which lead to absorption in distinct bands in the 1.4- to 1.52-μm wavelength range. High-temperature anneal steps have been shown to reduce this factor in waveguide loss. Best values for Si-based dielectric waveguide losses are in the 0.3-dB/cm range [47], a value not very shy of that achievable with III-V materials. Silicon semiconductor waveguides, using either a Si/SiGe/Si sandwich [48] or epitaxially grown Si on oxide [49], have been developed to similar levels of quality, the structures showing loss figures in the range of 0.5 dB/cm.

8.5 LOSS CHARACTERIZATION

The engineer will begin to characterize a waveguide by determining the number and types of modes that propagate and their attenuation, or loss. The former is usually simply accomplished through the use of a magnifying objective and a video camera, but the latter is often a more involved measurement. Although numerous techniques for the determination of waveguide losses have been demonstrated and widely used, such as multiple prism coupling or scattered light measurement [50], we will focus on the two that are most broadly applied to semiconductor structures. We note that both of these techniques provide a single loss value for the waveguide, irrespective of the loss mechanism. To understand the origin of the measured attenuation and to evaluate the relative contributions of scattering or absorption, a carefully designed experiment using a variety of structures that differ only by a single, controlled parameter must be employed.

8.5.1 Cut-Back Technique

Conceptually, the simplest approach to measuring the optical loss in a waveguide is by measuring the amount of light transmitted through sections of varying length for a constant input power. In practice, this is accomplished by measuring the output power of a single waveguide repeatedly as it is gradually shortened by having a piece cut off the end of it—hence the name *cut back*.

For the output intensities $\mathcal{I}_1$ and $\mathcal{I}_2$ measured for waveguide lengths L_1 and L_2, the loss of the waveguide may then be found from

$$\alpha = \left(\frac{1}{L_1 - L_2}\right) \ln\left(\frac{\mathcal{I}_2}{\mathcal{I}_1}\right) \tag{8.34}$$

where L_1 is the longer length. A more precise variant involves measuring output intensities at numerous lengths and then plotting the $\ln(\mathcal{I}_{out}/\mathcal{I}_{in})$ as a function of L. The resulting linear plot has a slope of $-\alpha$, and the scatter of points from the line allows one to determine the expected error.

Although the cut-back measurement is simple and relatively quick to perform, its accuracy is limited by the reproducibility of input coupling. The technique relies on identical coupling conditions for every measurement, which is experimentally difficult to guarantee. In addition, for low-loss waveguides, the changes in intensity with length can become quite small and eventually inside the error margins of the measurement.

8.5.2 Fabry-Perot Resonances

A powerful approach for waveguide loss measurement employs the intensity modulation of the Fabry-Perot resonances in the waveguide cavity [51]. The cleaved facets that delineate a semiconductor waveguide form a Fabry-Perot cavity, as we will see in the Chapter 9 when we turn to lasers, and the transmission through this resonator is a function of the loss in between the mirrors. This measurement technique does require good-quality mirror facets, which may be difficult to achieve in noncrystalline waveguide materials.

As we will derive in Chapter 9, the optical intensity transmitted through a Fabry-Perot cavity, $\mathcal{I}_t$, is related to the incident intensity, $\mathcal{I}_0$, as

$$\frac{\mathcal{I}_t}{\mathcal{I}_0} = \frac{(1 - R \cdot e^{-\alpha L})^2}{(1 - R \cdot e^{-\alpha L})^2 + 4R \cdot e^{-\alpha L} \sin^2\left(\frac{\phi}{2}\right)} \tag{8.35}$$

for facet reflectivity R, loss α, cavity length L, and phase difference between adjacent waves in the resonator, ϕ. The transmission has a maximum value of $\mathcal{I}_{max}/\mathcal{I}_0 = 1$ when $\phi = 0$, and a minimum value, for $\phi = \pi$, of

$$\frac{\mathcal{I}_{min}}{\mathcal{I}_0} = \frac{(1 - R \cdot e^{-\alpha L})^2}{(1 - R \cdot e^{-\alpha L})^2 + 4R \cdot e^{-\alpha L}} \tag{8.36}$$

We can thus define the contrast ratio, C_R [dimensionless], the ratio of maximum to minimum intensity in the resonator, as

$$C_R = \frac{\mathcal{I}_{max}}{\mathcal{I}_{min}} = \frac{(1 + R \cdot e^{-\alpha L})^2}{(1 - R \cdot e^{-\alpha L})^2} \tag{8.37}$$

which we can directly solve for the waveguide loss,

$$\alpha = -\frac{1}{L}\ln\left[\frac{1}{R}\frac{\sqrt{C_R} - 1}{\sqrt{C_R} + 1}\right] \tag{8.38}$$

We see from Equation 8.38 that the loss may be evaluated from knowledge of only the Fabry-Perot resonator contrast, the facet reflectivity, and the cavity length; the value is independent of the input or output coupling.

In practice, the Fabry-Perot technique involves coupling an external laser source, following suitable optical isolation, into the waveguide as shown in Figure 8.14. A detector measures the output intensity, and lock-in techniques are usually employed to enhance the signal-to-noise ratio. The trickiest aspect to the measurement is sweeping the waveguide Fabry-Perot cavity through a few maxima and minima to determine C_R. One typical approach involves tuning the wavelength of the external light source; a wavelength shift of several tenths of a nanometer is usually sufficient to sweep through several resonances, and this tuning can usually be accomplished by a small change in laser temperature (< 1°C). Alternatively, the temperature of the waveguide may be slightly varied, causing a shift in the effective resonator cavity length; this method often results in a physical shift of the waveguide as well and thus affects the input coupling of the light source.

The Fabry-Perot technique is especially useful for the characterization of low-loss waveguides, $\alpha < 1$ dB/cm, where the cut-back approach becomes unreliable. Because the contrast ratio, and thus the difference between minima and maxima, increases as the loss decreases, this method actually becomes more accurate as losses

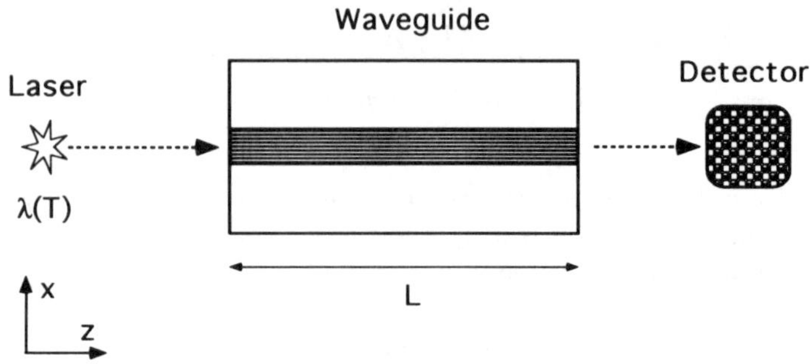

Figure 8.14 Schematic representation of the Fabry-Perot resonance technique for measuring waveguide loss. The external laser, whose wavelength can be tuned by temperature, is coupled into the waveguide using an end-fire or fiber-coupling setup.

diminish. A possible disadvantage is the need for monomode waveguides; a multimode structure will show the superposition of separate resonance peaks, of varying period, for each mode, so a distinct evaluation of the contrast ratio is not always possible.

8.6 PASSIVE WAVEGUIDE STRUCTURES

The waveguide configurations we discussed in the previous sections form the basis for many optoelectronic devices, including the lasers, modulators, and detectors that will make their appearance in the next three chapters. Waveguides on their own, without further processing, may also be used as passive elements for various functions (passive in this case implies the absence of any electrical function).

8.6.1 Curves

The flexibility of two-dimensional waveguide structures is maximized when these can be routed at will across the surface of the wafer. The ability to change the direction of the guided optical mode is essential, and this is often accomplished by the use of a waveguide curve.

As we see from the sketch of Figure 8.15(a), a curve makes a jog L [m] in the lateral $\mathbf{x}$ direction requiring a distance of T [m] to do so. Two forms are commonly used for the curve, namely a simple circle, for which

$$x = \sqrt{R^2 - z^2} \tag{8.39}$$

for radius R [m], or a raised-cosine curve, given by

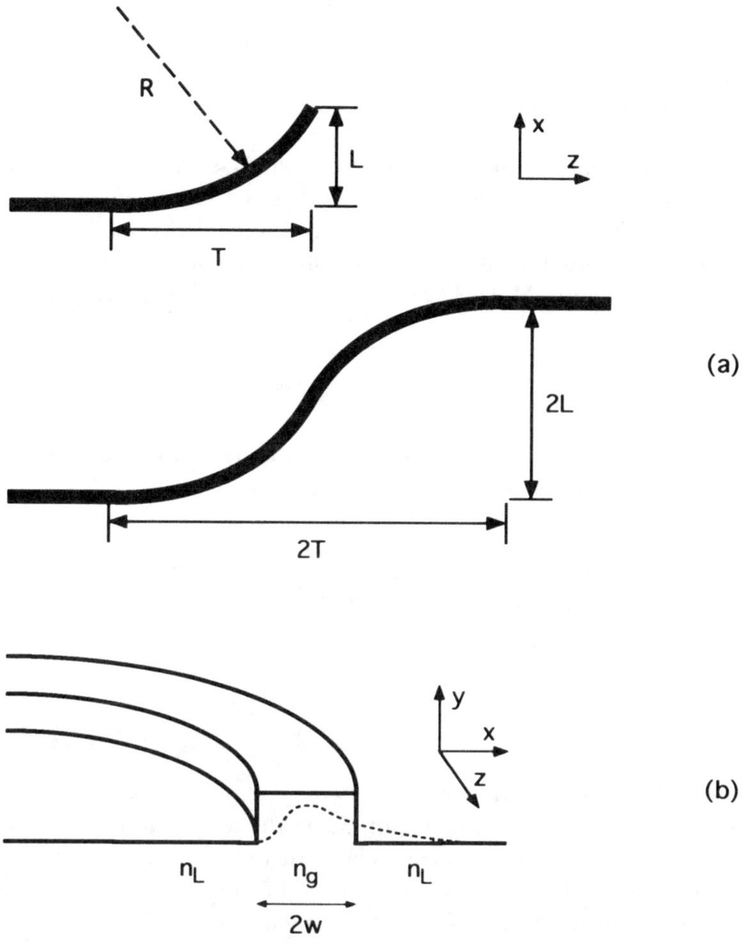

Figure 8.15 Waveguide curves: (a) the lateral jog, L, and the longitudinal displacement, T; curves can be circular or cosine-shaped. For typical radii, $T \gg L$, so that the sketches are not at all to scale. (b) Curved waveguide cross section showing the distortion of the mode that loses energy to radiation.

$$x = L \cos\left(\frac{\pi z}{2T}\right) \tag{8.40}$$

Important to note is that the radii employed are usually in the several hundred micron to millimeter range, as we will see. For that reason, the longitudinal displacement, T, is usually considerably larger than the lateral jog, L. For a circular curve, for example, the two are related as

$$T = \sqrt{2RL - L^2} \tag{8.41}$$

meaning that for a jog of $L = 10$ μm and a radius of 1 mm, a distance of T ≈ 100 μm is required.

The need for a large radius of curvature is based on loss considerations: as the waveguide changes direction, the mode is pulled outwards, shown by the schematically distorted mode of Figure 8.15(b). As a result, the optical field loses energy to radiation. Based on considerations that the entire optical field, for all R, must rotate with the same $d\theta/dt$, Marcatili and Miller [52] have shown that the curvature radiation losses are of the form

$$\alpha_{\text{curve}} = C_1 e^{-C_2 R} \tag{8.42}$$

from which it is clearly seen that radius, R, must be as small as possible. In Equation 8.42, the constants are related to the waveguide structure as

$$C_1 = \frac{\cos^2(k_{xg}w)\lambda_0 e^{2k_{xL}W}}{4k_{xL}n_L w^2\left[w + \frac{1}{2k_{xg}}\sin(2wk_{xg}) + \frac{1}{k_{xL}}\cos^2(wk_{xg})\right]} \tag{8.43}$$

$$C_2 = 2k_{xL}\left(\frac{\lambda_0\beta}{2\pi n_L} - 1\right) \tag{8.44}$$

where the various parameters come from a two-dimensional solution of the waveguide; β is the **z**-directed propagation constant, k_{xg} is the **x**-directed propagation constant in the waveguide, and k_{xL} is the decay constant outside the rib, which has a width $2w$. The exponential relationship between α and R implies a sensitive relationship of loss to radius, hence the large values of R typically used. The loss is also a critical function of k_{xL}, which stipulates good lateral confinement and thus small lateral decay constant, for low curve losses. These two parameters, radius and confinement, can be traded off against one another; strongly guided (e.g., deeply etched) waveguides can be bent at smaller radii for equivalent losses.

Through optimization of waveguide design for curves, losses below 1 dB/bend for a 90° curve have been demonstrated in GaAs/AlGaAs for radii in the millimeter regime [53] and silicon-based [54] systems for hundred-micron range radii.

Example

Using a two-dimensional solution of a curved waveguide, the radiation losses can be calculated using Equations 8.42–8.44. Using the same $Al_{0.3}Ga_{0.7}As$ core ($n_g = 3.35$) and $Al_{0.8}Ga_{0.2}As$ lower cladding ($n_s = 3.12$) as in Example 8.2.1, we remember that the 500-nm-thick waveguiding layer is etched 150 nm and that we solve the waveguide to yield n_{effL} = 3.247, k_{xL} = 2.49 μm^{-1}, k_{xg} = 2.89 μm^{-1}, and β = 24.13 μm^{-1} for a 1 μm wide waveguide. Plugging those values into our radiation loss expressions, and assuming a radius of 200 μm, we find a curvature loss of 7.2 dB/cm. Since a 200-μm radius curve makes a 90° transition in $\pi R/2$ = 314 μm, we can alternatively express the loss as 0.226 dB/90°.

Let's try this again with a more deeply etched version of this waveguide so that we can see the effect of better lateral confinement. Taking an etch depth of 200 nm, the waveguide can be solved again: we find this time that n_{effL} = 3.225, k_{xL} = 3.40 μm^{-1}, k_{xg} = 3.29 μm^{-1}, and β = 24.08 μm^{-1}. The increase in etch depth by a mere 50 nm has significantly raised the value of k_{xL}, implying a faster decay of the mode outside the waveguide. Indeed, using Equations 8.42–8.44, we find that the loss is 0.001 dB/cm for a 200-μm radius, implying $3 \cdot 10^{-5}$ dB/90°, a considerably smaller value than for the shallower waveguide. Thus, relatively minor changes in lateral optical confinement can lead to significant changes in curvature loss.

8.6.2 Couplers

If two waveguides are manufactured in close physical proximity, the evanescent field from one can excite an optical mode in the other; through coupling of the evanescent fields, power may be transferred between two waveguides. Coupling between vertically or laterally adjacent waveguides has an extensive literature, from which we extract a rudimentary sketch to illustrate the utility of waveguide couplers.

The two identical rib waveguides separated by a spacing of s [m] as shown in Figure 8.16, are assumed to have propagating optical fields of the form

$$(A,B) = (A_0(x,y),B_0(x,y)) \cdot e^{j\beta z}e^{j\omega t} \tag{8.45}$$

which may be related by the coupling equations [55]

$$\frac{dA_0}{dz} = \kappa B_0 \tag{8.46}$$

$$\frac{dB_0}{dz} = -\kappa A_0 \tag{8.47}$$

To write Equations 8.46 and 8.47 in this simple form, we have assumed codirectional coupling, phase matching between the modes, no waveguide attenuation, and a purely real coupling coefficient, κ [m^{-1}]. If we now assume an incident field of amplitude C in one guide only, $A_0(z=0) = 0$ and $B_0(z=0) = C$, these equations have solutions of the form

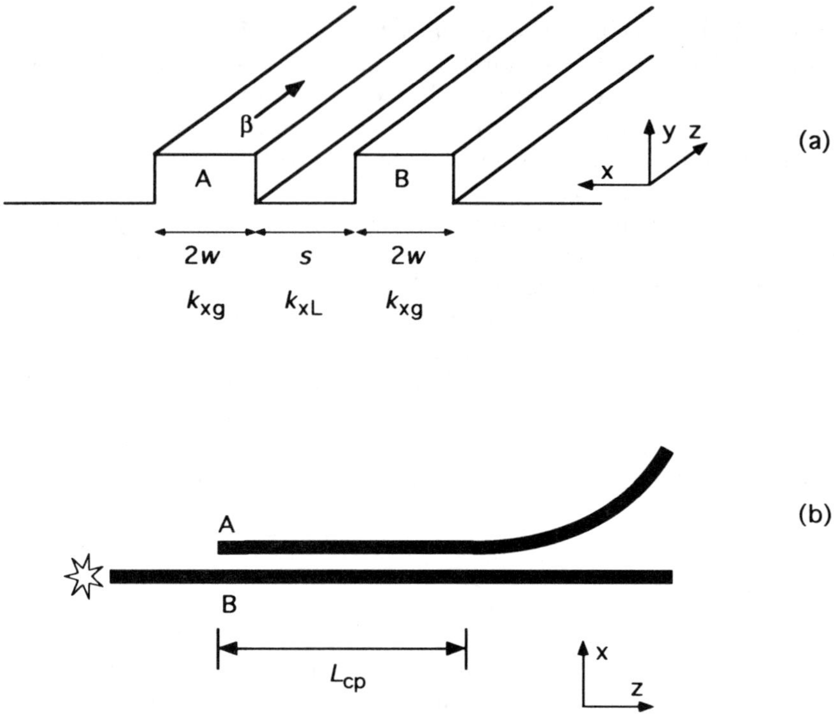

Figure 8.16 (a) Two waveguides, A and B, with mode coupling between them. Both waveguides are assumed identical. (b) Waveguide coupler layout; for a coupling length of $L_{cp} = L_\pi$, all the energy input into waveguide B will be transferred to A.

$$A_0(t) = C \sin(\kappa z) \qquad (8.48)$$
$$B_0(t) = C \cos(\kappa z)$$

which corresponds to two traveling waves, π, out of phase.

Equation 8.48 shows that an optical field incident in one guide will give rise to a field in the neighboring guide, and the energy will transfer between them sinusoidally with distance as the mode moves down the pair of waveguides. We see that complete transfer of energy from one guide to the other takes place periodically in **z** with a period given by the coupling length L_π [m], where $L_\pi = \mathrm{m}\pi/2\kappa$ for integer values of m. Thus, a typical coupler structure is seen in Figure 8.16(b), where the physical overlap distance, L_{cp} [m], may be varied in accordance with the application. For a length $L_{cp} = L_\pi = \pi/2\kappa$, the coupler will transfer all energy from guide B to A; for a length $L_{cp} = L_\pi/2 = \pi/4\kappa$, half of the energy will be transferred to A. Splitting the optical field evenly between the two guides results in a structure known as a 3-dB coupler.

The calculation of the coupling between waveguides often centers on an evaluation of coupling coefficient κ. It has been shown [56] that for two dielectric waveguides with good modal confinement, the coupling coefficient may be expressed as

$$\kappa = \frac{2k_{xg}^2 k_{xL} e^{-k_{xL}s}}{2\beta w(k_{xL}^2 + k_{xg}^2)} \tag{8.49}$$

where the k_{xg} and k_{xL} terms are the transverse propagation and decay constants as usual, β is the **z**-directed propagation constant, and the waveguide spacing is s. We see from this equation that the strongest dependence is on waveguide spacing and transverse decay constant, and thus mode confinement, as both waveguide spacing and transverse decay constant should be small for energy transfer within a reasonable coupling distance, L_{cp}. Clearly, widely spaced waveguides with small evanescent field components will couple poorly. Good couplers usually require shallow ridge or rib-etch depth, a condition that runs contrary to that required for small curvature radius. Thus, the design of curved and coupling structures will not only require detailed analysis and modeling to determine the etch depths, spacing, and coupling lengths, but will also usually represent a compromise.

Example

Since we have solved an interesting waveguide structure in two-dimensions for two different etch depths and thus lateral confinement in the previous example, we can directly employ those results and calculate the efficiency of a coupler based on those structures. For the shallow etch, 150 nm, by applying Equation 8.49, we find for two waveguides with a width of 1 μm that are separated by a spacing, s, of 1 μm, a coupling coefficient of $\kappa = 0.01\ \mu\text{m}^{-1}$. This leads to a coupling length of $L_\pi = 160$ μm. For the more deeply etched waveguide, with depth 200 nm, we find a smaller $\kappa = 0.005\ \mu\text{m}^{-1}$ and thus a longer $L_\pi = 345\ \mu$m.

From this and the last example, it is clear that the etch depth finally used for a particular application will prove to be a compromise between curvature losses and low coupling lengths. In addition, a more deeply etched waveguide might be more susceptible to scattering losses at etched sidewalls, whereas a shallow rib might result in an increased mode size.

8.6.3 Y-Junctions

A slightly less process- and design-sensitive coupler arrangement is the Y-junction of Figure 8.17. Its function seems intuitively clear: a mode entering from the left is

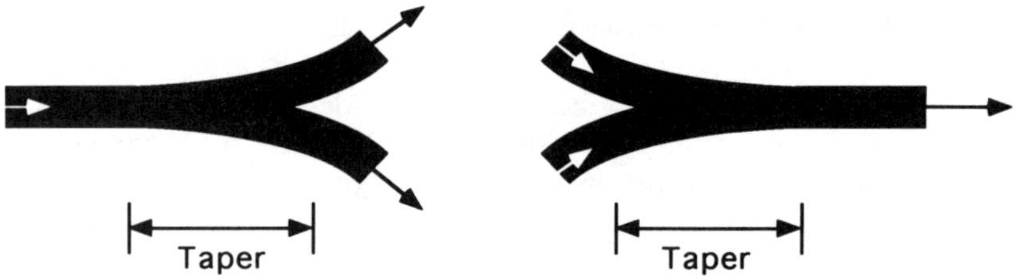

Figure 8.17 A waveguide Y-junction that divides the optical field into two output waveguides. The inverse structure, shown on the right, is used to combine the fields from two input waveguides.

split into two portions, one in either arm of the output. Conversely, the junction on the right combines the modes from the two input arms into a single output. Y-junctions with a virtually arbitrary number of arms may be employed to split or combine waveguide modes.

Design considerations become apparent when we consider what happens as the mode is split or two modes are combined in a Y-junction [57]. In the taper region between the straight waveguide and the point where the diverging guides begin to separate, the width of the structure is usually sufficient to allow higher order modes to propagate. Thus, single-mode input excites a multimode spectrum in the taper, which in turn excites the two single-mode output waveguides. Some of the energy from these higher order modes can be radiated into the substrate after the junction because the single-mode output waveguides will not support these modes. Similarly, for the combiner shown at the right of the figure, higher order modes are excited in the taper and cutoff at the output; for the case of optical input from only one arm, this will imply a 3-dB loss at the output because only one of the excited modes will emerge in the monomode output waveguide. Thus, a fraction of the split or combined power from the junction input is always lost to the substrate in this structure, unlike for the coupler discussed in Section 8.6.2.

The abrupt end of the multimode taper also gives rise to radiation losses, which can be reduced by minimizing the separation angle, ϕ, of the output waveguides. Typical values for ϕ are less than 4°; significant radiation losses may be avoided if this splitting angle is kept below [58]

$$\phi < \cos\left(\frac{n_L}{n_g}\right) - \tan^{-1}\left(\frac{k_{xg}}{\beta}\right) \tag{8.50}$$

where a larger index step between core and cladding clearly allows a larger splitting angle. Adequate control of photolithographic definition and waveguide etch technology must be assured to fabricate a smooth coupler transition.

REFERENCES

[1] Ferlinghetti, L., "Special Clearance Sale of Famous Masterpieces," *Starting from San Francisco,* New York: New Directions Books, 1959.

[2] Garmire, E., "Optical Waveguides in Single Layers of $Ga_{1-x}Al_xAs$ Grown on GaAs Substrates," *Applied Physics Letters,* Vol. 23, 1973, pp. 403–404.

[3] Takeuchi, H., and K. Oe, "Very Low Loss GaAs/AlGaAs Miniature Bending Waveguide with Curvature Radii Less than 1 mm," *Applied Physics Letters,* Vol. 54, 1989, pp. 87–89.

[4] Reinhart, F. K., R. A. Logan, and T. P. Lee, "Transmission Properties of Rib Waveguides Formed by Anodization of Epitaxial GaAs on $Al_xGa_{1-x}As$ Layers," *Applied Physics Letters,* Vol. 24, 1974, pp. 270–272.

[5] Austin, M. W., "Theoretical and Experimental Investigation of GaAs/AlGaAs and n/n^+ GaAs Rib Waveguides," *IEEE Journal of Lightwave Technology,* Vol. LT-2, 1984, pp. 688–694.

[6] Blum, F. A., D. W. Shaw, and W. C. Holton,, "Optical Striplines for Integrated Circuits in Epitaxial GaAs," *Applied Physics Letters,* Vol. 25, 1974, pp. 116–118.

[7] Kapon, E., and R. Bhat, "Low-Loss Single-Mode GaAs/AlGaAs Optical Waveguides Grown by Organometallic Vapor Phase Epitaxy," *Applied Physics Letters,* Vol. 50, 1987, pp. 1,628–1,630.

[8] Tracy, J. C., W. Wiegman, R. A. Logan, and F. K. Reinhart, "Three-Dimensional Light Guides in Single-Crystal GaAs-$Al_xGa_{1-x}As$," *Applied Physics Letters,* Vol. 22, 1973, pp. 511–512.

[9] Dutta, N. K., J. Lopata, R. Logan, and T. Tanbun-Ek, "Integrated Distributed Feedback Laser and Optical Amplifier," *Applied Physics Letters,* Vol. 59, 1991, pp. 1,676–1,677.

[10] Caracci, S. J., F. A. Kish, N. Holonyak, S. A. Maranowski, S. C. Smith, and R. D. Burnham, "High-Performance Planar Native-Oxide Buried-Mesa Index-Guided AlGaAs-GaAs Quantum Well Heterostructure Lasers," *Applied Physics Letters,* Vol. 61, 1992, pp. 321–323.

[11] Garmire, E., H. Stoll, A. Yariv, and R. G. Hunsperger, "Optical Waveguiding in Proton-Implanted GaAs," *Applied Physics Letters,* Vol. 21, 1972, pp. 87–88.

[12] Martin, W. E., and D. B. Hall, "Optical Waveguides by Diffusion in II-VI Compounds," *Applied Physics Letters,* Vol. 21, 1972, pp. 325-327.

[13] Kumar, M., V. Gupta, G. N. DeBrabander, P. Chen, J. T. Boyd, A. J. Steckl, A. G. Choo, H. E. Jackson, R. D. Burnham, and S. C. Smith, "Optical Channel Waveguides in AlGaAs Multiple-Quantum-Well Structures Formed by Focused Ion-Beam-Induced Compositional Mixing," *IEEE Photonics Technology Letters,* Vol. 4, 1993, pp. 435–438.

[14] Yu, L. S., Z. F. Guan, W. Xia, Q. Z. Liu, F. Deng, S. A. Pappert, P. K. L. Yu, S.S. Lau, L.T. Florez, and J.P. Harbison, "Photoelastic Waveguides Formed by Interfacial Reactions," *Applied Physics Letters,* Vol. 62, 1993, pp. 2,944–2,946.

[15] Yamada, Y., M. Yamada, H. Terui, and M. Kobayashi, "Optical Interconnections Using a Silica-Based Waveguide on a Silicon Substrate," *Optical Engineering,* Vol. 28, 1989, pp. 1,281–1,287.

[16] Valette, S., J. P. Jadot, P. Gidon, S. Renard, A. Fournier, A. M. Grouillet, H. Denis, P. Philippe, and E. Desgranges, "Si-Based Integrated Optics Technologies," *Solid State Technology,* 1989, pp. 69–74.

[17] Nagata, T., T. Tanaka, K. Miyake, H. Kurotaki, S.Yokoyama, and M. Koyanagi, "Micron-Size Optical Waveguide for Optoelectronic Integrated Circuits," *Japanese Journal of Applied Physics,* Vol. 33, 1994, pp. 822–826.

[18] Takato, N., K. Jinguji, M. Yasu, H. Toba, and M. Kawachi, "Silica-Based Single-Mode Waveguides on Silicon and Their Application to Guided-Wave Interferometers," *IEEE Journal of Lightwave Technology,* Vol. 6, 1988, pp. 1,003–1,010.

[19] Wagner, C., and P. P. Deimel, "Optical Radiation from Tapered Rib-Waveguides," *IEEE Journal of Lightwave Technology,* Vol. 11, 1993, pp. 1,769–1,773.

[20] Fischer, U., T. Zinke, B. Schüppert, and K. Petermann, "Singlemode Optical Switches Based on SOI Waveguides with Large Cross-Section," *Electronics Letters,* Vol. 30, 1994, p.406–408.

[21] Rickman, A. G., G. T. Reed, and F. Namavar, "Silicon-On-Insulator Optical Rib Waveguide Loss and Mode Characteristics," *IEEE Journal of Lightwave Technology*, Vol. 12, 1994, pp. 1,771–1,776.
[22] Soref, R. A., "All-Silicon Active and Passive Guided Wave Components for $\lambda = 1.3$ and 1.6 μm," *IEEE Journal of Quantum Electronics*, Vol. QE-22, 1986, pp. 873–879.
[23] Splett, A., and K. Petermann, "Low Loss Single-Mode Optical Waveguides with Large Cross-Section in Standard Epitaxial Silicon," *IEEE Photonics Technology Letters*, Vol. 6, 1994, pp. 425–427.
[24] Soref, R. A., F. Namavar, and J. P. Lorenzo, "Optical Waveguiding in a Single-Crystal Layer of Germanium Silicon Grown on Silicon," *Optics Letters*, Vol. 15, 1990, pp. 270–272.
[25] Schmidtchen, J., B. Schüppert, and K. Petermann, "Passive Integrated-Optical Waveguide Structures by Ge-diffusion in Silicon," *IEEE Journal of Lightwave Technology*, Vol. 12, 1994, pp. 842–848.
[26] Suzuki, S., M. Yanagisawa, Y. Hibino, and K. Oda, "High-Density Integrated Planar Waveguide Circuits Using SiO_2-GeO_2 Waveguides with a High Refractive Index Difference," *IEEE Journal of Lightwave Technology*, Vol. 12, 1994, pp. 790–796.
[27] Henry, C. H., G. E. Blonder, and R. F. Kazarinov, "Glass Waveguides on Silicon for Hybrid Optical Packaging," *IEEE Journal of Lightwave Technology*, Vol. 7, 1989, pp. 1,530–1,539.
[28] Kimble, T. C., M. D. Himel, and K. H. Guenther, "Optical Waveguide Characterization of Dielectric Films Deposited by Reactive Low-Voltage Ion Plating," *Applied Optics*, Vol. 32, 1993, pp. 5,640–5,644.
[29] Kogelnik, H., "Theory of Optical Waveguides," in *Guided-Wave Optoelectronics*, Tamir, T. (ed.), Berlin: Springer-Verlag, 1988, Section 2.5.7.
[30] Nishihara, H., M. Haruna, and T. Suhara, *Optical Integrated Circuits*, New York: McGraw Hill, 1989, Section 2.3.
[31] Marcatili, E. A. J., "Dielectric Rectangular Waveguide and Directional Coupler for Integrated Optics," *Bell System Technical Journal*, Vol. 48, 1969, pp. 2,071–2,102.
[32] Furuta, H., H. Noda, and A. Ihaya, "Novel Optical Waveguide for Integrated Optics," *Applied Optics*, Vol. 13, 1974, pp. 322–326.
[33] Hunsperger, R. G., A. Yariv, and A. Lee, "Parallel End-Butt Coupling for Optical Integrated Circuits," *Applied Optics*, Vol. 16, 1977, pp. 1,026–1,032.
[34] Emkey, W. L., "Optical Coupling Between Single-Mode Semiconductor Lasers and Strip Waveguides," *IEEE Journal of Lightwave Technology*, Vol. LT-1, 1983, pp. 436–443.
[35] Zappe, H. P., M. Shen, and H. E. G. Arnot, "Hybrid Coupling and Monolithic Integration for Integrated Optical Sensors," *Proceedings of the 6th European Conference on Integrated Optics*, Neuchâtel, Switzerland, April 18–20, 1993, pp. 11.8–11.9.
[36] Kuwahara, H., M. Sasaki, and N. Tokoyo, "Efficient Coupling from Semiconductor Lasers into Single-Mode Fibers with Tapered Hemispherical Ends," *Applied Optics*, Vol. 19, 1980, pp. 2,578–2,583.
[37] Suhara, T., and H. Nishihara, "Integrated Optics Components and Devices Using Periodic Structures," *IEEE Journal of Quantum Electronics*, Vol. QE-22, 1986, pp. 845–867.
[38] Hunsperger, R. G., *Integrated Optics: Theory and Technology*, 3rd Ed., Berlin: Springer-Verlag, 1991, p. 97.
[39] Moss, T. S., and T. D. F. Hawkins, "Infrared Absorption in Gallium Arsenide," *Infrared Physics*, Vol. 1, 1961, pp. 111–115.
[40] Hunsperger, op.cit., p. 78.
[41] Fan, H. Y., "Effects of Free Carriers on the Optical Properties," in *Semiconductors and Semimetals*, Willardson, R. K., and A. C. Beer (eds.), Vol. 3, New York: Academic Press, 1967, p. 405.
[42] Ridley, B. K., *Quantum Processes in Semiconductors*, Oxford, England: Clarendon, 1988, Chapter 5.2.
[43] Hall, D. G., "In-Plane Scattering in Planar Optical Waveguides: Refractive Index Fluctuations and Surface Roughness," *Journal of the Optical Society of America*, A, Vol. 2, 1985, pp. 747–752.
[44] Tien, P. K., "Light Waves in Thin Films and Integrated Optics," *Applied Optics*, Vol. 10, 1971, pp. 2,395–2,413.

[45] Deri, R. J., and E. Kapon, "Low-Loss III-V Semiconductor Optical Waveguides," *IEEE Journal of Quantum Electronics,* Vol. 27, 1991, pp. 626–640.
[46] Hibbs-Brenner, M. K., and C.T. Sullivan, "Low-Loss AlGaAs Optical Rectangular Waveguides at 830 nm," *Applied Physics Letters,* Vol. 56, 1990, pp. 1,529–1,531.
[47] Henry, C. H., R. F. Kazarinov, H. J. Lee, K. J. Orlowsky, and L. E. Katz, "Low Loss Si_3N_4-SiO_2 Optical Waveguides on Si," *Applied Optics,* Vol. 26, 1987, pp. 2,621–2,624.
[48] Pesarcik, S. F., G. V. Treyz, S. S. Iyer, and J. M. Halbout, "Silicon Germanium Optical Waveguides with 0.5 dB/cm Losses for Singlemode Fibre Optic Systems," *Electronics Letters,* Vol. 28, 1992, pp. 159–160.
[49] Schmidtchen, J., A. Splett, B. Schüppert, K. Peterman, and G. Burbach, "Low Loss Singlemode Optical Waveguides with Large Cross-Section in Silicon-On-Insulator," *Electronics Letters,* Vol. 27, 1991, pp. 1,486–1,488.
[50] Najafi, S. I., *Introduction to Glass Integrated Optics,* Norwood, MA: Artech House, 1992, Chapter 5.5.
[51] Walker, R. G., "Simple and Accurate Loss Measurement Technique for Semiconductor Waveguides," *Electronics Letters,* Vol. 21, 1985, pp. 581–583; Erratum, *Electronics Letters,* Vol. 21, 1985, p. 714.
[52] Marcatili, E. A. J., and S. E. Miller, "Improved Relations Describing Directional Control in Electromagnetic Wave Guidance," *Bell System Technical Journal,* Vol. 48, 1969, pp. 2,161–2,188.
[53] Deri, R. J., E. Kapon, and M. Schiavone, "Bend Losses in GaAs/AlGaAs Optical Waveguides," *Electronics Letters,* Vol. 23, 1987, pp. 845–847.
[54] Pennings, E. C. M., G. H. Manhoudt, and M. K. Smit, "Low-Loss Bends in Planar Optical Ridge Waveguides," *Electronics Letters,* Vol. 24, 1988, pp. 998–999.
[55] Yariv, A., "Coupled-Mode Theory for Guided-Wave Optics," *IEEE Journal of Quantum Electronics,* Vol. QE-9, 1973, pp. 919–933.
[56] Somekh, S., E. Garmire, A. Yariv, H. L. Garvin, and R.G. Hunsperger, "Channel Optical Waveguide Directional Couplers," *Applied Physics Letters,* Vol. 22, 1973, pp. 46–47.
[57] Izutsu, M., Y. Nakai, and T. Sueta, "Operation Mechanism of the Single-Mode Optical-Waveguide Y Junction," *Optics Letters,* Vol. 7, 1982, pp. 136–138.
[58] Kuznetsov, M., "Radiation Losses in Dielectric Waveguide Y-Branch Structures," *IEEE Journal of Lightwave Technology,* Vol. LT-3, 1985, pp. 674–677.

Chapter 9

Lasers

Had my eyes dazzled by the shimmering track of beams.
Walt Whitman, 1856 [1]

Lasers have become mundane. The track of red beams emanating from the supermarket checkout counter dusted off some of their Buck Rogers aura, and when semiconductor lasers supplanted the phonograph needle, the microscopic devices became as much an established fixture in the household as the toaster. Key to this development was fabrication technology: structurally, most semiconductor lasers derive from the waveguide configurations we discussed in Chapter 8, so the capabilities of semiconductor manufacturing can be brought to bear on these optoelectronic devices. Thus, laser performance has become affordable, which we can see clearly when considering that the semiconductor laser found in compact disc players can be had for pocket change. For the consumer, the laser has become an over-the-counter item.

Despite its myriad secular applications, the laser in general is both the subject of considerable ongoing research and employed as an optical tool. Shortly after its invention in 1958 [2], the unique features of this light source became apparent: light emission is monochromatic and coherent, and hence can be easily collimated. Output is intense and generated with a high efficiency, implying low-power consumption. In addition, it later became clear that lasers can be modulated at high speeds, which is essential for telecommunications [3] or computer data transmission [4]. Lasers can also emit extremely short optical pulses [5], which renders them useful as a research tool in numerous fields.

The semiconductor laser retains all of these benefits despite its small physical size and manufacturability. Following considerable early theoretical and experimental work on semiconductors as potential laser materials,[1] the efficient electroluminescence from GaAs pn junctions led to the first semiconductor injection laser demonstrations in 1962 [6–8]. Once room-temperature operation was achieved [9] by use of the GaAs/AlGaAs double heterostructure [10] in 1970, the device had become a focus

[1]For a historical overview, see Dupuis, R. D., "An Introduction to the Development of the Semiconductor Laser," *IEEE Journal of Quantum Electronics*, Vol. QE-23, 1987, pp. 651–657, and other articles in this special issue.

of considerable research, which continues, with the addition of numerous new material systems, unabated today.

This chapter peruses the basics of laser operation and semiconductor laser structure and behavior. Befitting such a developed field, there are numerous excellent texts that discuss many of these subjects in greater detail.[2] We begin with a discussion of spontaneous and stimulated light emission and combine this with the idea of optical amplification to derive the concept of optical gain. Continuing with an analysis of the feedback generated by optical oscillation in a resonant cavity, we are led quite naturally to the laser *per se.* From there we turn our attention to manufacturable semiconductor device structures and outline the basic measurable electrical and optical characteristics. Finally, we address the use of quantum wells, take a glance at LEDs, and finish with a discussion of distributed feedback and surface-emitting lasers.

9.1 LASER OPERATION—A QUICK HEURISTIC DISCUSSION

We begin with a material that can be coerced into emitting light under the proper external stimulus: light and, more practically, current are suitable stimuli. Electron transitions from the conduction to the valence band generate photons by either spontaneous or stimulated emission, the latter occurring when a photon propagating through the semiconductor incites an electron to undergo a transition and emit a second photon identical to the first. If the conduction band is heavily populated with electrons, the probability of stimulated emission is high, and a single generated photon can give rise to significant numbers of clones of itself. Thus, an optical field propagating through the material will grow in intensity, a phenomenon known as *gain.* The overpopulation of the conduction band necessary for gain to exist, known as population inversion, can be achieved by heavy current injection, particularly across a semiconductor pn junction.

We now take this piece of material into which we are injecting a current, and which exhibits gain, and place it inside an optically resonant cavity. The resonator forces the field to traverse the material repeatedly, so that the intensity grows. At the same time, the cavity permits only a distinct set of resonant frequencies to oscillate, and these increase in intensity until virtually all the energy of the system is found in one resonant mode. At that point, the device begins to lase and zaps the reflecting surface of the compact disc. In the rest of this chapter, we will see how to design and fabricate a device that permits such zapping.

[2]See, for example: Casey, H. C., and M. B. Panish, *Heterostructure Lasers,* Vols. A and B, San Diego: Academic Press, 1978; or Agrawal, G. P. and N. K. Dutta, *Long-Wavelength Semiconductor Lasers,* New York: Van Nostrand Reinhold, 1986.

9.2 PHOTON EMISSION IN SEMICONDUCTORS

The basis of any electrically pumped laser or LED is the conversion of electrical power to optical power, and thus the emission of photons from the material. In semiconductors, optical emission results from the transitions of electrons from higher to lower energy levels, so that the energy band diagrams we discussed in Chapter 2 again become relevant. Transitions can either be spontaneous (i.e., the electron decides one fine picosecond to leap to its demise) or stimulated, in which the presence of a photon causes an electronic transition that generates a further photon. We thus begin our analysis of the laser with a study of optical emission.

9.2.1 Spontaneous and Stimulated Emission

The semiconductor from which we are to fabricate our laser can be modeled by the two level electronic system shown in Figure 9.1. This model is equivalent to the simplified version of a semiconductor bandgap we derived in Figure 2.8, where we ignored the variation of energy, E [J], with momentum, k [kg m/s^2], and simply drew two energy levels uniform in k and real space. The two energy levels are denoted E_1 and E_2, with electron densities n_1 and n_2 [m^{-3}]; these energy levels may represent the valence and conduction bands, respectively. Because we ignore the k variation, the transitions we are about to discuss are all direct; i.e., $\Delta k = 0$.

Several electronic transitions are now possible. For spontaneous emission, electrons from level 2 drop to level 1 at a rate [in m^{-3}s^{-1}] given by

$$\left.\frac{dn_2}{dt}\right|_{\text{spon}} = -A_{21}n_2 = \frac{1}{\tau_{21}}n_2 \tag{9.1}$$

where $A_{21} = 1/\tau_{21}$ [s^{-1}] represents a transition rate per electron and thus τ_{21} [s] a transition time or lifetime. This spontaneous transition, which occurs without any external stimulus, generates photons of energy $E_\nu = E_2 - E_1 = h\nu$. The rate is clearly related to the population density of the upper level, n_2.

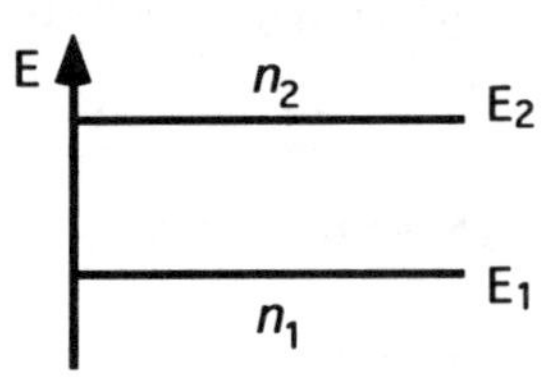

Figure 9.1 A two-level model system for electronic transitions in a semiconductor. The two energy levels are E_1 and E_2, with electron densities n_1 and n_2, representing, for example, the valence and conduction bands, respectively.

In the presence of an optical field $\Phi(\nu)$ [Js/m^3], representing a photon density per unit frequency, ν [Hz], and per unit volume, both absorption and stimulated emission may take place. Absorption of photons with energy $E_\nu = E_2 - E_1$ leads to upward transitions at a rate

$$\frac{dn_1}{dt} = -B_{12} n_1 \Phi(\nu) \tag{9.2}$$

corresponding to the inverse of spontaneous emission. The net effect is to increase the population of the upper level. This absorption rate is defined by a constant B [m^3/Js^2].

The optical field also gives rise to stimulated emission, in which a photon from the exciting field incites an electron transition from level 2 to 1, resulting in the generation of a photon with the same energy, polarization, and direction as the first. The number of electrons lost from level 2 per unit time due to stimulated emission is

$$\left.\frac{dn_2}{dt}\right|_{\text{stim}} = -B_{21} n_2 \Phi(\nu) \tag{9.3}$$

so that the total rate of electron loss from level 2 to 1 is given by the sum of the stimulated (Equation 9.3) and spontaneous (Equation 9.1) components

$$-\frac{dn_2}{dt} = A_{21} n_2 + B_{21} n_2 \Phi(\nu) \tag{9.4}$$

In thermal equilibrium, the net rates of upward and downward transitions must be equal, $dn_1/dt = dn_2/dt$, leaving us with a characteristic equation for this system,

$$A_{21} n_2 + B_{21} n_2 \Phi(\nu) = B_{12} n_1 \Phi(\nu) \tag{9.5}$$

The rate constants A and B are known collectively as the Einstein A and B coefficients [11], whose values we will determine presently.

9.2.2 Electron Distribution and Transition Rates

The electronic transitions we have just described have been defined without making any assumptions concerning the electron populations n_1 and n_2. By adding to our model the electron distribution in energy, we can find values for the Einstein coefficients. In thermal equilibrium, electron density as a function of energy far from the Fermi level is approximated by a Boltzmann function, namely

$$\frac{n_2}{n_1} = \exp\left[-\frac{(E_2 - E_1)}{k_B T}\right] = \exp\left[-\frac{\mathrm{h}\nu}{k_B T}\right] \tag{9.6}$$

since we are considering a photon energy $\mathrm{h}\nu$ where h is Planck's constant [Js] and ν the frequency [$\mathrm{Hz} \equiv \mathrm{s}^{-1}$]. Note that Equation 9.6 implies that, in thermal equilibrium, n_2 is always smaller than n_1.

The induced transitions are characterized by the constants B_{12} and B_{21}, which are in turn characteristic of the material. For a system in thermal equilibrium at temperature T [K], the optical field $\Phi(\nu)$ takes the form of a black body (thermal) radiation spectrum

$$\Phi(\nu) = \frac{\Phi_0}{(e^{(\mathrm{h}\nu/k_B T)} - 1)} \tag{9.7}$$

where Φ_0 [$\mathrm{Js/m^3} \equiv \mathrm{J/m^3Hz}$] is given by

$$\Phi_0 = \frac{8\pi \mathrm{h}}{\lambda^3} \tag{9.8}$$

for wavelength in the material, λ [m]. For convenience, we will continue to use both λ and ν, remembering that $\lambda\nu = c$. Combining Equations 9.7 and 9.8 with 9.5 allows us to write

$$\frac{n_2}{n_1} = \frac{B_{12}\Phi(\nu)}{A_{21} + B_{21}\Phi(\nu)} = \exp\left(-\frac{\mathrm{h}\nu}{k_B T}\right) \tag{9.9}$$

which we can simplify to yield

$$B_{12}\Phi_0 - B_{21}\Phi_0 e^{-\mathrm{h}\nu/k_B T} = A_{21}(1 - e^{-\mathrm{h}\nu/k_B T}) \tag{9.10}$$

By looking at Equation 9.10 for a moment, we realize that two conditions are necessary for a solution: (1) $B_{12} = B_{21}$, and (2) $A_{21} = B_{12}\Phi_0$. The first of these conditions implies that the induced upward and downward transition rates are equal, or in other words, a photon is as likely to transfer an electron from E_1 to E_2 as from E_2 to E_1. We can now simply define $B_{12} = B_{21} \equiv B$ [$\mathrm{m^3/Js^2}$] and $A_{21} \equiv A$ [$\mathrm{s^{-1}}$], and take advantage of Equation 9.1 in order to relate

$$A = \frac{1}{\tau_{21}} = \frac{1}{\tau_{sp}} \tag{9.11}$$

and

$$B = \frac{1}{\Phi_0 \tau_{sp}} \tag{9.12}$$

where τ_{sp} [s] is known as the spontaneous lifetime of an electron state.

With these definitions, we can already derive two hints for conditions that will be necessary for lasing. Stimulated emission, as we noted in our quick heuristic discussion, is central to laser action. We can see from the results above that the ratio of stimulated emission to absorption is given by

$$\frac{\text{stimulated}}{\text{absorption}} = \frac{n_2}{n_1} \tag{9.13}$$

and stimulated to spontaneous emission by

$$\frac{\text{stimulated}}{\text{spontaneous}} = \frac{B\Phi(\nu)}{A} \tag{9.14}$$

It is clear from these last two expressions that a dominance of stimulated emission from a laser medium is achieved through an excess population of electrons in level 2 (a condition known as population inversion, which implies that the system is not in thermal equilibrium) and through the presence of an intense optical field, $\Phi(\nu)$. We will see in the next section how we go about achieving these conditions.

Example

From Equation 9.14, we see that stimulated and spontaneous emission are equal when $\Phi = A/B$. Since we know A and B from Equations 9.11 and 9.12, we can easily write down the optical field necessary for equal stimulated and spontaneous components, namely $\Phi = \Phi_0 = 8\pi h/\lambda^3$. We can convert this to an optical intensity by writing $\mathcal{I}_\nu = c\Phi/g(\nu) \approx c\Phi\Delta\nu$, for speed of light, c, and linewidth, $\Delta\nu$. If we consider the case of an optical field propagating through a piece of GaAs, $\lambda_0 = 850$ nm, n = 3.45, and taking $\Delta\nu = 10^{14}$ Hz, we obtain $\mathcal{I}_\nu = 9.7 \cdot 10^9$ W/m^2, recalling that $\lambda = \lambda_0/n$ and $c = c_0/n$. For a typical semiconductor laser cross-sectional area, say 5×0.2 μm^2, this corresponds to a power of about 10 mW. This calculation is very approximate and we should not take the numbers too seriously; it is more useful to illustrate qualitatively the effect of an optical field on the ratio of stimulated to spontaneous emission.

9.2.3 Lineshape Function

The variation of the optical field with frequency is also important for a detailed calculation of the interaction of photons with a two-level system, such as our semiconductor model. If all existent photons are generated solely by this dual-level system, we might imagine that the optical field is a delta function at the gap energy, i.e. $\Phi(\nu) = \Phi \cdot \delta(\nu - \nu_0)$ where $\nu_0 = (E_2 - E_1)/\text{h}$, meaning of course that the field is zero for all other frequencies. For various reasons, however, the transition is not truly monochromatic but has a finite width and a characteristic shape. This latter quality is defined by the lineshape function.

We define the lineshape function, $g(\nu)$ [Hz^{-1}] as simply the distribution of photon intensity in frequency (or energy). The function is chosen so that we may normalize it as

$$\int_{-\infty}^{\infty} g(\nu)d\nu = 1 \tag{9.15}$$

and it is identical for photon emission or absorption. Thus $g(\nu)$ defines the spectral shape in energy of the optical field, which we may then write as

$$\Phi(\nu) = \Phi_\nu g(\nu) \tag{9.16}$$

where the constant Φ_ν [J/m^3] is the energy density per volume of the optical field, leading to $\Phi(\nu)$ as an energy density per frequency and per volume when multiplied with the lineshape function.

We can now return to our expressions for stimulated emission (Equation 9.3) and include the shape of the optical field. The number of photons emitted per unit time and per unit volume is, similar to before,

$$-\left.\frac{dn_2}{dt}\right|_{\text{stim}} = \frac{n_2}{\Phi_0 \tau_{\text{sp}}}\Phi_\nu g(\nu) = \frac{n_2 \lambda^3 \Phi_\nu}{8\pi \text{h} \tau_{\text{sp}}} g(\nu) \tag{9.17}$$

We can use this expression to define a transition rate *per atom*, W_i [s^{-1}] as

$$W_i = \frac{\lambda^3 \Phi_\nu}{8\pi \text{h} \tau_{\text{sp}}} g(\nu) \tag{9.18}$$

which, using the Einstein coefficients, is of the form $W_i = B\Phi(\nu)$. The emission rate parameter, W_i, defines the number of photons being absorbed or emitted by stimulated emission per second per atom. Note that the rate is the same per atom for transitions in either direction. By extension, $W_i(n_2 - n_1)$ is the rate of photon emission per second per volume so that $W_i(n_2 - n_1)\text{h}\nu$ becomes the emitted energy per second per volume or the intensity per volume. We see thus that because it is integral to the definition of W_i, the lineshape function is central to an analysis of laser emission.

We briefly make note of alternative notation that is occasionally employed. The transition rate is often written as

$$W_i = \rho\sigma(\nu) = \rho S g(\nu) \tag{9.19}$$

where σ is defined as a transition cross section [m^2] and S as an oscillator strength [m^2/s], the terminology deriving from an electronic transition model that represents the atom as an electron oscillator. The term ρ [$m^{-2}s^{-1}$] is then the photon flux, per unit time and area. We will not use these definitions further, but the reader may stumble across these elsewhere.

9.2.4 Spectral Shape and Broadening

Having seen how the lineshape function of an atomic transition directly affects the transition rate, we need to discuss spectral shape and width in greater detail. As we mentioned previously, the width of the transition is not infinitely narrow and numerous factors contribute to a spectral broadening of $g(\nu)$.

One fundamental source of spectral broadening is the finite value of the lifetime, τ_{sp}. To see this, we can use a damped oscillator model[3] for the decay of an atom leading to an emitted photon. The atom is thus characterized by an electric field, $\mathcal{E}$,

$$\mathcal{E}(t) = e^{-t/\tau_{sp}} e^{j2\pi\nu_0 t} \tag{9.20}$$

for resonant frequency ν_0 with a Fourier transform

$$\mathcal{E}(\nu) \propto \frac{1}{1 + j2\pi(\nu - \nu_0)\tau_{sp}} \tag{9.21}$$

Realizing that the intensity of a transition is proportional to $|\mathcal{E}(\nu)|^2$, we can write

$$|\mathcal{E}(\nu)|^2 \propto \frac{1}{(\nu - \nu_0)^2 + \tau_{sp}^{-2}} \tag{9.22}$$

for frequencies near resonance. Because the normalization condition of Equation 9.15 must hold for the lineshape function and by defining the linewidth as $\Delta\nu = 1/\pi\tau_{sp}$, we can derive from this last expression the spectral envelope, $g(\nu)$,

$$g(\nu) = \frac{\Delta\nu}{2\pi\left[(\nu - \nu_0)^2 + \left(\frac{\Delta\nu}{2}\right)^2\right]} \tag{9.23}$$

[3]This derivation is based on a nice discussion in Yariv, A., *Optical Electronics*, 4th Ed., Philadelphia: Saunders, 1991, Chapter 5.

This form for $g(\nu)$ is known as Lorentzian and has the shape shown schematically in Figure 9.2; a Lorentzian spectrum is typical of a linearly damped system such as the one we have used for our finite-lifetime model. The linewidth, $\Delta\nu$, is defined as the full width at half maximum (FWHM), the point at which $g(\nu)$ has decayed to half its peak value; the peak is centered at ν_0. This spectral form for photon emission from atomic systems is indeed what is measured experimentally.

Numerous material factors can contribute to a further spectral broadening of the lineshape function [12]. We have used the finite (nonzero) value for spontaneous lifetime as one source that contributes to homogeneous broadening, implying that all atoms have the same $g(\nu)$. Phonon collisions, causing a random shift in the phase of the atomic eigenfunction and collisions between atoms, the latter predominantly in gasses and negligible in semiconductors, also contribute to broadening in a homogeneous medium.

In addition, inhomogeneous broadening in semiconductors due to crystalline defects, material inhomogeneities, and strain, also causes a widening of the emission or absorption spectrum. In an inhomogeneous system, different atoms or groups of atoms can have different lineshape functions, so that $g(\nu)$ for the system is a weighted average of the individual lineshapes of the constituent atoms. Doppler broadening, derived from the motion of atoms in a gaseous laser material, is a well known source of inhomogenous broadening, albeit one that is practically irrelevant for semiconductor systems.

9.3 OPTICAL GAIN

The photons generated in a semiconductor by the processes we have just studied propagate through the material and continue to interact with atoms as they do so.

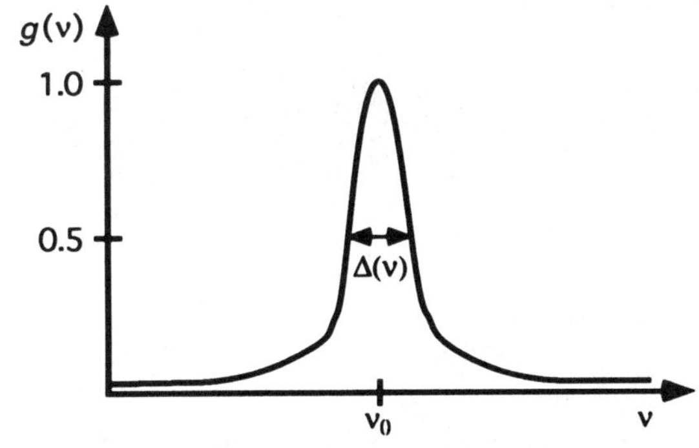

Figure 9.2 The Lorentzian form of the lineshape function, $g(\nu)$; its width, $\Delta\nu$, is defined as the full width at half maximum, where the value has decreased to half the peak value at ν_0.

If no external energy is input into this system, the photons will typically be absorbed, resulting in an attenuation of the optical field. This is the material absorption we saw in earlier chapters. A laser, however, requires amplification of light, and we shall now see that it is possible to increase the number of photons as the optical field propagates and thereby achieve optical gain.

9.3.1 Amplification and Attenuation

We have drawn schematically the amplification and attenuation of an optical field in Figure 9.3. Both processes are mediated by electronic transitions, which for a transition rate of W_i comprise $W_i n_2$ stimulated photon emissions and $W_i n_1$ photon absorptions per unit time and unit volume, still referring to our two-level model of Figure 9.1. Since the former process increases the number of photons and the latter decreases it, we can define a net gain in photons per volume as $(n_2 - n_1)W_i$ or gain in optical power per volume as $(n_2 - n_1)W_i h\nu$.

How do these processes affect the intensity of a wave propagating through the material in the **z** direction as shown in the figure? We can define the energy intensity (energy per area per unit time or equivalently power per area) of the propagating field, $\mathcal{I}_\nu$ [J/m^{2}s], as

$$\mathcal{I}_\nu = \lambda\nu\Phi_\nu = \frac{c}{\mathrm{n}}\Phi_\nu \tag{9.24}$$

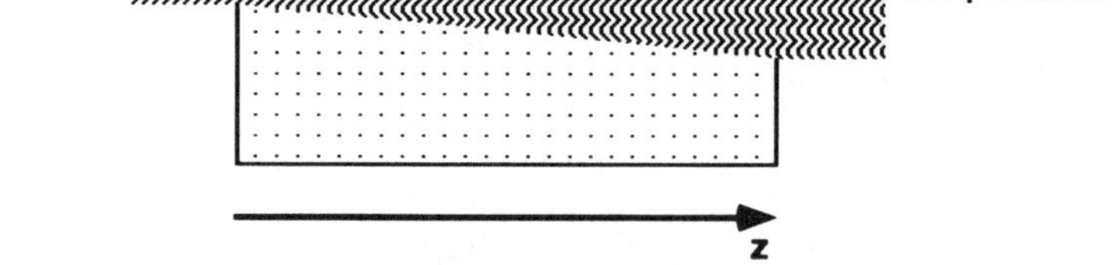

Figure 9.3 Amplification and attenuation in a semiconductor medium. The upper trace shows, schematically, the decay of the optical field in a region where more photons are absorbed than generated by stimulated emission (gain, γ, is negative) and the lower an amplified field if stimulated emission dominates (γ positive).

for optical field energy density, Φ_ν; refractive index, n; and speed of light, c. We saw above that power per unit volume is given by $W_i(n_2 - n_1)\mathrm{h}\nu$, so that we can write the change in intensity with propagation as

$$\mathrm{d}\mathcal{I}_\nu = (n_2 - n_1)W_i\mathrm{h}\nu\,\mathrm{d}z \tag{9.25}$$

Using the definition of W_i (Equation 9.18) and that of intensity (Equation 9.24), we can rewrite this expression as a differential equation in intensity

$$\mathrm{d}\mathcal{I}_\nu = \frac{\lambda^2}{8\pi\tau_{sp}}(n_2 - n_1)\mathcal{I}_\nu g(\nu)\,\mathrm{d}z \tag{9.26}$$

or

$$\frac{\mathrm{d}\mathcal{I}_\nu}{\mathrm{d}z} = \frac{\lambda^2}{8\pi\tau_{sp}}(n_2 - n_1)g(\nu)\mathcal{I}_\nu \tag{9.27}$$

Equation 9.27 clearly has a solution of the form

$$\mathcal{I}_\nu(z) = \mathcal{I}_\nu(0) \cdot e^{\gamma z} \tag{9.28}$$

where the factor γ [traditionally cm^{-1}] is given by

$$\gamma = \frac{\lambda^2}{8\pi\tau_{sp}}(n_2 - n_1)g(\nu) \tag{9.29}$$

and is referred to as the gain of the material.

We can see immediately from Equations 9.28 and 9.29 that the sign of γ is important for deciding whether an optical field will grow or decay as it propagates, and the sign is given by the relative values of n_1 and n_2. If $n_1 > n_2$, the case for thermal equilibrium where the lower energy level is more populated, $\gamma < 0$ and the optical wave is attenuated. On the other hand, if $n_2 > n_1$, a nonequilibrium situation in which the upper level has more electrons, $\gamma > 0$ and the semiconductor has gain. This latter case requires an external energy source to populate the upper level, a process known as pumping. If we can achieve this, however, we have fulfilled one necessary condition for a laser, namely a medium that amplifies the photon field.

9.3.2 Semiconductor Gain

The existence of gain is essential for laser operation, but the variation of gain with photon energy and carrier density is also important in the design. These variations, particularly the former, are known collectively as the gain profile. We saw in Equation 9.29 that the gain is directly proportional to the difference between the electron populations in the upper and lower energy levels. In a semiconductor, this implies the difference between electron concentrations in the conduction and valence bands, respectively. Calculations of the gain profile are in general quite involved [13], but we can write an approximate expression for gain in a semiconductor as [14]

$$\gamma(\nu) = \frac{\lambda^2}{8\pi^2\tau_{sp}}\left[\frac{2}{\hbar}\frac{m_c^* m_v^*}{(m_c^* + m_v^*)}\right]^{3/2}\left(2\pi\nu - \frac{E_g}{\hbar}\right)^{1/2}[f_c(\nu) - f_v(\nu)] \tag{9.30}$$

The terms m_c^* and m_v^* are the conduction and valence band effective masses, respectively, and E_g is the energy gap. f_c and f_v represent the occupation probabilities of the conduction and valence bands, respectively, due to the Fermi-Dirac distribution of carriers,

$$f_c(\nu) = \left[\exp\left(\frac{h\nu - E_{Fc}}{k_B T}\right) + 1\right]^{-1} \tag{9.31}$$

$$f_v(\nu) = \left[\exp\left(\frac{h\nu - E_{Fv}}{k_B T}\right) + 1\right]^{-1} \tag{9.32}$$

The Fermionic nature of electrons in the semiconductor system contrasts with the Maxwell-Boltzmann statistics, which apply to most other laser systems. This is primarily due to the low spatial density of carriers in the various levels of the latter.

An analysis of Equation 9.30 shows that $\gamma(\nu)$ is only positive (meaning that there is gain rather than loss in the material) for a narrow range of photon energies. It is first required that $h\nu > E_g$, because no photons can contribute to stimulated emission for energies below the bandgap. Once this is given, the sum of the Fermi factors, $[f_c(\nu) - f_v(\nu)]$, is the term that defines population inversion. The condition $[f_c(\nu) - f_v(\nu)] > 0$ is equivalent to $n_2 > n_1$ in Equation 9.29, implying the existence of the inverted carrier distribution necessary for gain. We can easily show, from Equations 9.31 and 9.32, that $[f_c(\nu) - f_v(\nu)] > 0$ is identical to the condition $h\nu < E_{fc} - E_{fv}$, the quasi-Fermi level spacing.

Thus, the photon energy range in which we expect positive gain is given by

$$E_g < h\nu < E_{fc} - E_{fv} \tag{9.33}$$

As is shown schematically in Figure 9.4, photons of energy below the bandgap do not interact with the semiconductor (i.e., the material is transparent). For photons of energy larger than the quasi-Fermi level spacing, the semiconductor becomes

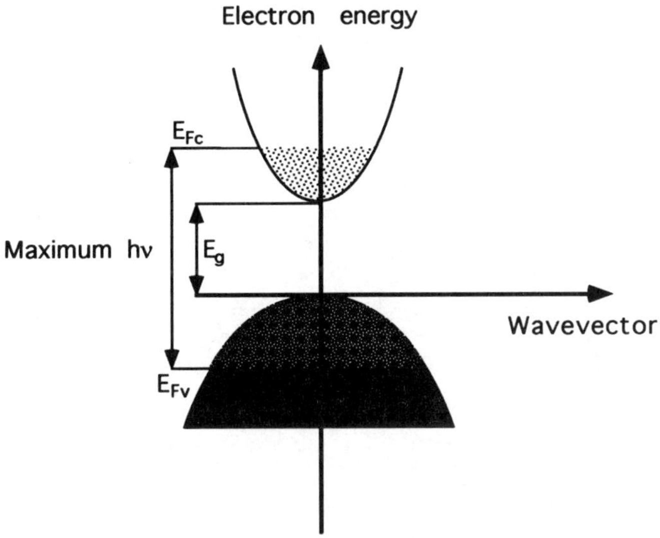

Figure 9.4 Electronic transitions give rise to lasing action. The photons must have an energy, $h\nu$, greater than the bandgap (E_g) but less than the quasi-Fermi level spacing ($f_c(\nu) - f_v(\nu)$), to experience positive gain.

absorbing, as no electron transitions resulting from stimulated emission are possible. We recall that the top of the inverted electron population is defined by the quasi-Fermi level.

It should be clear from this discussion that a semiconductor in thermal equilibrium, for which $[f_c(\nu) - f_v(\nu)] < 0$, will never exhibit gain and thus never lase. The nonequilibrium situation, the population inversion necessary for gain, is most easily achieved in semiconductors by the use of a pn junction. Through injection across the junction, very high local carrier densities, and thus inversion, can be achieved. The ability to fabricate pn junctions is key to the success of an electrically pumped semiconductor laser.

We can use Equation 9.30 to calculate the dependence of gain in GaAs on photon energy, using the electron density, or pump intensity, as a parameter. As we can see in Figure 9.5, gain is always negative (the semiconductor thus absorbing) for carrier densities below about 10^{18} cm^{-3}. As we pump the material harder and raise the injected carrier density above a threshold value, the peak gain becomes positive, and the spectral region for which $\gamma(\nu) > 0$ grows. The threshold carrier concentration is typically referred to as the transparency density; at this carrier density, the semiconductor neither absorbs nor amplifies, hence the name. Figure 9.6 depicts the peak gain as a function of inversion carrier density, showing that the

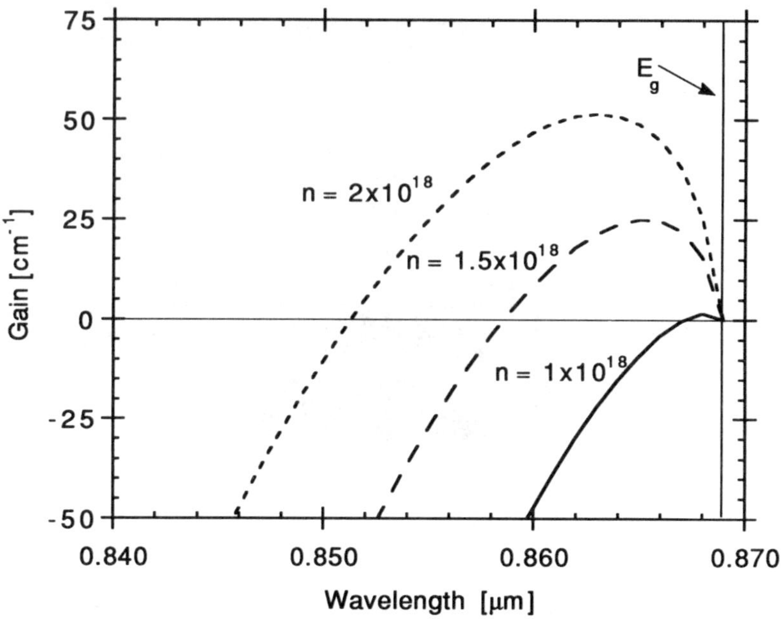

Figure 9.5 Gain as a function of photon wavelength for different electron concentrations as calculated by Equation 9.30. Increased pumping is achieved through increased current injection, thus raising the carrier concentration. Above the transparency value of 1×10^{18} cm^{-3}, gain becomes positive for an increasing wavelength range.

transparency density is on the order of 1×10^{18} cm^{-3} for GaAs and that peak gain increases approximately linearly with injected carrier density above threshold.

The linear relation between peak gain and inversion carrier concentration shown in Figure 9.6 also implies a linear increase with current. We may thus write the peak gain as

$$\gamma_{\text{peak}} = \gamma_D(n - n_{tr}) \propto \gamma_D(J - J_{tr}) \tag{9.34}$$

for transparency density, n_{tr} [cm^{-3}]; transparency current density, J_{tr} [A/cm^2]; and differential gain, γ_D [cm^2]. From the figure, we derive a value of $\gamma_D \approx 2 \cdot 10^{-16}$ cm^2, implying that once transparency is reached, very high gain can be achieved in semiconductors with little additional pumping. The increase in peak gain with device current [15] at room temperature results in nominal gain coefficients in the range of 0.044 cm/A.

9.3.3 Gain Saturation

The simple model for optical gain we have just discussed ignores a number of phenomena that are nevertheless important for understanding amplification in semiconduc-

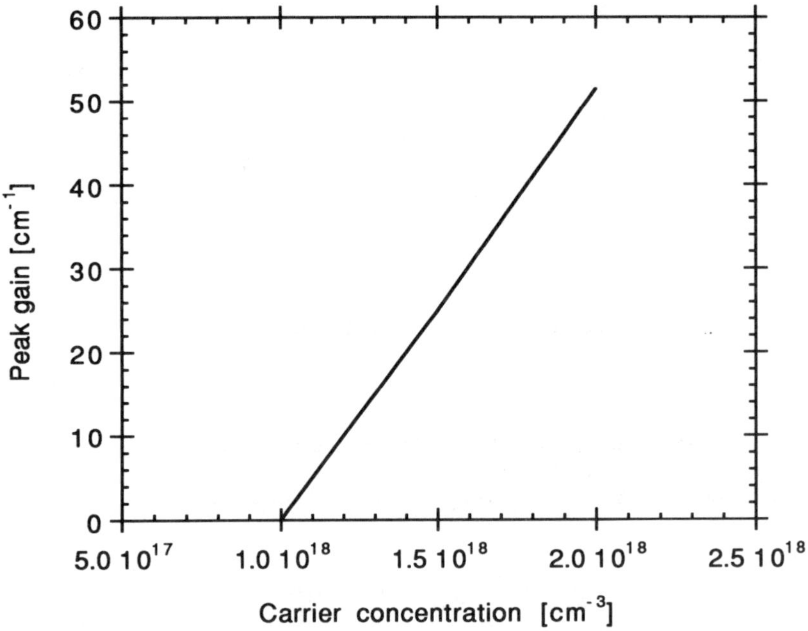

Figure 9.6 Calculated peak gain as a function of injected carrier density for the same system as in Figure 9.5. The approximately linear relationship permits the definition of differential gain as the slope of the characteristic (~ $2 \cdot 10^{-16}$ cm^{-2}), and the intercept defines the transparency carrier concentration (~ 10^{18} cm^{-3}).

tors. We have seen that amplification requires a population inversion, the dominance of electrons in a higher energy level achieved by pumping. As the transition rate per atom, W_i, given by Equation 9.18, is the same for upward and downward transitions, a higher population of the upper level ($n_2 > n_1$) implies that there are more downward transitions than upward ones. Because downward transitions reduce the population of the upper level, while of course contributing to the optical field, the gain begins to saturate as photon generation increases. If we recall that W_i is directly proportional to the intensity of the optical field, Φ_v, we see that gain should begin to decrease as the density of photons increases.

We can show the effect of gain saturation in a homogeneous medium by considering the effects of pumping and transitions in the more realistic three-level model of Figure 9.7. Electrons are raised to higher energy levels by the pump rates R_1 and R_2 [cm^{-3}sec^{-1}] and decay back to lower energies by spontaneous emission, characterized by τ_{sp} or by nonradiative processes, τ_{20} and τ_{10}. The laser transition between levels 1 and 2 is characterized by W_i. If we now write down the rate equations that account for the changes in carrier concentration in these levels, we obtain

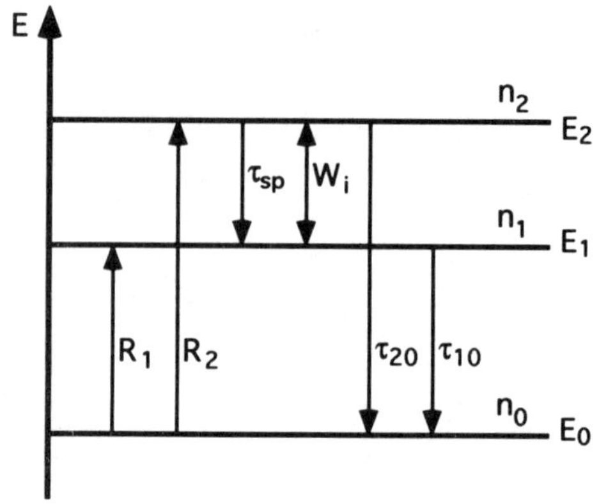

Figure 9.7 A three-level model system used to derive the concept of gain saturation. The pumping rates given by *R* raise electrons into the higher levels, from which they decay with time constants, τ, or contribute to laser emission through the induced transition rate, W_i.

$$\frac{dn_1}{dt} = R_1 + \frac{n_2}{\tau_{sp}} + n_2 W_i - n_1 W_i - \frac{n_1}{\tau_{10}} \tag{9.35}$$

and

$$\frac{dn_2}{dt} = R_2 - \frac{n_2}{\tau_{sp}} - n_2 W_i + n_1 W_i - \frac{n_2}{\tau_{20}} \tag{9.36}$$

At steady-state pumping, we have $dn_1/dt = dn_2/dt = 0$, so that we can solve Equations 9.35 and 9.36 for carrier concentration to yield the ominous

$$n_2 - n_1 = \frac{R_2\left(\frac{1}{\tau_{spon}} + \frac{1}{\tau_{20}}\right)^{-1}\left(1 - \frac{\tau_{10}}{\tau_{spon}}\right) - R_1\tau_{10}}{1 + \left\{\left(\frac{1}{\tau_{spon}} + \frac{1}{\tau_{20}}\right)^{-1} + \tau_{10}\left[1 - \frac{1}{\tau_{spon}}\left(\frac{1}{\tau_{spon}} + \frac{1}{\tau_{20}}\right)^{-1}\right]\right\}W_i} \tag{9.37}$$

which becomes considerably less threatening if we realize that it can be written in the form

$$n_2 - n_1 = \frac{\Delta n_0}{1 + \tau_{sat} W_i} \propto \gamma \tag{9.38}$$

In this latter expression, Δn_0 is the inverted population, $(n_2 - n_1)$, in the absence of an optical field, and τ_{sat} [s] is a saturation time constant. Since $(n_2 - n_1)$ is proportional to the gain, γ, and W_i is proportional to the optical field, Φ, Equation 9.38 predicts the expected decrease of gain with increasing photon density as shown in Figure 9.8. The saturation optical field, Φ_{sat}, is defined as the field at which the gain has dropped to half its original value, which occurs when $W_i = 1/\tau_{sat}$.

Although an intense optical field can reduce the gain of a laser system to a smaller value, it should be noted that the sign of γ is given by $(n_2 - n_1)$ and is independent of the optical field. Thus, if population inversion has not been achieved, gain will be negative (i.e., the system will absorb), and an increase in the optical field will not make it positive. The only way to achieve gain in a system is to pump more carriers into the upper level.

9.3.4 Hole Burning

A further nonideality in the behavior of laser gain is related to gain saturation, but on an energetically or spatially restricted level [16]. If, for example, an excessively large photon density is found at a particular wavelength, λ_h, the gain at that wavelength will be subject to homogeneous saturation. If the remaining wavelengths

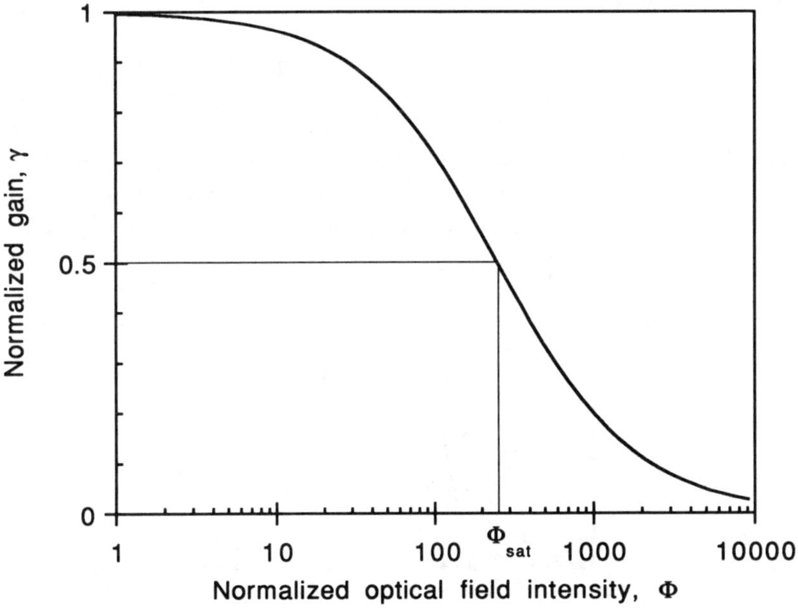

Figure 9.8 Gain saturation as a function of optical field intensity, Φ. The saturation value, Φ_{sat}, is defined as that which reduces gain to half of its unsaturated value.

contributing to lasing are less saturated, the emitted laser spectrum will have a sharp dip, or hole, as seen in Figure 9.9. Such an effect is known as spectral hole burning.

The same phenomenon can take place spatially; if a region of the laser is subject to an enhanced photon density, gain can saturate locally, reducing the emission as a result of spatial hole burning. This can occur with a broadband photon density and can degrade laser action, particularly through spatial mode instabilities.

9.4 LASER OSCILLATION

In the previous sections, we have seen how a material, when subject to external pumping, can reach a nonequilibrium state in which an optical field propagating through it can be amplified. This is not yet sufficient to make a laser; we still require a means to generate an intense coherent optical field of a single wavelength. A resonator lends itself naturally to this function, as it forces the field to traverse the amplifying medium repeatedly and selects a distinct set of resonant wavelengths. We thus begin with an analysis of the Fabry-Perot resonator, and then we'll see what happens when we put a material with gain into it.

9.4.1 Fabry-Perot Etalon

The Fabry-Perot etalon, as shown in Figure 9.10, is a conceptually simple resonator, which forms the basis for almost all laser structures. The etalon consists of two parallel mirrors separated by a distance, L [m], of transmission coefficient, t, and reflection coefficient, r. If we imagine a wave of amplitude, $\mathcal{E}_0$, incident from the left, the resonance condition will be based on a phase matching between waves after each successive round trip in the resonator. For propagation constant $k = 2\pi/\lambda$, where

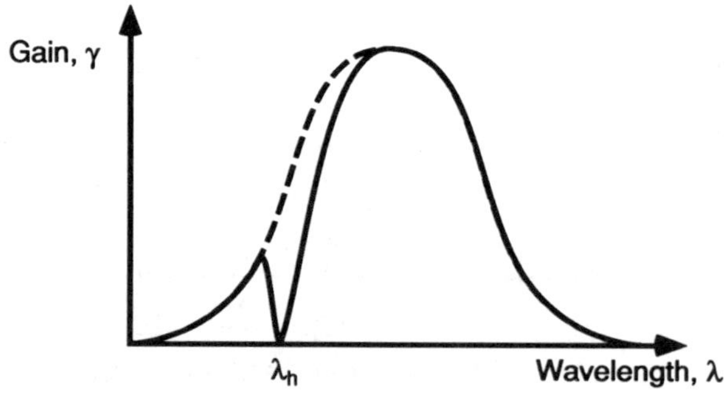

Figure 9.9 Spectral hole burning in a laser medium with an enhanced photon density at the wavelength λ_h. A higher optical field at that wavelength leads to gain saturation and a decrease in γ.

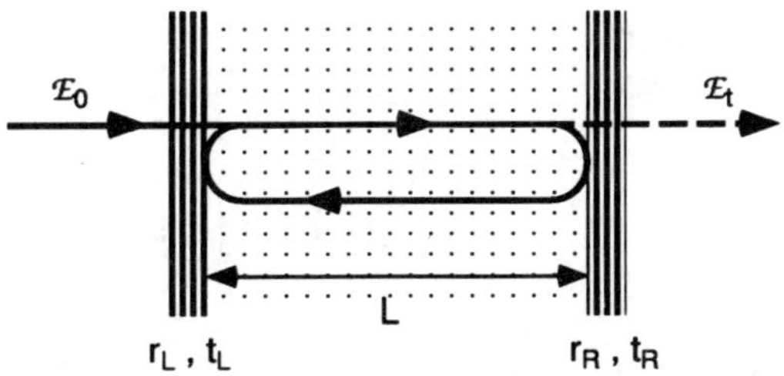

Figure 9.10 The Fabry Perot etalon consists of two parallel mirrors separated by a distance, *L*. An incident wave is internally reflected numerous times, but transmission is significant only if the phase shift between successive round trips is a multiple of 2π.

$\lambda = \lambda_0/n$ for refractive index n between the mirrors, the phase after one complete round trip will be given by $\theta = 2kL$. The total transmitted amplitude will then be given as the sum of the portions transmitted after each round trip, or

$$\begin{aligned}\mathcal{E}_t &= \mathcal{E}_0(t_Lt_R + t_Lt_Rr_Lr_Re^{j\theta} + t_Lt_Rr_L^2r_R^2e^{j2\theta} + \ldots) \\ &= \mathcal{E}_0t^2(1 + r^2e^{j\theta} + r^4e^{j2\theta} + \ldots)\end{aligned} \tag{9.39}$$

the latter for $r_R = r_L = r$ and $t_R = t_L = t$. Because $t^2 = T$ (transmittance) and $r^2 = R$ (reflectance), we can express this sum as

$$\mathcal{E}_t = \frac{\mathcal{E}_0 T}{1 - R \cdot e^{j\theta}} \tag{9.40}$$

A more useful expression results if we develop this relation in terms of optical intensity rather than amplitude. Remembering that intensity, $\mathcal{I}$, is given by $|\mathcal{E}|^2$, and that the reflectance and transmittance are related as T = 1 − R, we can write the ratio of transmitted to incident intensity through the etalon as

$$\frac{\mathcal{I}_t}{\mathcal{I}_0} = \frac{(1 - R)^2}{(1 - R)^2 + 4R\sin^2\left(\frac{\theta}{2}\right)} \tag{9.41}$$

This expression for etalon transmission can be simplified further by using the definition of a parameter, $\mathcal{F}$,

$$\mathcal{F} = \frac{\pi\sqrt{R}}{1 - R} \tag{9.42}$$

such that the transmission ratio becomes

$$\frac{\mathcal{I}_t}{\mathcal{I}_0} = \frac{1}{1 + \left(\frac{2\mathcal{F}}{\pi}\right)^2 \sin^2\left(\frac{\theta}{2}\right)} \tag{9.43}$$

$\mathcal{F}$ is known as the Fabry-Perot cavity finesse and, as we shall see presently, is a measure of how sharp the resonance is. A high value for $\mathcal{F}$ implies a narrow-band transmission through the etalon at a specific wavelength.

We see from this last equation that the resonator transmission has a maximum value of unity if the relative phase between waves that have undergone two complete reflections (one round trip in the etalon) is given by $2m\pi$ for integer m. Since the phase was determined to be related to wavelength and cavity length as $\theta = 2kL = 4\pi L/\lambda$, maximum transmission occurs for a mirror spacing of $L = m\lambda/2$. Minimum transmission is reached if the waves are π out of phase, so that $\theta = m\pi$ for odd m, which is achieved if the mirror spacing is $m\lambda/4$, again for m odd. The transmission ratio has a value $(1 - R)^2/(1 + R)^2$ at its minimum.

Converting to oscillation frequency, ν, recalling that $\nu = cn/\lambda_0$, for index of refraction, n, and speed of light, c, the above discussion implies that we have a resonance for the Fabry-Perot etalon at frequencies

$$\nu = \frac{cm}{2Ln} \tag{9.44}$$

for integer m. The spacing between resonances is then

$$\Delta\nu_{FP} = \frac{c}{2Ln} \tag{9.45}$$

where $\Delta\nu_{FP}$ $[s^{-1}]$ is also often referred to as the free spectral range; note that $\Delta\nu$, the linewidth, should not be confused with $\Delta\nu_{FP}$, the mode spacing or free spectral range.

Finally, very close to a resonance frequency, ν_m, for which $\sin(\theta) \approx \theta$, the transmission ratio can be approximated as

$$\frac{\mathcal{I}_t}{\mathcal{I}_0} \approx \frac{1}{1 + \left(\frac{2\mathcal{F}}{\pi}\right)^2 \left(\frac{\theta}{2}\right)^2} \tag{9.46}$$

so that $\mathcal{I}_t/\mathcal{I}_0$ drops to a value 1/2 for $\mathcal{F}\theta/\pi = 1$ or $\theta = \pi/\mathcal{F}$. We can thus define the spectral width as the full-width at half-maximum, FWHM $= 2\theta = 2\pi/\mathcal{F}$, and because

2π corresponds to the resonance spacing $\Delta\nu_{FP}$, we can finally write down the width of the Fabry-Perot resonance as

$$\text{FWHM} = \frac{\Delta\nu_{FP}}{\mathcal{F}} \tag{9.47}$$

The preceding bit of calculation and definition should become clearer after taking a look at the plotted transmission characteristics of a Fabry-Perot etalon in Figure 9.11. At the resonance frequencies $\nu_m = 2\pi$ and $\nu_{m+1} = 4\pi$, transmission reaches unity; the sharpness of this resonance is given by the finesse, $\mathcal{F}$. It is clear that for a high-finesse cavity, the linewidths are narrow and the resonances are sharp. Poorer cavities, with smaller values for $\mathcal{F}$, tend to have considerable etalon transmission between resonance frequencies.

Example

Calculating the finesse of a Fabry-Perot cavity made from the uncoated, cleaved facets of a GaAs laser (n = 3.45), where $R = 0.31$ yields $\mathcal{F} = 2.54$; this is a rather

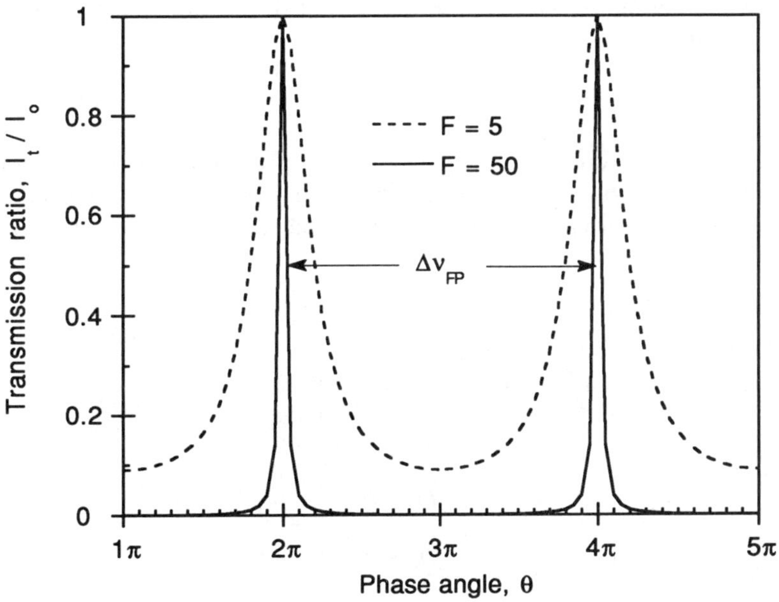

Figure 9.11 Transmission characteristics of a Fabry-Perot etalon showing resonances at phase shifts of 2π, 4π, ... The free spectral range is $\Delta\nu_{FP}$. A cavity with a high finesse ($\mathcal{F} = 50$) has a considerably sharper resonance than one with a lower value, $\mathcal{F} = 5$.

low value. For the finesse values used in Figure 9.11, we require $R = 0.54$ for $\mathcal{F} = 5$ and $R = 0.94$ for $\mathcal{F} = 50$; achieving these finesse values would require high-reflectivity coatings on a normal GaAs-based laser.

9.4.2 The Etalon with Optical Gain

All of the pieces of the laser puzzle are now in place. The Fabry-Perot etalon forms a resonator in which an intense optical field can interact with a material exhibiting optical gain. As we shall see, by overlapping the resonant frequency of the cavity with the point of maximum gain, lasing at that resonant wavelength becomes possible. In this section, then, we complete the last step and combine the concepts of gain with those of the resonant cavity.

We begin by remembering that our propagation constant is a complex value given by $k = \beta + j\alpha/2$, where $\beta \equiv k_z$, implying that the electric field propagating through a material varies as (Equation 3.50)

$$\mathcal{E} = \mathcal{E}_0 \cdot e^{jkz} \cdot e^{j\omega t} = \mathcal{E}_0 \cdot e^{j\beta z} \cdot e^{-\alpha z/2} \cdot e^{j\omega t} \tag{9.48}$$

with position and time. We can thus write the intensity variation of the wave, ignoring the time dependence, as

$$\mathcal{I} = \mathcal{I}_0 \cdot e^{j2\beta z} \cdot e^{-\alpha z} \tag{9.49}$$

The real part of the propagation constant, β, is based on the material real refractive index and can reflect changes in the index due to changes in the free-carrier concentration. Recalling that the real part of the dielectric constant is related to the susceptibility (Equation 3.40) as

$$\epsilon_R = 1 + \chi_{eR}(\omega) \tag{9.50}$$

where χ_{eR} represents the real component of susceptibility due to the presence of electrons, we can write the real part of the propagation constant as

$$\begin{aligned} \beta &= \beta_0 + \beta_0\left(\frac{\chi_{eR}}{2n^2}\right) \\ &= \beta_0 + \Delta\beta \end{aligned} \tag{9.51}$$

for real refractive index, n. $\Delta\beta$ is primarily due to the pump electrons and is negative (Equation 3.32 or Equation 8.3).

Similarly, the imaginary term α is proportional to the imaginary susceptibility χ_{eI}, and includes any residual material absorption and any gain. The ratio of real to imaginary susceptibilities, χ_{eR}/χ_{eI}, is referred to as the linewidth enhancement factor; this factor defines how much the real refractive index of a material changes with a given change in gain since, as we saw earlier, negative absorption implies gain. When a laser is subject to high-speed modulation, a change in the real refractive index due to carrier injection can increase the transition linewidth. This is due to small, quick shifts in frequency, known as chirp; hence the name linewidth enhancement factor.

We can write the loss term most simply as $\alpha = \alpha_0 - \gamma$, where α_0 is background loss due to scattering or free carrier absorption, and γ is our friend the optical gain. Thus, the field intensity (Equation 9.49) may now be expressed as

$$\mathscr{I} = \mathscr{I}_0 \cdot e^{j2(\beta+\Delta\beta)z} \cdot e^{-(\alpha_0-\gamma)z} \tag{9.52}$$

and we are ready to pop this wave into the resonator. Returning to Equation 9.40, we see that if the $R \cdot e^{j\theta}$ term becomes unity, the etalon reaches resonance, and we have unity loop gain between the mirrors. For an intervening material with loss or gain, the phase θ is the argument of the exponential in Equation 9.52, so that the condition for unity loop gain becomes

$$R \cdot e^{j2(\beta+\Delta\beta)z} \cdot e^{-(\alpha_0-\gamma)z} = 1 \tag{9.53}$$

The implications of Equation 9.53 are best analyzed if we consider the amplitude and phase terms separately. The amplitude restriction is simply

$$R \cdot e^{-(\alpha_0-\gamma)z} = 1 \tag{9.54}$$

which we can rewrite as

$$\gamma = \alpha_0 + \frac{1}{L}\ln\left(\frac{1}{R}\right) \equiv \gamma_{th} \tag{9.55}$$

where we have defined the gain necessary to fulfill the condition of Equation 9.54 as the threshold gain, γ_{th}. We can interpret this result as γ_{th} being the gain necessary to overcome the residual material losses, α_0, and the mirror losses due to nonunity R. This mirror "loss" is of course the useful light emitted from the laser.

The phase condition implied by Equation 9.53 is,

$$R \cdot e^{j2(\beta+\Delta\beta)z} = 1 \tag{9.56}$$

which restricts the argument of the exponential to

$$2(\beta+\Delta\beta)L = 2m\pi \tag{9.57}$$

Since the $(\beta+\Delta\beta)$ term is the real part of the propagation constant, k, and $k = 2\pi n/\lambda_0$ for refractive index, n, this last relationship implies that the allowed wavelengths in the resonator that fulfill the phase requirements are

$$\lambda_0 = \frac{2nL}{m} \tag{9.58}$$

for integer index m. Equation 9.58 defines the allowed cavity wavelengths, or modes, and thus the longitudinal modes of the laser, and is equivalent to Equation 9.44, the resonance condition we derived previously.

We can extend this analysis to a calculation of the mode spacing. Taking the differential of Equation 9.58,

$$d(m\lambda_0) = md\lambda_0 + \lambda_0 dm = d(2Ln) = 2Ldn \tag{9.59}$$

and considering the case for spacing between two adjacent modes, so that $|dm| = 1$, the mode spacing $\Delta\lambda_0$ can easily be shown to be

$$|\Delta\lambda_0| = \frac{\lambda_0^2}{2Ln - 2L\lambda_0\left(\frac{dn}{d\lambda_0}\right)} \tag{9.60}$$

The second term in the denominator is proportional to the dispersion, $dn/d\lambda_0$, which defines how much the laser medium refractive index changes with wavelength. In many cases, this term may be neglected so that the wavelength separation takes on a simpler form. $|\Delta\lambda_0|$ is then the distance in wavelength between adjacent modes.

Finally, the mode spacing in wavelength can also easily be written as a spacing in frequency: from $\lambda = c/\nu$, and thus $d\lambda = d(c/\nu)$, we find that $\Delta\lambda = (\lambda^2/c)\Delta\nu_{FP}$.

Example

We can determine the expected mode spacing of an AlGaAs Fabry-Perot cavity using Equation 9.60. Assuming that the dispersion term is negligible, and taking $n = 3.35$, $L = 500\ \mu m$, and $\lambda_0 = 820$ nm, we find that $\Delta\nu_{FP} = 90$ GHz or equivalently in wavelength $\Delta\lambda_0 = 0.2$ nm. Compare this calculated value with the sidemode spacing of Figure 9.21. We can also easily determine the resonance width of this cavity, taking the value of finesse as $\mathcal{F} = 2.54$; we then find the FWHM from Equation 9.47 to be FWHM $= \Delta\nu_{FP}/\mathcal{F} = 35$ GHz, which we find from $\Delta\lambda = (\lambda^2/c)\Delta\nu_{FP}$ to be equivalent to 0.08 nm.

9.4.3 Laser Resonance

A Fabry-Perot resonator with an intermediate laser medium can achieve a resonance when the loop gain for a field oscillating between the mirrors reaches unity and, at the same time, has a set of discrete allowed resonator modes spaced $\Delta\lambda_0$ apart. We have just seen this. Since the background loss, α_0, does not vary appreciably with wavelength, but gain, γ, is a strong function of λ, it seems clear that there will be a region of wavelengths for which the cavity experiences unity loop gain *and* one or more modes will be allowed. Phenomenologically, as current is pumped into a laser, the material emits spontaneous and eventually stimulated emission. For a pumping level that raises the gain above the background loss—the laser threshold—most of the optical energy will suddenly be confined to one or several resonant modes, and the device will begin to emit coherent laser radiation.

We can see this clearly in Figure 9.12(a), which schematically shows the loss, gain, and allowed modes as a function of wavelength. For increasing current injection, the gain will eventually rise above the background loss at a point known as lasing threshold. Ideally, at that point all the modes for which $\gamma > \alpha_0$ could lase. In practice, however, inhomogeneities in the material or defects on the laser mirrors will cause a single mode to dominate, so that a typical output spectrum will be as shown in Figure 9.12(b). This figure shows the measured spectrum of a monomode laser both below and above threshold, clearly demonstrating how the broad spontaneous emission evolves into a narrow peak at the onset of lasing. Thus, we have theoretically assembled our laser and even operated a real one; we will now turn to the fabrication of practical semiconductor laser structures.

9.5 SEMICONDUCTOR LASER STRUCTURES

The discussion in the first half of this chapter has made apparent that we need several items to fabricate a semiconductor laser: a material which exhibits optical gain at the wavelength of interest, a means to confine the pump electrons with the optical field, and a resonant cavity to select a particular wavelength. In general, a direct bandgap semiconductor material, predominantly one of the III-Vs, forms the basis for the device; the double heterostructure serves to confine electrons and photons in the active region; and the cleaved facets of a waveguide-based design form the mirrors of the Fabry-Perot cavity.

9.5.1 Materials and Heterostructures

The appropriate choice of semiconductor laser material is primarily determined by the desired wavelength. An essential requirement is, of course, efficient photon emission at this wavelength, which implies that direct bandgap materials are almost exclusively

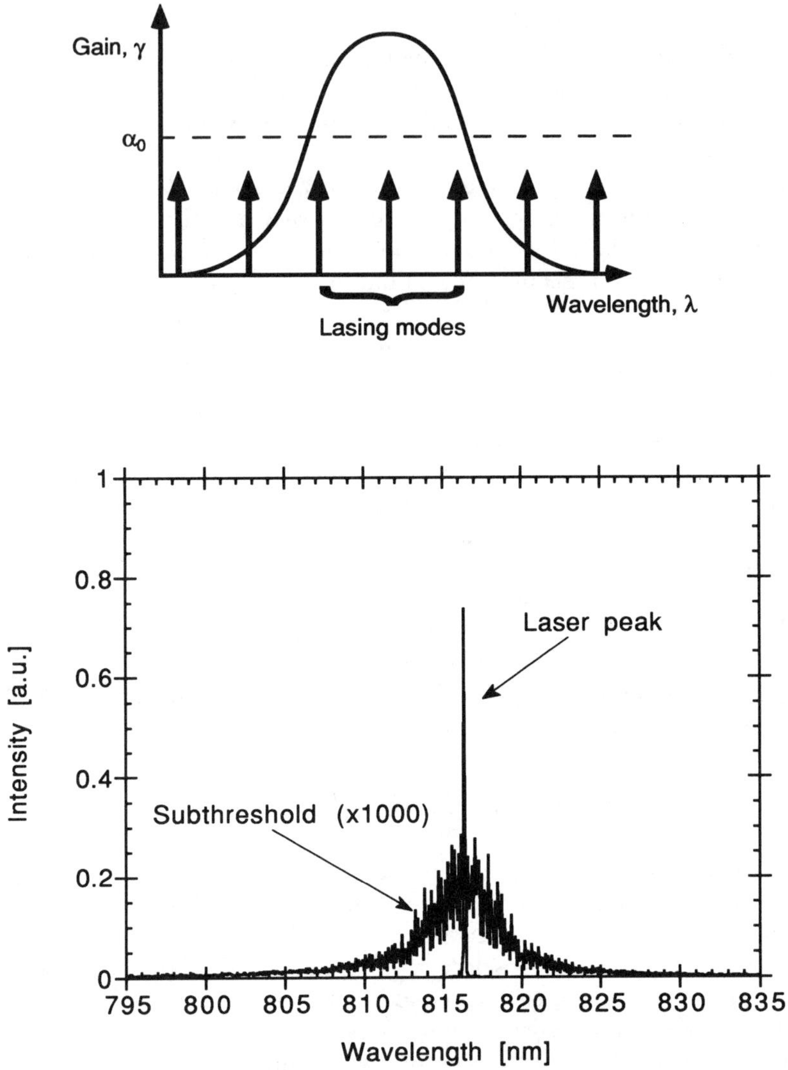

Figure 9.12 (a) A schematic plot of gain, background loss, and allowed Fabry-Perot modes as a function of wavelength. For the pumping level shown, gain exceeds loss for the center three modes, implying that these three can lase. (b) A measured output spectrum, on a linear scale, for a monomode Fabry-Perot laser both below and above threshold. The difference between spontaneous emission (which has been multiplied by 1,000) and the laser spectrum is clearly seen. The device employed a 3-μm-wide, 500-μm-long Fabry-Perot cavity with a threshold current of 12 mA; spectra were taken at bias currents of 11 mA and 30 mA.

employed. As we discussed in Chapter 4, the GaAs/AlGaAs system is the workhorse for lasers in the 780- to 850-nm wavelength range; this material system has been widely exploited because GaAs growth and fabrication are both very well developed. Recent work on ZrF_4-BaF_2-LaF_3-AlF_3-NaF (ZBLAN) fibers [17], which are suitable for use in the near-infrared regime covered by the GaAs/AlGaAs system, may enhance the palette of applications for these lasers. The use of InGaAs grown on GaAs is useful for wavelengths in the region of 880 to 1100 nm. For telecommunications, $In_{1-x}Ga_xAs_yP_{1-y}$ grown on InP [18] has been the most attractive due to wavelengths that vary from 920 nm (InP) to 1.6 μm ($Ga_{0.47}In_{0.53}As$). This band includes the dispersion minimum for silica optical fibers, $\lambda = 1.3$ μm, a wavelength reached by $In_{1-x}Ga_xAs_yP_{1-y}$ with $x = 0.28$ and $y = 0.6$. The residual loss in an optical fiber at 1.3 μm is still on the order of 0.5 dB/km and reaches a minimum at 1.55 μm, another relevant wavelength for InGaAsP lasers. In the latter case, the dispersion problem is generally attacked by reducing the linewidth of the laser source. For applications requiring wavelengths further into the infrared, either the III-V antimonide (2 to 5 μm) or the lead chalcogenide, PbSe/PbEuSe (2.5 to 12.5 μm), systems are useful, albeit the latter only at liquid nitrogen temperatures.

Many applications benefit from shorter wavelengths, particularly in the visible spectrum. The quaternary AlGaInP system [19] has been particularly successful for lasers in the 650- to 690-nm range (red). Reliable lasers at shorter wavelengths are still elusive creatures [20]. Great progress has been made in the II-VI materials, and ZnSe-based lasers are at present the most promising, emitting at 470 nm [21]. Currently operating only in pulsed mode at room temperature, the lifetime of these diodes is still measured in hours. This material system is, in addition to GaN and SiC, a very active research area [22]. A more extensive discussion of material issues was the subject of Chapter 4.

The emission wavelength is determined by the dominant optical transitions in the material, usually the conduction to valence band transition. Recall that a semiconductor laser uses a pn junction to create a locally enhanced carrier density that leads to spontaneous and stimulated emission; this inversion population is spatially limited to a small volume we term the "active region." A fundamental advantage of the materials we have outlined is the ability to fabricate a heterostructure, a layer sequence that confines both electrons and holes, thus facilitating the generation of an inversion population.

We discussed heterostructures in Section 2.7.1—the double heterostructure (DH), as the reader certainly recalls, is a sandwich of lattice-matched material (such as GaAs and AlGaAs, for all aluminum mole fractions up to AlAs) where the center of the sandwich has a lower energy gap than the sides (see Figure 2.20). Thus, electrons injected into a double heterostructure will be confined to the region of lower E_g, where they recombine, thus defining the active region of the laser. Fortuitously, the center region with lower energy gap also typically has the higher refractive index, which, as we saw in Chapters 7 and 8, forms a slab waveguide and confines the generated light. This leads to the nomenclature "separate confinement double heterostructure," implying a layer sequence that confines both electrons and photons. For quantum-well lasers, due to the small optical confinement factor of the well, electrons and photons are truly separately confined, the former by the quantum well and the latter by the waveguide core.

The heterostructure laser [23] thus typically employs a layer structure similar to those of the examples shown in Figure 9.13. The $Al_{0.8}Ga_{0.2}As$ cladding and $Al_{0.3}Ga_{0.7}As$ core are not coincidentally quite similar to the waveguides we considered in Chapters 7 and 8. In the absence of quantum wells, which we will discuss in the

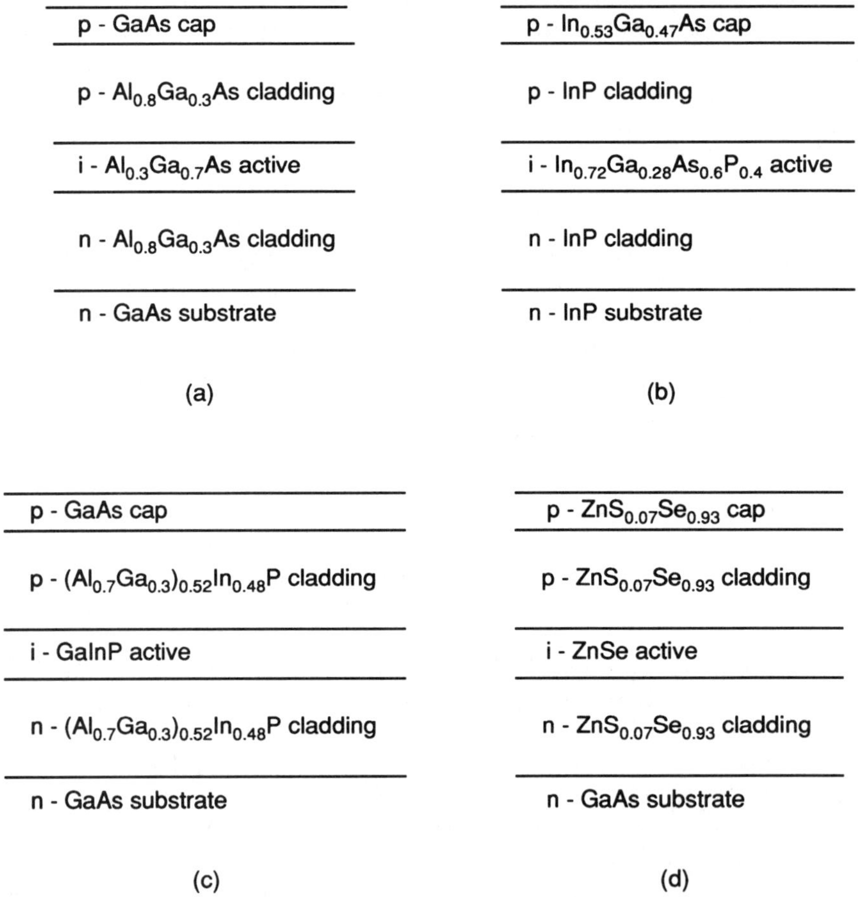

Figure 9.13 Model double heterostructure layer sequences such as those typically employed for lasers in the (a) AlGaAs/GaAs, (b) InGaAsP/InP, (c) GaInP/GaAs, and (d) ZnSe systems.

next section, the emission wavelength is defined by the energy gap of the core material, typically an undoped layer between the p and n doped cladding.

9.5.2 Quantum-Well Lasers

We saw in section 2.7.2 that the quantum well is a natural extension of the double heterostructure, namely one in which the center region is so thin that quantum-confinement effects become relevant. The use of a quantum well as the active region of a laser, first demonstrated in 1975 [24], has numerous advantages. The most important benefits are the decrease of active region volume and thus threshold current, coupled with the ability to accurately tune the emission wavelength to a transition between two quantum-well subbands. Many currently fabricated semiconductor lasers employ quantum wells [25] and take advantage of these effects.

Gain calculations for a quantum-well laser need to reflect the step-like density of states (Figure 2.22) for a two-dimensional system. As a result, a plot of γ as a function of wavelength is more complicated than for the bulk case and also shows a step behavior. The peak gain is lower in magnitude than in bulk, but also narrower in energy, and saturates due to the constant (as opposed to increasing as $\sqrt{E}$) density of states in a given miniband. The required threshold gain for lasing also tends to be higher because the narrow width of the quantum well implies a low confinement factor.

However, in a quantum well, a higher gain can be achieved at low electron densities, leading to a reduction in the threshold current density as well as a reduced real refractive index change with pumping (i.e., a lower linewidth enhancement factor). The latter implies reduced chirp and decreased linewidth with respect to a standard double heterostructure device. One can take advantage of strain effects in quantum wells to tune polarization or wavelength and in addition profit from the relatively low temperature sensitivity of threshold current.

The quantum-well wavelength is defined by transitions between the two lowest subbands, whose spacing is determined by the quantum-well width. For emission wavelengths in the range 800 to 850 nm, a typical $Al_{0.3}Ga_{0.7}As$ core with a GaAs quantum well requires well thicknesses of 4 to 10 nm. Because the quantum well itself is too narrow to adequately guide light (a 5-nm well has $\Gamma = 0.024$, for example), a double heterostructure is still employed to confine the optical field. Several configurations enjoy popularity: the stair-step structure of Figure 9.14(a) just pops the quantum well into the center of a DH layer sequence, but the graded-index separate confinement heterostructure (GRINSCH) of Figure 9.14(b) and (c) employs a sloped transition to the waveguide core, which is designed to enhance electron capture and increase confinement factor for a given active region width [26]. Both linear and parabolic gradings have been employed.

If one quantum well is good, multiple quantum wells are often better [27]. Using several quantum wells (three to five is popular) increases the overall gain, but

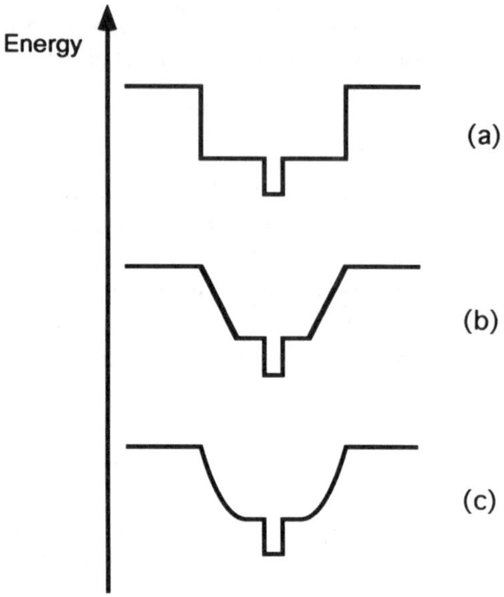

Figure 9.14 Heterostructure configurations employing quantum wells: (a) stair step, (b) linear graded-index separate confinement heterostructure (GRINSCH), and (c) parabolic GRINSCH.

the increase is sublinear with the number of wells. Performance is superior for multiquantum well structures at high operating currents, and multiquantum well structures can have advantages for integrated modulators as we shall discuss in Chapter 11. Finally, the clever use of quantum structures, including quantum wires, dots, and sophisticated two-dimensional layer structures, is a very active research area and has led to numerous novel device structures.[4]

9.5.3 Typical Laser Designs

The simplest type of semiconductor laser cavity is shown in Figure 9.15; a large metal contact on a slab of semiconductor serves to inject the pump electrons, and the two opposing cleaved facets form the mirrors of the Fabry-Perot cavity. The active region is a pn junction in a double heterostructure layer sequence. Typical cavity lengths are 200 to 500 μm. In this so-called broad-area laser, optical power, in a spatially broad band of laser radiation, is emitted from both facets because there is no confinement in the lateral direction. The large active region area also results in high total current levels and high power dissipation. The wide lateral structure

[4]For example: Faist, J., F. Capasso, D. Sivco, C. Sirtori, A. L. Hutchinson, and A. Y. Cho, "Quantum Cascade Laser," *Science*, Vol. 264, 1994, pp. 553–556.

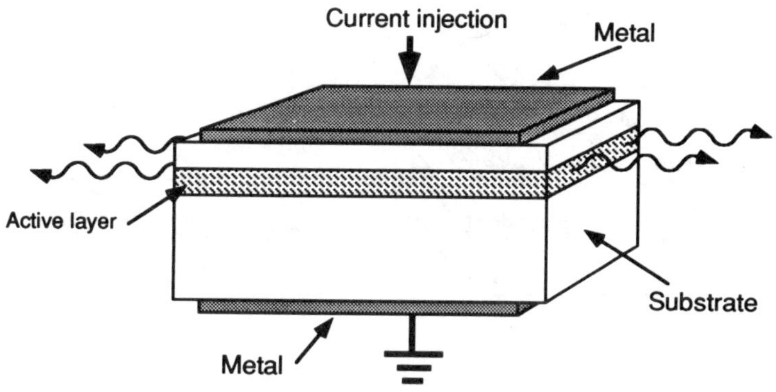

Figure 9.15 A broad-area laser employs a cleaved optical cavity with no lateral confinement. Current injection is from the top metal contact; the backside metal is grounded. Spatially broad emission is from both facets and many lateral modes emit simultaneously.

implies that this laser emits in many lateral modes, so that monomode behavior is, of course, never attained. Despite these limitations, the broad-area laser is a useful development tool because the threshold current density, J_{th}, is lower than for a stripe laser; the large ratio of pumped area to perimeter implies that current leakage is minimized. Thus, a laser designer will often fabricate a broad-area laser from a freshly grown layer structure to evaluate its quality and emission wavelength.

Significant improvement in laser performance is achieved by moving to one of many stripe laser configurations, a selection of which is shown in Figure 9.16. Many stripe laser arrangements are based on the passive waveguides we studied in Chapter 8. In these structures, lasing is confined to a narrow stripe with typical lateral dimensions of 2 to 5 μm, and, as with the waveguides, there are numerous means for accomplishing the lateral structuring. The stripe laser can be made to emit in a single mode; it can be driven continuously, which is known as continuous wave (CW) operation; it has a lower threshold current (but not threshold current *density*) than a broad-area laser; and it benefits from potentially low capacitance so that high-speed operation is possible.

Two fundamental mechanisms are at work for defining the lasing region in the structures of Figure 9.16. The first three (a–c) have no index guiding and might not, in the absence of current injection, function as passive waveguides. Each structure limits the lateral extent of current flow, either by (a) patterning the contact, (b) limiting injection by only doping a selected region, or (c) confining the current by implantation to generate high-resistivity regions. The lateral extent of lasing is then limited to the volume into which the current is injected, and where gain is therefore positive. We thus have a local enhancement of the imaginary refractive index, which leads to gain-guiding.

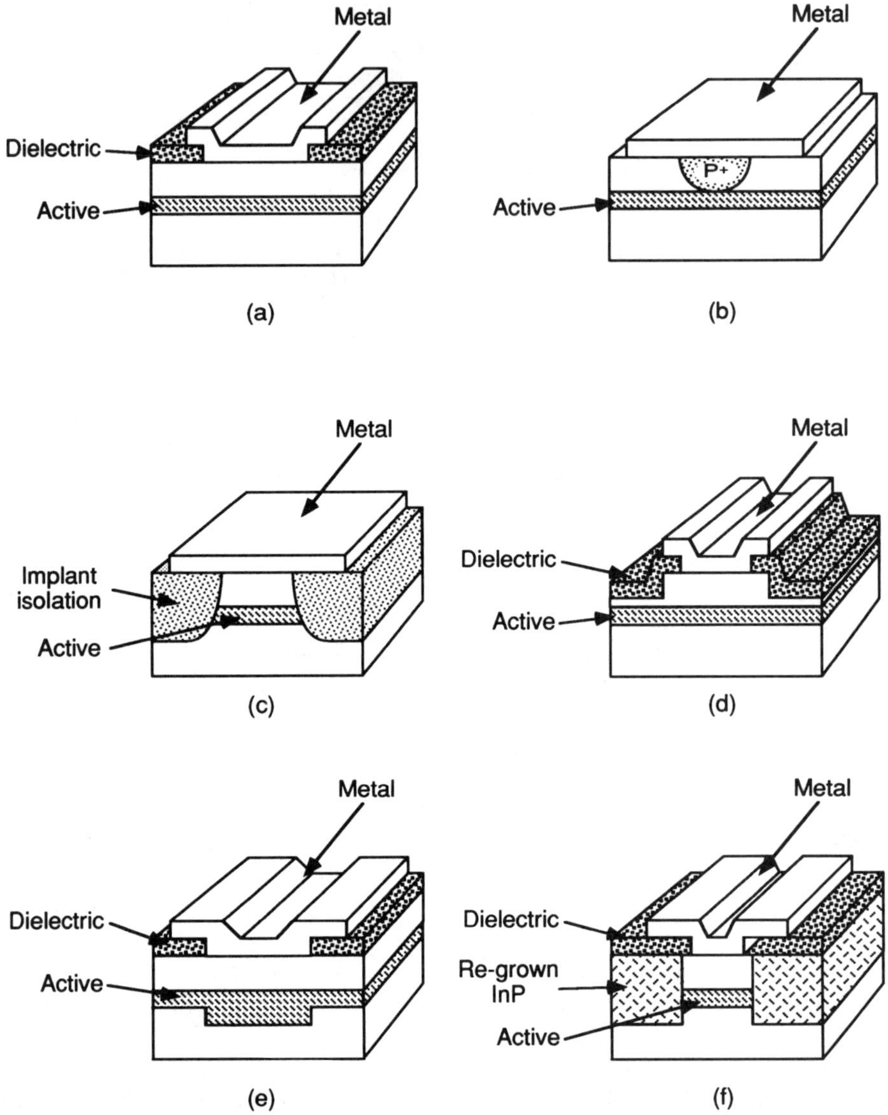

Figure 9.16 Some examples for stripe laser designs: (a) structured metal contact, (b) selective contact doping, (c) isolation implantation, (d) strip loaded, (e) inverted rib, and (f) etched ridge and regrowth. The backside metal contacts are not shown.

Index guiding, as we saw it for passive waveguides, provides stronger confinement, independent of injection conditions. The strip loaded structure (d) is the simplest technologically, where a deeper etch depth can also be used to generate a ridge. An inverted rib (e) is similar conceptually, but requires regrowth of the upper layers for confinement and electrical contact. Finally, etch and regrowth (f), where the use of semi-insulating Fe-doped InP is popular for the regrown lateral cladding layer, is commonly used in the InGaAsP system.

While gain-guided stripe lasers are relatively easy to fabricate, the poor lateral confinement implies that the emission spectra are spatially broad when compared to index-guided structures. In addition, required injection currents are higher due to excess leakage at the edges of the gain-guided region. Also, in gain-guided devices, modal instabilities often arise at high injection levels, the far field usually consists of two lobes, and spatial hole burning effects can become troublesome. For these reasons, even a small amount of index guiding is usually desirable, and some variations on the themes presented in Figure 9.16 (d–f) are most often employed. Advances in microfabrication technology, however, have permitted the design and fabrication of a veritable menagerie of other shrewd laser structures, many incorporating advanced quantum and optical effects [28].

9.5.4 Reliability and Lifetime

Semiconductor processing of the structures shown in Figure 9.16, the subject of Chapter 5, is essentially the same as for waveguides. A crucial element in the successful manufacture of a semiconductor laser, however, is the generation of the mirrors. For (100) oriented GaAs substrates, cleaving the {011} planes is the simplest means to form a facet. However, dry-etched mirrors [29] are a useful alternative, particularly when monolithic laser integration with other components becomes relevant. The facets, cleaved or etched, tend to be one of the weak points in laser reliability, primarily due to the very intense optical field on the mirrors [30]. At high optical power levels, slight imperfections in the crystal on the facet can lead to local heating and thus catastrophic optical damage (COD), the sudden demise of the laser due to destruction of the resonant Fabry-Perot cavity. Passivation, high- or anti-reflection coatings on the facets can help to alleviate these difficulties, but facet degradation remains a serious reliability issue.

The bulk semiconductor material itself can also lead to laser degradation, particularly material defects that appear during growth or during thermal processing. These defects tend to migrate, forming networks known as dark-line defects in the bulk, which can cause excessive leakage current or localized heating and lead to laser degradation. The pn junction, as in a transistor, is sensitive to current spikes and can easily be destroyed (by shorting) through static discharge. Such device aspects are the subject of considerable study, particularly by industry, as laser lifetimes on the order of 10^5 to 10^6 hours are required for a reliable product.

9.6 ELECTRICAL AND OPTICAL CHARACTERIZATION

Laser diode characterization is not unlike that for waveguides. The laser can be placed on a temperature-controlled chuck; electrical contact is made to the surface by micromanipulator probers and to the backside by the chuck. Alternatively, the device may be soldered to a machined mount and contacted by bond wires. The optical emission is gathered either by a lens or a fiber for measurement of intensity and spectrum. A standard set of electrical and optical measurements is generally performed using continuous wave (CW) signals (i.e., dc); high-speed lasers require not only specialized device structures [31] but also sophisticated measurement techniques, both of which are beyond the scope of our present discussion.

A current source is used to bias the laser, its level ramped to the threshold current, at which point lasing commences and the output intensity suddenly jumps several orders of magnitude. Further increase of current bias leads to a linear increase in optical output intensity. For CW measurements, a stable operating point is maintained for subsequent measurement of the spectrum. Essential parameters are the threshold current, output power and spatial distribution, efficiency, spectrum, and temperature sensitivity.

9.6.1 Threshold Current

As current is injected into the laser, spontaneous (LED) emission is generated below the onset of lasing, as we saw in Figure 9.12(b). The transition to lasing is an abrupt one and takes place at an injection current equal to the threshold current, I_{th}. Returning to the gain model we derived above, we see that a number of things happen at threshold: for current levels below I_{th}, gain increases linearly with injection as we can see in the schematic plot of Figure 9.17. At threshold, however, lasing begins and

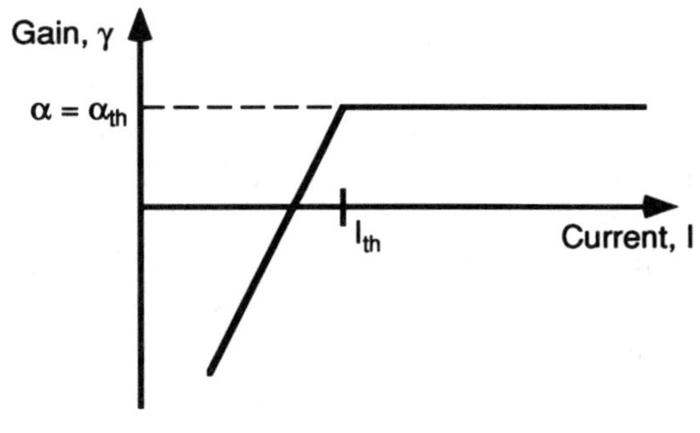

Figure 9.17 The increase of optical gain with injected current below and above lasing threshold. At threshold, the resultant optical field gives rise to gain saturation, reducing γ to the level of the background plus mirror losses, α. For higher pumping levels, the gain remains saturated at that level.

the intense optical field immediately leads to gain saturation, reducing the maximum gain to the threshold value equal to the sum of background and mirror losses. Beyond threshold, the optical output power then increases linearly with injected current.

An expression for threshold current can easily be derived from our definition of gain, Equation 9.29, and the relationship between threshold gain and the material and mirror losses, Equation 9.55. If we include the confinement factor of the laser medium, Γ, in the latter expression, we have

$$\Gamma \gamma_{th} = \alpha_0 + \frac{1}{L}\ln\left(\frac{1}{R}\right) \tag{9.61}$$

Equation 9.29 can be rewritten as [32]

$$\gamma = \frac{\lambda^2}{8\pi}\left(\frac{J}{qt}\right)\left(\frac{1}{\Delta\nu}\right) \tag{9.62}$$

which derives from the relationship between excess carrier concentration and current density J [A/m^2], $(n_2 - n_1)/\tau_{sp} = J/qt$, for active area thickness t [m], and the approximation of linewidth, $1/\Delta\nu$, for the lineshape function. From the previous two expressions, we can then directly write down the threshold current density

$$J_{th} = \frac{8\pi}{\Gamma\lambda^2}(qt\Delta\nu)\left(\alpha_0 + \frac{1}{L}\ln\frac{1}{R}\right) \tag{9.63}$$

From this last equation, we see that a plot of threshold current versus $(1/L \ln 1/R)$ allows us to determine the slope (involving linewidth and confinement factor) and the loss α_0. The device length, L, and active area thickness, t, typically a quantum well or heterostructure, are known; confinement factor, Γ, and material background loss, α_0, can be calculated from knowledge of the material and its layer structure. Finally, values for $\Delta\nu$ are typically in the range of 10^{14} s^{-1}. Acceptable values for J_{th} are in the range of several hundred A/cm^2, with good lasers giving thresholds on the order of < 50 A/cm^2, in the units traditionally employed.

A major aspect of any new laser design and development is the reduction of threshold current; a lower J_{th} implies less total current flow and thus less thermal generation and greater overall efficiency. In Equation 9.63, we can find a number of structural parameters that permit threshold optimization. A reduction of the active layer thickness decreases the pumped volume and hence the current necessary to achieve population inversion. Likewise, a high confinement factor, achieved by a large index step in a heterostructure layer sequence, confines the field to the active region, enhancing stimulated laser emission. The effects of these variations can be seen in Figure 9.18, which shows the variation of threshold current density with t

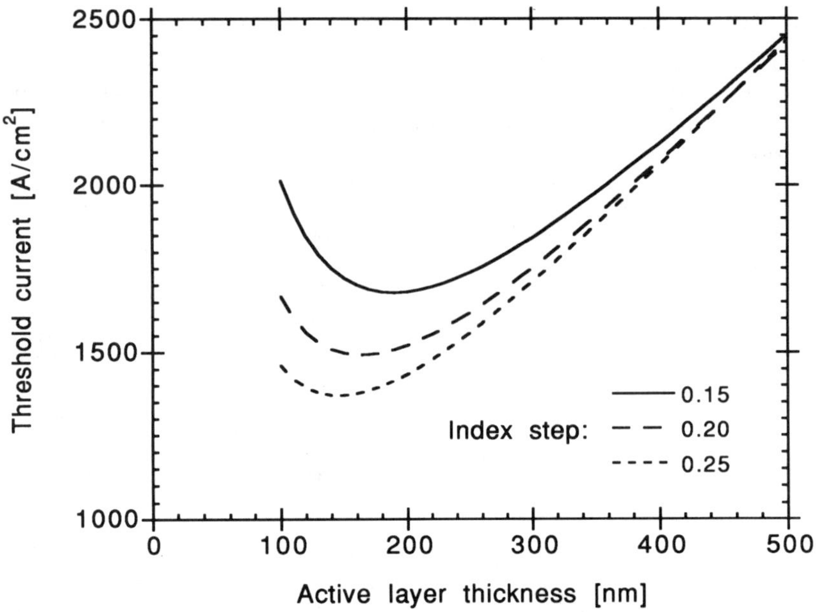

Figure 9.18 Threshold current density as a function of active layer thickness, t, with the index step between active and cladding for a double heterostructure as a parameter. The curves were calculated for a device 500 μm long, with $R = 0.31$ and background loss of $\alpha_0 = 10$ cm^{-1}. Small active volume and large index step, implying high Γ, lead to reduced threshold current density.

and Δn, the index step in a heterostructure. The overall decrease in J_{th} with decreasing active region thickness is due to the reduced active volume, and the subsequent increase for small values of t results from a reduction in Γ for small thicknesses.

Example

The expected threshold current for a GaAs laser emitting at 850 nm can be approximated by Equation 9.61. By taking a mirror reflectivity of $R = 0.31$ and a cavity length of 200 μm, the mirror loss becomes 59 cm^{-1}. Assuming a confinement factor of $\Gamma = 0.6$ (basing this on Figure 7.10 and an active region thickness of 200 nm) and a residual background loss of 10 cm^{-1}, we derive a threshold gain value of $\gamma_{th} = 114$ cm^{-1}. Using Figure 9.6, or the differential gain constant of $\gamma_D = 2 \cdot 10^{-16}$ cm^2, we find a necessary carrier concentration of about $\Delta n = 3.2 \cdot 10^{18}$ cm^{-3}. We can relate this to the necessary injection current by use of the recombination lifetime, which we take as $\tau = 7$ ns, by $q \cdot \Delta n/\tau = I/\text{volume} = J/t$ for current density J and active layer thickness $t = 200$ nm. From this we determine directly a threshold current density $J_{th} = 1.46$ kA/cm^2. This value agrees with the one we determine from

Figure 9.18 (where the index step is 0.2), which was calculated based on Equation 9.63. Note that these calculations depend on accurate knowledge of recombination lifetimes, which are a critical function of material type and quality, and may thus be a significant source of uncertainty.

9.6.2 Output Power and Efficiency

The lasing threshold is an abrupt transition, which is manifested in a change from the exponential increase in optical output power characteristic of spontaneous emission (such as that for an LED) below threshold to the linear power output once lasing has been achieved. A typical optical power/current plot, often referred to as a PI or LI (power-current or light-current) characteristic, is presented in Figure 9.19. The transition to lasing is also accompanied by a significant jump in overall output power, often a factor 100 higher for the laser than the subthreshold LED output.

The power output of the laser is related to the various structural and electrical device parameters simply as

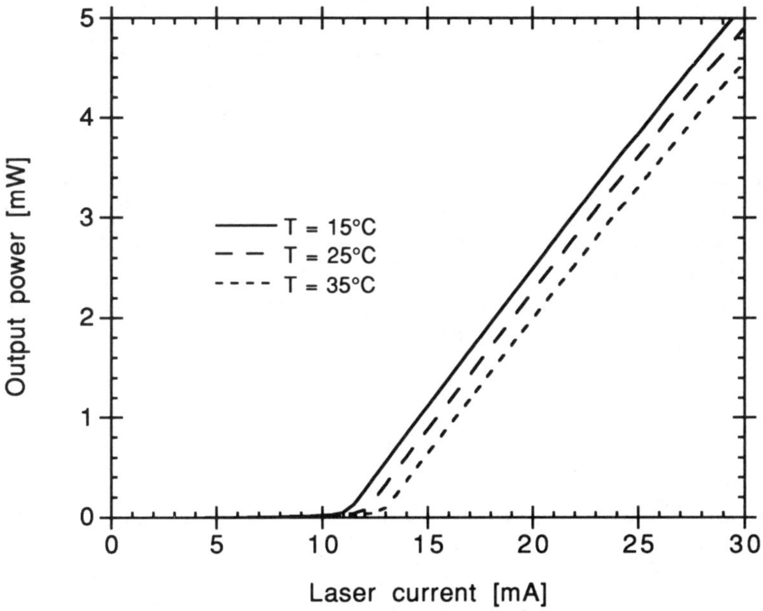

Figure 9.19 The measured optical output power of a laser as a function of drive current for different laser temperatures. The transition to a linear region is the unmistakable indication of lasing. The intercept of the characteristic with the current axis defines the threshold current, and the slope leads to the efficiency.

$$P_{out} = \frac{1}{2}\eta_i \left[\frac{\frac{1}{L}\ln\frac{1}{R}}{\alpha_0 + \frac{1}{L}\ln\frac{1}{R}} \right] \left(\frac{h\nu}{q}\right)(I - I_{th}) \quad (9.64)$$

The factor 1/2 is required because we are considering the output *per facet* of the laser. We see then that the threshold current can easily be determined from the PI characteristic by extrapolating the linear output back to an intersect on the current axis. The slope of the linear region allows us to determine the laser efficiency, where several types of efficiency are commonly quoted.

The term η_i [dimensionless] in Equation 9.64 is the internal quantum efficiency, defined as the ratio of radiative transitions to all electronic transitions. η_i is typically very close to unity, although its value may be reduced if defects in the material give rise to a large nonradiative recombination component or if a significant fraction of the current is lost to leakage outside the laser region. More generally applied is the differential efficiency, η_d [dimensionless], defined as

$$\eta_d = \frac{1}{2}\eta_i \left[\frac{\frac{1}{L}\ln\frac{1}{R}}{\alpha_0 + \frac{1}{L}\ln\frac{1}{R}} \right] = \frac{1}{2}\eta_i \left[\frac{\alpha_m}{\alpha_0 + \alpha_m} \right] = \frac{1}{2}\eta_i\eta_e \quad (9.65)$$

where α_0 is the material loss and α_m is the loss through the mirrors. The term in the square brackets of Equation 9.65 relates how many photons escape through the mirrors as useful output power to all photons lost, which is often referred to as the emission efficiency, η_e. If we take the quantum efficiency $\eta_i = 1$, then η_d defines the ratio of photons escaping from the laser to all photons generated. Differential efficiency is easily related to the slope of the PI curve, dP_{out}/dI, as

$$\eta_d = \frac{q}{h\nu}\left(\frac{dP_{out}}{dI}\right) \quad (9.66)$$

recalling again that we are considering the output and efficiency per facet.

Finally, we are often interested in the overall efficiency of the laser with respect to the total electrical power we pump into it. Since considerable power may be dissipated in the resistive regions of the pn diode, with total resistance R_{lase}, a "wall plug" efficiency η_{wp} [dimensionless] can be defined as

$$\eta_{wp} = \frac{P_{optical}}{P_{electrical}} = \frac{P_{out}}{IR_{lase}} \quad (9.67)$$

Numerous factors can lead to output curves that deviate from the expected linear behavior of Figure 9.19. As we show schematically in Figure 9.20, the device heating that results from higher drive powers can cause the PI curve to bend over, reducing the differential efficiency; the effect is based on an increase of the threshold current with temperature. An increase of leakage current or internal loss with higher currents can give rise to the same phenomenon. In addition, higher currents can allow more modes to lase, promote instabilities in the already lasing modes, or lead to hole burning, all of which can generate kinks in the output characteristic, as also seen in the figure.

Example

Let us examine the efficiency of some ideal and real laser structures. For a 500-μm-long cavity, reflectance, $R = 0.31$, and background loss, $\alpha_0 = 10$ cm^{-1}, we find from Equation 9.65 a differential efficiency of $\eta_d = 0.35$ per facet, assuming unity quantum efficiency, $\eta_i = 1$. In this case, the differential efficiency is equal to the emission efficiency, $\eta_d = \eta_e$. The mirror loss, which goes into useful output power, is 23 cm^{-1} for this structure.

If we now analyze the laser of Figure 9.19, we determine the output slope to be 0.27 W/A at $\lambda = 820$ nm, which we can plug into Equation 9.66 to yield a differential efficiency of $\eta_d = 0.18$ per facet. Relating back to Equation 9.65, this value for η_d implies a background loss of $\alpha_0 \approx 107$ cm^{-1}, an order of magnitude higher than what we assumed earlier. Clearly scattering and absorption losses in this structure are significant. Continuing to evaluation of the wall plug efficiency, we find from the PI characteristic that 5 mW are emitted at 29.5 mA, which gives a value $\eta_{wp} = 17\%$, using Equation 9.67 and a laser resistance of about 1 Ω. While these figures might be typical, better laser efficiencies can be achieved.

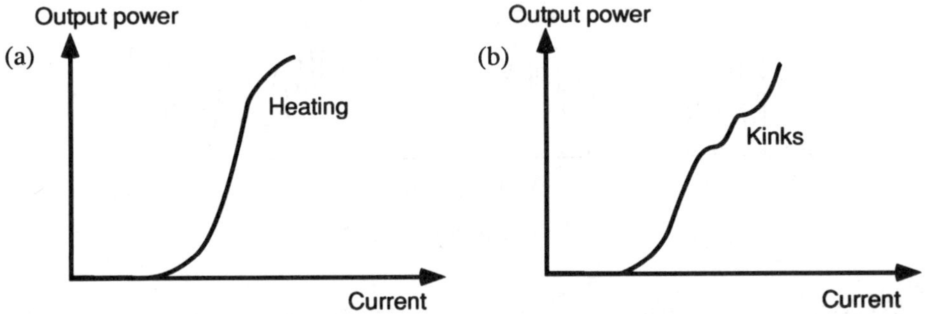

Figure 9.20 (a) At higher drive currents, heating or enhanced leakage can cause the slope of the PI curve to decrease, and (b) mode instabilities or hole burning may result in kinks in the output power.

9.6.3 Spectrum

We already received a hint as to the form of the laser output spectrum in Figure 9.12, where the lasing line appeared in sharp contrast to the broad spontaneous emission. In Figure 9.21, we see a detail of a monomode laser spectrum plotted in logarithmic units (dB), as is commonly done with intensity. The sidemodes, with a spacing of $\Delta\lambda_0$, are clearly seen about –30 dB below the lasing peak, implying a thousand-fold lower intensity.

The intensity ratio between the primary and sidemodes is referred to as the side mode suppression ratio (SMSR). Values of 20 to 30 dB are considered adequate. Fabry-Perot lasers, however, will usually exhibit several simultaneously lasing modes and are thus referred to as multimode lasers; pumping to higher current levels above threshold will almost always generate an output spectrum with several modes. Because many laser structures employ an optical waveguide configuration, such as those we studied in Chapter 8, multiple lateral modes are also permitted if the waveguide is sufficiently wide. These lateral modes lase at distinct, but closely spaced, wavelengths. Lasers that emit in a single longitudinal mode are often desirable devices [33], but

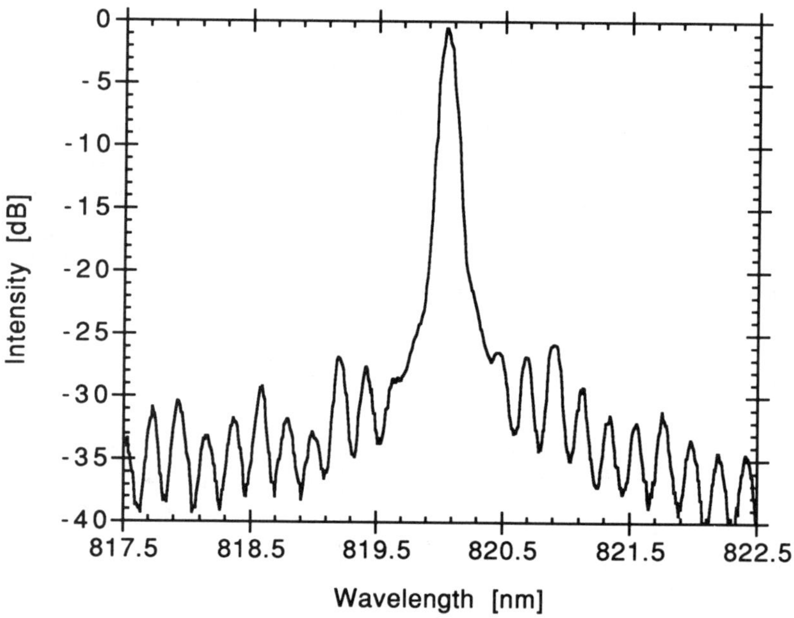

Figure 9.21 Output spectrum of a Fabry-Perot laser diode using the commonly employed logarithmic vertical axis giving output in dB. We see a single mode with sidemodes at about –30 dB lower intensity. This device had a Fabry-Perot cavity length of 500 μm and width of 3 μm; bias current was 30 mA.

a distributed feedback structure, which we will discuss in Section 9.8, is usually required to guarantee monomode output.

Output wavelength in Fabry-Perot structures can shift with both operating current and device temperature. We see an example of the former in Figure 9.22, which plots wavelength against device current for the laser whose spectrum we plotted in Figure 9.21. At certain currents, the wavelength jumps to the next value allowed by the Fabry-Perot cavity; these jumps are known as mode hops and are inherent features of this type of laser. Note that the wavelength increases slowly between the mode hops, indicating a limited available range of wavelength tuning.

9.6.4 Spatial Intensity Distribution

Although the study of laser structures themselves is interesting enough, most applications require that we couple the emitted laser radiation into some kind of an optical system, fiber, waveguide, or lens. For that reason, the spatial form of the emitted beam is of interest. The intensity distribution of the optical field directly at the emitting facet is known as the near-field pattern, and it is essentially the same as

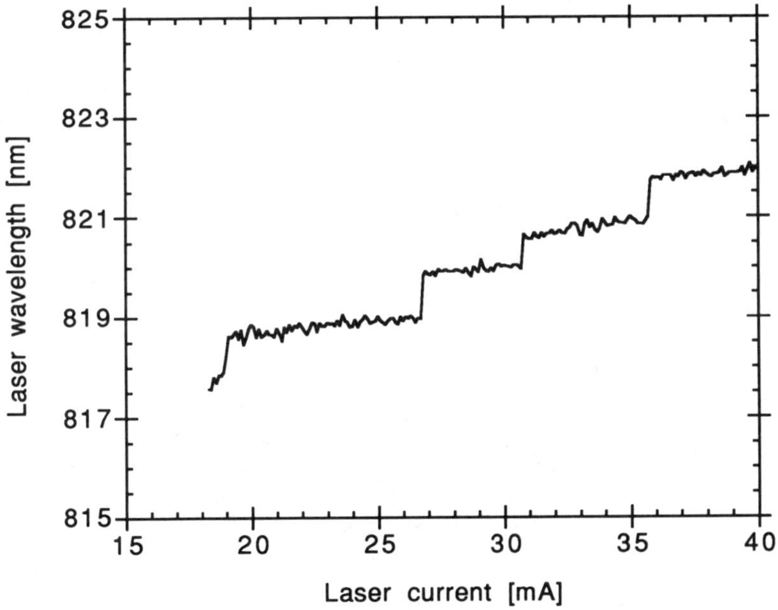

Figure 9.22 Laser output wavelength as a function of current for a Fabry-Perot laser with cavity length of 500 μm and a width of 3 μm. Note mode hops at roughly 19, 22, 31, and 35 mA. The jitter in the signal is measurement system noise.

the intensity distributions we calculated for the semiconductor waveguide structures of the last chapter. The emitted modes are Gaussian in shape, and they have a width that is a critical function of the layer sequence and guiding mechanisms employed to fabricate the laser. For index-guided structures, the lateral mode width, defined as FWHM, is typically on the order of 2 μm; gain-guided lasers have considerably wider lateral mode widths. The transverse width is a function of the heterostructure design and increases in width as t decreases due to a reduction of the transverse confinement factor.

At some distance from the laser facet, the far-field pattern becomes important; this is usually the relevant parameter for determining coupling efficiencies to, say, optical fibers, since the fiber is always at some finite distance from the laser. The beam diverges significantly with distance from the facet, and it does so differently in the planes normal and parallel to the junction. As we can see in Figure 9.23, the lateral distribution (parallel to the junction) for an index-guided laser typically diverges with an angle on the order of $\theta_{\parallel} = 10°$ (again FWHM). Strong confinement by the heterostructure adversely affects the far field in the transverse direction, perpendicular to the junction; divergence in that plane is typically $\theta_{\perp} = 30°$ or more.

9.6.5 Temperature Variations

Earlier, we hinted briefly that the temperature of the laser can strongly affect the optical and electrical characteristics. The most prominent effects are an increase in threshold current and a shift of lasing wavelength with increasing temperature. The

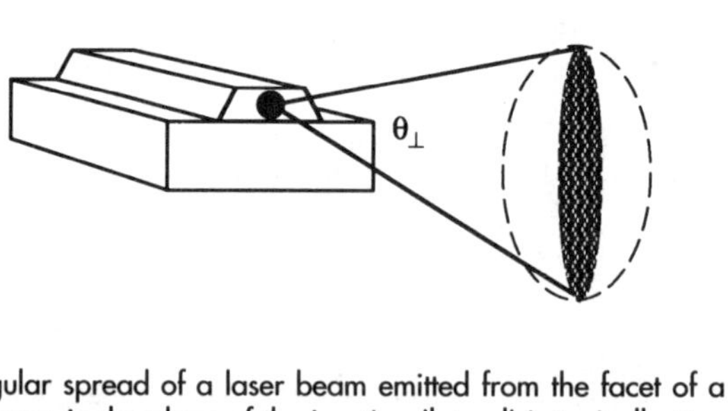

Figure 9.23 Far-field angular spread of a laser beam emitted from the facet of a double heterostructure laser. Divergence in the plane of the junction (lateral) is typically $\theta_{\parallel} = 10°$ and normal to the junction (transverse) about $\theta_{\perp} = 30°$.

former is based on an increase in background loss, α_0, and the widening of the electron and hole distributions as device temperature increases. The rise in I_{th} with temperature is an exponential one, varying as

$$I_{th} = I_{th0} e^{T/T_0} \tag{9.68}$$

T_0 is an experimentally determined parameter, found by a fit to threshold current data as the laser temperature is varied on a heated chuck; I_{th0} is the threshold current at 0° C. Typical values for commonly employed laser structures are in the range $T_0 =$ 100° to 150° C; the laser of Figure 9.19, for example, has $T_0 = 140°$ C. In quantum-well lasers, GaAs devices characteristically have $T_0 = 200°$, whereas for InP structures, $T_0 = 100°$ is typical due to the shallower wells. The change in threshold due to device heating can often be directly seen through a deviation in the slope of output power from its linear rise; the curves will tend to bend over for high current levels, which have led to a significant rise in junction temperature, as we saw schematically in Figure 9.20.

9.7 LIGHT-EMITTING DIODES

While a semiconductor laser diode is, if we submit to the tyranny of semantics, strictly also a light-emitting diode, the designation LED refers to emitters of incoherent, spectrally broad optical radiation.[5] In contrast to lasers, LEDs emit light at relatively low efficiency (typically less than several percent), with a very broad spectrum (many tens of nanometers) and reduced directionality. However, since almost any semiconductor will emit light if sufficiently pumped, and since a high-quality resonant cavity is not required, LEDs are considerably easier to manufacture than lasers. They can often be made from materials from which lasers cannot (or cannot yet) be fabricated. As a result, light-emitting diodes can be mass fabricated at low cost. The vast majority are employed as indicators or in displays, with a wide automotive market likely in the future. These consumer-oriented uses are coupled with a broad range of further applications including sensors [34], communications [35], and displays.

Because the requirements for incoherent emission are less severe than for lasers, both direct and indirect bandgap materials can be employed for the fabrication of LEDs. The indirect materials often take advantage of impurities as recombination centers to enhance electroluminescence efficiency. In Si, for example, the use of Er doping has allowed the fabrication of light emitters with 1% efficiency from a sharp line at a wavelength of 1.54 μm [36]. The III-V materials, however, due to the high-efficiency of emission, form the bulk of LED devices. In addition, because displays

[5]See, for example: Bergh, A. A., and P. J. Dean, *Light-Emitting Diodes*, Oxford, England: Clarendon, 1976; or: Williams, E. W., and R. Hall, *Luminescence and the Light Emitting Diode*, Oxford, England: Pergamon, 1978.

are a dominant market, the visible wavelengths are central to much LED development. For example, $GaAs_{1-x}P_x$, for $x < 0.45$, is direct bandgap in the wavelength range 871 to 627 nm, the latter corresponding to yellow. GaInP is direct to a wavelength of 564 nm, which is green, and GaP has also been used for efficient green emission [37]. High-brightness LEDs have been fabricated using AlGaInP wafer bonded onto a transparent GaP substrate (GaAs is absorbing) in the 600- to 615-nm wavelength regime [38]. Emission in the blue, down to 466 nm, requires the use of the II-VI system, ZnSe [39] and SiC [40], both being successful electroluminescence candidates; GaAlInN is employed for green to UV wavelengths.

Structurally, LEDs are simpler than lasers. Two representative configurations are shown in Figure 9.24; as in the laser, a pn junction is employed, along with similar double heterostructures. Most devices are either surface [41] or edge [42] emitters, the former providing the ability to emit vertically from a large area and the latter showing less beam divergence in the direction perpendicular to the junction plane. The designs are usually optimized for maximum output coupling of the generated light, where internal absorption in the substrate and reflection at the surfaces,

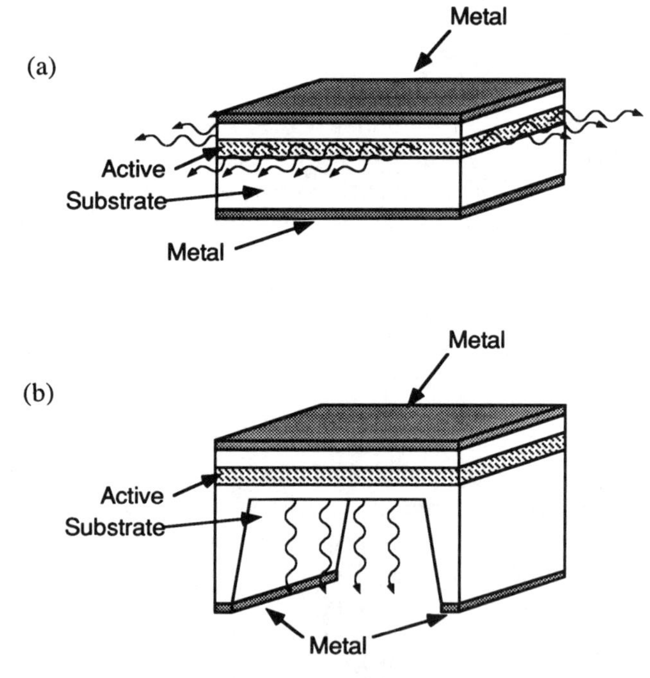

Figure 9.24 Two typical LED structures, both employing the same type of pn-junction heterostructure for light generation: (a) the surface emitter relies on an etch removal of the substrate to optimize output, and (b) the edge emitter is often structurally similar to the broad-area or stripe laser.

both due to critical angle and standard Fresnel reflectivity, are the primary adverse factors. Efficient coupling of LED emission into fibers is often an important design criterion [43], where shaping of the substrates or the use of lens structures is often required to optimize coupling.

For specialized LED structures with an intense optical field plus an adequate overlap of pump electrons and generated photons, stimulated emission can begin to dominate the optical output. If feedback is suppressed (i.e., the resonant cavity is missing), the electroluminescent output will reach a high-intensity state known as superluminesence [44]. Using material and device structures that would lead to a laser if the resonant cavity were established, superluminescent LEDs have been employed in numerous areas, including optical coherence domain reflectometry [45] and optical fiber gyroscopes [46].

9.8 DISTRIBUTED FEEDBACK LASERS

Semiconductor lasers based on a Fabry-Perot cavity are relatively easy to fabricate, as the cleaved mirror facets essentially come for free to those with a supple wrist, and they are certainly suitable for a very wide range of uses. However, as we hinted briefly when discussing the output spectrum, it is generally difficult to achieve true monomode operation using this type of structure. The problem derives from the cavity itself, which permits a large number of closely spaced wavelengths that are only selected by the spectrally broad gain spectrum.

This problem can be circumvented by the use of a wavelength-selective mirror, a function for which a grating is well suited. Gratings, as we saw briefly in Section 8.3.3, are periodic structures etched into or deposited on a semiconductor, giving rise to a periodic modulation of refractive index. Bragg reflection from the grating is a function of wavelength, so a grating can act as a distributed mirror for one specific frequency. Study of the distributed-feedback mechanism in laser material led to the first grating-based laser in 1973 [47], and these devices are at present most lucrative for coherent optical communications systems [48] in which monomode sources are essential. In addition, since a cleaved facet is unnecessary, a distributed-feedback laser is suitable for monolithic integration with an integrated optical circuit, as we will discuss in greater detail in Chapter 12.

9.8.1 Distributed Feedback

Distributed-feedback lasers are of two generic types, as outlined in Figure 9.25: the distributed Bragg reflector (DBR) laser employs separate pumped and grating regions, whereas the distributed-feedback laser (DFB) employs a periodic modulation in the pumped active layer of the device itself. In both cases, the corrugation periodically modulates the refractive index in the waveguide, which forms the active layer of the

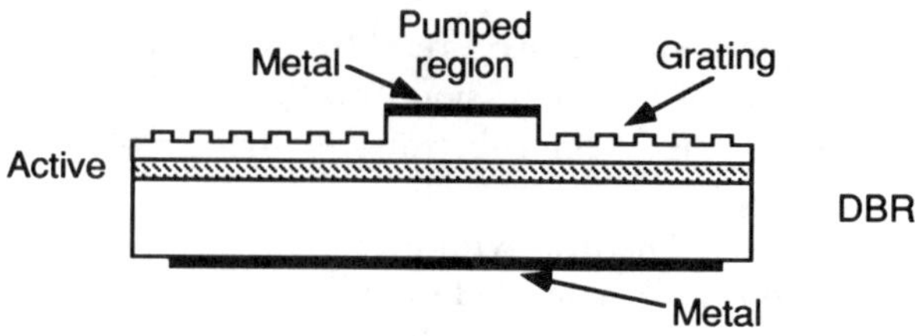

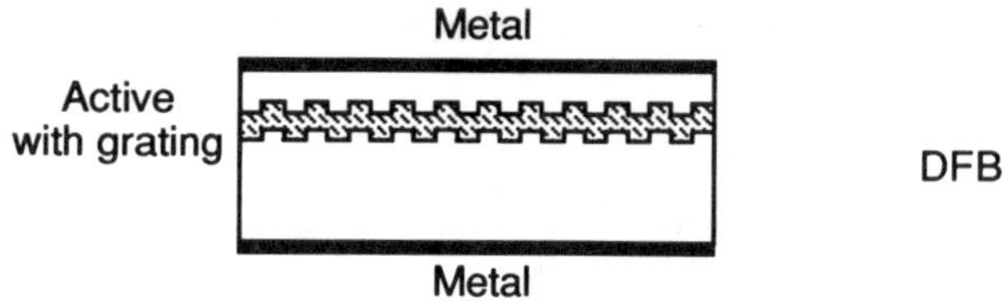

Figure 9.25 Schematic cross section of generic (a) DBR and (b) DFB lasers. The DBR has separate pumped and grating reflector regions and can use a cleaved facet on one end for simplicity. The DFB has a corrugation in the pumped region itself, so that the mirror coexists with the active layer.

laser, either by a longitudinal patterning of the waveguiding layer itself or by an etch into the cladding. The latter thus periodically modulates the effective waveguide index. The period of the grating is chosen so that a significant backward reflection is achieved, giving rise to the feedback necessary for a resonant cavity. For grating period, Λ [m], and active waveguide effective index, N, the required grating period may be found from

$$m\lambda_{DBR} = 2N\Lambda \tag{9.69}$$

for a DBR laser emission wavelength λ_{DBR}. The integer index, m, is the grating order, whose value determines the efficiency and direction of reflected light.

Although the thorough treatment of the coupled wave theory [49], which is required to fully analyze the behavior of a distributed-feedback laser, is outside the scope of this chapter,[6] we can write down a few equations to help demonstrate the mechanisms at work. For a **z**-directed active region waveguide modulated by a grating

[6]For an in-depth discussion, see: Suematsu, Y., K. Kishino, S. Arai, and F. Koyama, "Dynamic Single-Mode Semiconductor Lasers with a Distributed Reflector," in *Semiconductors and Semimetals*, Tsang, W. T., (ed.), Vol. 22B, San Diego: Academic Press, 1985, pp. 205–255.

with period Λ, the total field, $\mathcal{E}_z$, is given as the sum of a forward, $\mathcal{E}_f$, and a backward, $\mathcal{E}_r$, traveling wave,[7]

$$\mathcal{E}_z = \mathcal{E}_f + \mathcal{E}_r \tag{9.70}$$

This total field can be shown upon solution of the coupled wave equations

$$\frac{d\mathcal{E}_f}{dz} = j\Delta\beta\mathcal{E}_f + j\kappa\mathcal{E}_r \tag{9.71}$$

$$-\frac{d\mathcal{E}_r}{dz} = j\kappa\mathcal{E}_f + j\Delta\beta\mathcal{E}_r \tag{9.72}$$

to be of the form

$$\mathcal{E}_f(z) = \mathcal{E}_{f0} \cdot e^{j\beta_B z} + r \cdot \mathcal{E}_{r0} \cdot e^{-j\beta_B z} \tag{9.73}$$

$$\mathcal{E}_r(z) = r \cdot \mathcal{E}_{f0} \cdot e^{j\beta_B z} + \mathcal{E}_{r0} \cdot e^{-j\beta_B z} \tag{9.74}$$

In these relations, $\Delta\beta$ is the difference between the propagation constant in absence of a grating, β, and the Bragg wave number, $\Delta\beta = \beta - m\pi/\Lambda$; κ is the coupling constant; r is a complex reflection coefficient, which is a function of κ; and β_B is the propagation constant of the coupled waves, an eigenvalue whose value is determined by the boundary conditions at the laser facets.

From these last two expressions, it is clear that for wavelengths near the Bragg wavelength, forward and backward traveling waves are coupled. The forward component has a portion arising from reflection of the backward component and vice versa. The extent of this interaction is mediated by the coupling coefficient, κ, which defines how strongly the grating affects the mode propagating in the waveguide and is typically found by calculating an overlap integral of the mode intensity with the grating profile [50]. Determination of κ is generally nontrivial and usually requires numerical simulation; typical values are in the range 20–80 cm^{-1}. The magnitude of the coupling constant is a critical function of grating order, m, and shape, where square gratings, for example, lead to higher κ values than sinusoidal profiles. The corrugation depth and grating distance from the active region are also important parameters in the calculation of coupling constant.

The grating order, m in Equation 9.69, critically defines the nature of coupling between the modes. For m = 1, a first-order grating, only forward and backward modes are allowed, as shown in Figure 9.26. A second-order grating also couples

[7]We follow the treatment of Agrawal, G. P,. and N. K. Dutta, *Long-Wavelength Semiconductor Lasers*, New York: Van Nostrand Reinhold, 1986, Chapter 7.

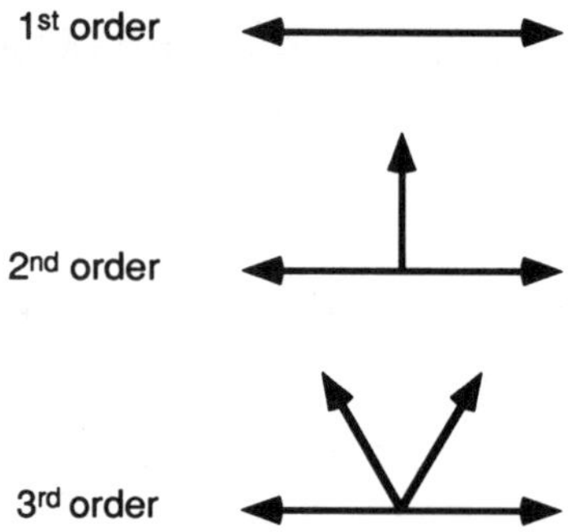

Figure 9.26 Mode coupling directions for gratings of increasing order: first order has only forward and backward transmission; second and third order have vertical reflection components.

energy vertically from the laser, which is useful for fabricating a surface emitter [51] but at the expense of considerably less power being returned into the laser cavity; thus r is low. A third-order grating also has surface emission, but couples a significant fraction back into the laser cavity. Higher order gratings are commonly used when wavelength structures require grating periods too small for available lithography capabilities.

9.8.2 DBR/DFB Fabrication

Active layer structure and device fabrication are similar to those for Fabry-Perot lasers, except, of course, that a grating must be generated. For the DBR structure, the reflector can be formed near the end of device processing, although a recess etch is frequently required to place the grating close to the waveguide that forms the active region [52]. The grating itself is formed by electron-beam or holographic exposure and subsequent etch; embossing techniques have also been employed for the softer materials of the lead salt lasers [53]. The required period, Λ, is wavelength dependent and becomes shorter for shorter laser wavelengths: first-order gratings can have grating periods of less than 100 nm for typical GaAs/AlGaAs lasers, which becomes challenging for most lithography techniques. For that reason, third-order gratings are often employed.

The DFB is more involved from the material growth aspect. Half the layer sequence is grown, stopping at or above the active layer of the laser. The grating is then exposed and etched into the material, whereupon the remaining layers are completed in a second growth step. This process requires adequate facilities for performing regrowth and requires accurate layer thickness control. Since in many designs the grating must be generated before the exact wavelength of the active region is known, a mismatch between grating and device wavelength is possible. Finally, etching the grating directly into the active layer, whether for the DFB or DBR, results in etch damage, so a separation of the corrugation from the core layers is often desirable. Advanced etching and regrowth techniques, however, have recently made the fabrication of active grating DFBs practical.

9.8.3 Operation and Special Features

Distributed feedback lasers have similar operating characteristics to their Fabry-Perot cousins, with the added advantage that monomode operation is assured due to the mode filter of the grating reflector. A typical spectrum of a DBR laser is shown in Figure 9.27. The DFB laser has a modal degeneracy at the reflection peak, meaning that two modes could lase. A $\lambda/4$ shift in the grating or an unpumped window at the end of the laser is typically used to force monomode operation.

Spectral linewidth is smaller for distributed feedback structures than for Fabry-Perot lasers, with values of < 1 MHz typical for DBRs, versus several GHz for cleaved mirrors. DBR linewidth can be reduced further by coating the facets to suppress any Fabry-Perot noise. Achieving SMSR values of greater than 15 dB is difficult with standard cleaved cavity lasers, although the DBR can routinely provide sidemode

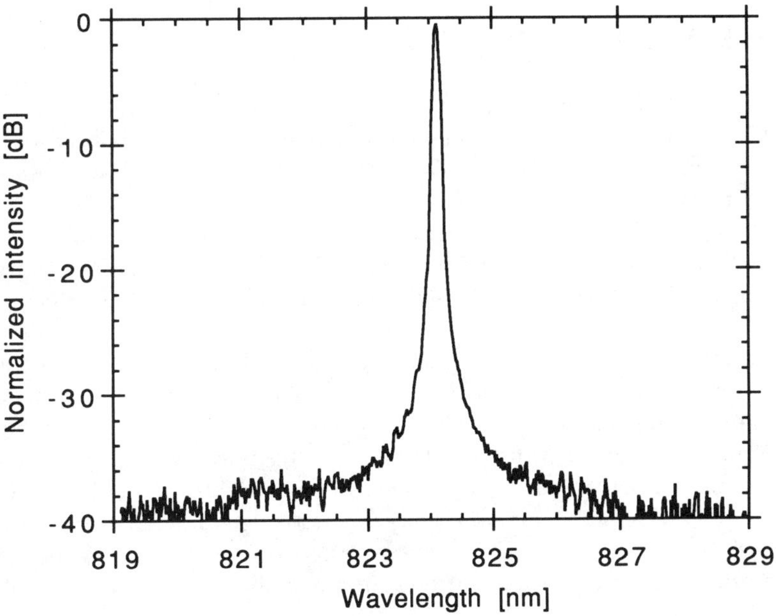

Figure 9.27 Spectral output of a DBR laser using an etched grating with a period Λ = 385 nm. The device had a pumped section 500 μm long and ridge width of 3 μm; drive current was 55 mA.

suppression of 30 dB or greater. In addition, under high-speed modulation, distributed-feedback lasers remain single mode.

The temperature dependence of wavelength in a DBR laser is due to a change in the index of the grating and shows a small shift with temperature; $\Delta\lambda/\Delta T \leq 0.1$ nm/°C is typical. In addition, due to the wavelength-selective reflection, the mode hops of Figure 9.23 are widely spaced in a DBR structure. For comparison, the Fabry-Perot laser will usually shift at a rate $\Delta\lambda/\Delta T \geq 0.5$ nm/°C between hops. Distributed-feedback lasers do tend to be less efficient than Fabry-Perots, primarily due to losses in the grating; both scattering and surface or substrate coupling, for $m \neq 1$, result in a loss of reflected light and thus reduced overall η. Threshold current densities can likewise be higher, although values < 1 kA/cm^2 can be reached.

Finally, DBR lasers can form the basis for electrically tunable lasers, where the laser wavelength can be continuously tuned over a significant range without mode hops [54]. A multisection device is usually required; the pumped active region, a phase control segment whose optical properties are modulated by current injection, and the grating, whose effective wavelength is also shifted by the refractive index, change due to current injection. Although technologically somewhat difficult to fabricate, the useful features of having an electrically tunable semiconductor light source have provided significant impetus in their development.

9.9 SURFACE EMITTING LASERS

A semiconductor laser area under significant development at present is concerned with the design and fabrication of surface-emitting devices. As opposed to the edge emitters on which we have focused above, surface-emitting lasers direct optical radiation normal to the wafer surface. If our goal is to fabricate an integrated optical circuit where optical operations are done on chip, this type of laser structure is less suited than a waveguide based configuration; nevertheless, many applications benefit greatly from surface emission and the use of these devices will likely increase in the future.[8]

Surface-emitting lasers have numerous features that render them attractive in some situations. The ability to easily fabricate two-dimensional laser arrays with individual addressing for free space data transmission, or coherent arrays for high-power output, are useful advantages. Device testing may be done on-chip, as the cleaving of a laser facet is no longer required; this can also simplify processing to some extent. Layer growth, however, may be more demanding. The optical beam quality is usually good, as surface emission often proves to have a narrow beam divergence and low astigmatism, and focusing or collimation of the optical field is sometimes possible.

[8]For an overview, see: Evans, G., and J. M. Hammer (eds.), *Surface Emitting Semiconductor Lasers and Arrays*, Boston: Academic Press, 1993.

Three representative surface-emitting laser structures are presented in Figure 9.28, each using a distinct mechanism for generating optical emission normal to the plane of the wafer. Using sophisticated dry-etch technology, the conceptually simplest approach etches a deflecting mirror behind a standard Fabry-Perot cavity. Alternatively, higher order gratings couple a fraction of reflected light toward the surface in a DBR structure. The vertical cavity surface emitting laser (VCSEL) turns the

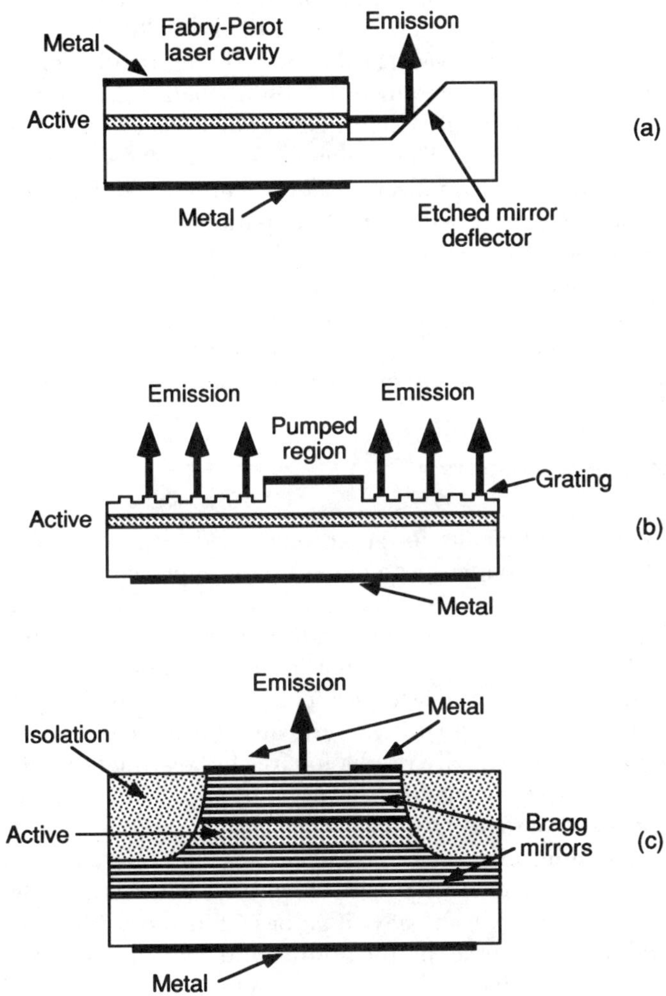

Figure 9.28 Cross-sectional views of three surface-emitting laser structures: (a) dry-etched mirror deflection, (b) higher order grating coupling from a DBR, and (c) the vertical cavity surface-emitting laser.

entire cavity on its side, using stacks of dielectric or semiconductor as Bragg mirrors to promote laser oscillation in the vertical direction. Each of these techniques has its own advantages and limitations, which we now discuss briefly.

9.9.1 Mirror Deflectors

Just as waveguide corner reflectors can be used to deflect light beams in the plane of a wafer, an etched mirror as that shown in Figure 9.28(a) may be employed to deflect the optical field in the vertical direction. Using tightly controlled dry-etch techniques, typically ion beam etching or CAIBE, the latter also successfully used for etching Fabry-Perot mirror facets, a high-quality mirror may be fabricated adjacent to a normal in-plane laser cavity. The technique is straightforward conceptually, but requires very high-quality mirror facets to assure a uniform and well-directed surface beam—flatness and angular orientation are of paramount importance. Through the use of curved or shaped mirror surfaces, beam shaping or collimation may be achieved and etched-lens structures may also be employed to spatially define the vertical emission. The mirror-reflector structure has the advantage that the laser characteristics may be optimized independently of the mechanism for achieving vertical emission, so that power conversion efficiency and spectral quality can be high.

9.9.2 Surface Grating Couplers

We saw in Figure 9.26 that the Bragg gratings used for generating feedback in a DBR or DFB laser have coupling components that emit normal to the surface, for second- or higher order gratings. Taking advantage of this phenomenon for the generation of a grating-coupled surface-emitting laser leads to a structure as shown in Figure 9.28(b). In this DBR laser, a portion of the optical field in the grating is directed back into the pumped laser section, giving rise to the necessary optical feedback, but a portion is also emitted vertically, for a second-order grating. In a sense, surface emission essentially results for free because higher order gratings are typically used for edge-emitting DBR devices anyway, as the demands on grating definition are less severe than for a first-order grating.

Using DBR gratings for surface emission has the advantage that the grating shape can be used to shape and direct the emitted optical field. By employing a focusing grating, the surface beam may either be collimated or focused at a particular distance. In addition, by contacting the grating and changing its effective refractive index electrically, the surface beam may be dynamically steered, allowing extra functionality for many applications. While grating design and fabrication for these functions is not trivial, the flexibility involved in the use of gratings for surface-emitting lasers often makes the effort worthwhile.

9.9.3 VCSELs

A related, but structurally very distinct, category of surface-emitting distributed-feedback laser is the vertical cavity surface-emitting laser. This device employs a multilayer stack of deposited dielectrics or epitaxially grown layers as a Bragg reflector, as we can see in the cross section of Figure 9.28(c). The laser cavity is thus oriented vertically and has quite different physical dimensions than the structures we have considered previously. The active region usually has a length on the order of several nanometers to a micron and the Bragg mirrors have a typical total thickness on the order of one micron; thus, the entire VCSEL laser cavity is only several microns in length.

The multilayer Bragg stacks are formed from alternating layers of either dielectrics, such as SiO_2/Si or SiO_2/TiO_2, or, more popularly, epitaxially grown semiconductors, such as AlAs/GaAs; 15 to 30 periods are routinely employed. The refractive index differences in the two constituent materials of the stack form a vertical Bragg reflector, and the relative layer thicknesses for each material must be accurately chosen for a particular oscillation wavelength. To achieve the high reflectivity necessary for a usable device ($R \geq 0.95$ is usually required), careful attention must be paid to the material indices involved as well as the operating temperature.

Once the Bragg mirrors surrounding the active region have been grown, VCSEL processing is relatively simple and the fabrication of surface emitting arrays is straightforward. The lateral definition of the device, typically into a circular emitter as seen from above, may be achieved by implantation (as shown in the figure), oxidation, or by etching of a mesa. Current injection into the active region is typically through the mirrors; the mirror electrical and thermal resistance must be kept low to prevent excessive power dissipation and thus a temperature rise, which critically affects the wavelength at which the Bragg mirrors reflect. If this stipulation is fulfilled, VCSELs can operate with low threshold current, due to the small active volume, large mode spacing, short cavity, and good spatial mode characteristics.

REFERENCES

[1] Whitman, W., *Crossing Brooklyn Ferry*, 1856.

[2] Schawlow, A. L., and C. H. Townes, "Infrared and Optical Masers," *Physical Review*, Vol. 112, 1958, pp. 1,940–1,948.

[3] Koch, T. L. and U. Koren, "Semiconductor Lasers for Coherent Optical Fiber Communications," *IEEE Journal of Lightwave Technology*, Vol. 8, 1990, pp. 274–293.

[4] Harder, C. S., B. J. Van Zeghbroeck, M. P. Kesler, H. P. Meier, P. Vettiger, D. J. Webb, and P. Wolf, "High-Speed GaAs/AlGaAs Optoelectronic Devices for Computer Applications," *IBM Journal of Research and Development*, Vol. 34, 1990, pp. 568–584.

[5] Arahira, S., S. Oshiba, Y. Matsui, T. Kunii, and Y. Ogawa, "Terahertz-Rate Optical Pulse Generation from a Passively Mode-Locked Semiconductor Laser Diode," *Optics Letters*, Vol. 19, 1994, pp. 834–836.

[6] Nathan, M. I., W. P. Dumke, G. Burns, F. H. Dills, and G. Lasher, "Stimulated Emission of Radiation from GaAs PN Junctions," *Applied Physics Letters*, Vol. 1, 1962, pp. 62–64.

[7] Quist, T. M., R. H. Rediker, R. J. Keyes, W. E. Krag, B. Lax, A. L. McWhorter, and H. J. Ziegler, "Semiconductor Maser of GaAs," *Applied Physics Letters,* Vol. 1, 1962, pp. 91–92.
[8] Hall, R. N., G. E. Fenner, J. D. Kingsley, T. J. Soltys, and R. O. Carlson, "Coherent Light Emission from GaAs Junctions," *Physical Review Letters,* Vol. 9, 1962, pp. 366–368.
[9] Hayashi, I., M. B. Panish, P. W. Foy, and S. Sumski, "Junction Lasers which Operate Continuously at Room Temperature," *Applied Physics Letters,* Vol. 17, 1970, pp. 109–111.
[10] Alferov, Z. I., V. M. Andreev, E. L. Portnoi, and M. K. Trukan, "AlAs-GaAs Heterojunction Injection Lasers with a Low Room – Temperature Threshold," *Soviet Physics—Semiconductors,* Vol. 3, 1970, pp. 1,107–1,110.
[11] Einstein, A., "Zur Quantentheorie der Strahlung," *Physikalische Zeitschrift,* Vol. 18, 1917, pp. 121–128 (in German).
[12] Saleh, B. E. H., and M. C. Teich, *Fundamentals of Photonics,* New York: John Wiley & Sons, 1991, Chapter 12.2D.
[13] Stern, F., "Calculated Spectral Dependence of Gain in Excited GaAs," *Journal of Applied Physics,* Vol. 47, 1976, pp. 5,382–5,386.
[14] Yariv, A., *Optical Electronics*, 4th Ed., Philadelphia: Saunders, 1991, p. 562.
[15] Stern, F., "Gain-Current Relation for GaAs Lasers with N-Type and Undoped Active Layers," *IEEE Journal of Quantum Electronics,* Vol. QE-9, 1973, pp. 290–294.
[16] Yamada, M., and Y. Suematsu, "Analysis of Gain Suppression in Undoped Injection Lasers," *Journal of Applied Physics,* Vol. 52, 1981, pp. 2,653–2,664.
[17] Chen, Y. H. and F. Auzel, "Photon Avalanche in Er:ZBLAN Fibre Pumped at 690 nm," *Electronics Letters,* Vol. 30, 1994, pp. 1,602–1,603.
[18] Shen, C. C., J. J. Hsieh, and T. A. Lind, "1500-h Continuous Operation of Double Heterostructure GaInAsP/InP Lasers," *Applied Physics Letters,* Vol. 30, 1977, pp. 353–354.
[19] Roentgen, P., W. Heuberger, G. L. Bona, and P. Unger, "MOVPE of AlGaInP/GaInP Heterostructures for Visible Lasers," *Journal of Crystal Growth,* Vol. 107, 1991, pp. 724–730.
[20] Gunshor, R. L., and A. V. Nurmikko, "Blue Lasers on the Horizon," *IEEE Spectrum,* May 1993, pp. 28–33.
[21] Haase, M. A., J. Qiu, J. M. DePuydt, and H. Cheng, "Blue-Green Laser Diodes," *Applied Physics Letters,* Vol. 59, 1991, pp. 1,272–1,274.
[22] Morkoç, H., S. Strite, G. B. Gao, M. E. Lin, B. Sverdlov, and M. Burns, "Large-Band-Gap SiC, III-V Nitride and II-VI Zn-Se-Based Semiconductor Device Technologies," *Journal of Applied Physics,* Vol. 76, 1994, pp. 1,363–1,398.
[23] Panish, M. B., "Heterostructure Injection Lasers," *IEEE Transactions on Microwave Theory and Techniques,* Vol. MTT-23, 1975, pp. 20–30.
[24] van der Ziel, J. P., R. Dingle, R. C. Miller, W. Wiegmann, and W. A. Nordland, "Laser Oscillations from Quantum States in Very Thin GaAs-$Al_{0.2}Ga_{0.8}As$ Multilayer Structures," *Applied Physics Letters,* Vol. 26, 1975, pp. 463–465.
[25] Zory, P. (ed.), *Quantum Well Lasers,* Boston: Academic Press, 1993.
[26] Tsang, W. T., "Quantum Confined Heterostructure Semiconductor Lasers" in Dingle, R. (ed.), *Semiconductors and Semimetals,* Vol. 24, San Diego: Academic Press, 1987, pp. 397–458.
[27] McIlroy, P. W. A., A. Kurobe, and Y. Uematsu, "Analysis and Application of Theoretical Gain Curves to the Design of Multi-Quantum Well Lasers," *IEEE Journal of Quantum Electronics,* Vol. QE-21, 1985, pp. 1,958–1,963.
[28] Gourley, P. L., "Microstructured Semiconductor Lasers for High-Speed Information Processing," *Nature,* Vol. 371, 1994, pp. 571–577.
[29] Behfar-Rad, A., and S. S. Wong, "Monolithic AlGaAs-GaAs Single Quantum Well Ridge Lasers Fabricated with Dry-Etched Facets and Ridges," *IEEE Journal of Quantum Electronics,* Vol. 28, 1992, pp. 1,227–1,231.
[30] Epperlein, P. W., P. Buchmann, and A. Jakubowicz, "Lattice Disorder, Facet Heating and Cata-

strophic Mirror Damage of AlGaAs Quantum Well Lasers," *Applied Physics Letters,* Vol. 62, 1992, pp. 455–457.
[31] Ralston, J. D., D. F. G. Gallagher, P. J. Tasker, H. P. Zappe, I. Esquivias, and J. Fleissner, "A Vertically Compact 15 GHz GaAs/AlGaAs Multiple Quantum Well Laser Grown by Molecular Beam Epitaxy," *Electronics Letters,* Vol. 27, 1991, pp. 1,720–1,721.
[32] Lasher, G. J., "Threshold Relations and Diffraction Loss for Injection Lasers," *IBM Journal of Research and Development,* Vol. 7, 1963, pp. 58–61.
[33] Agrawal, G. P., "Single-Longitudinal Mode Semiconductor Lasers," in *Progress in Optics,* Wolf, E. (ed.), Vol. XXVI, Elsevier, 1988, pp. 165–225.
[34] Gerges, A. S., F. Farahi, T. P. Newson, J. D. C. Jones, and D. A. Jackson, "An Interferometric Fiber Optic Sensor Using a Short Coherence Length Source," *Electronics Letters,* Vol. 23, 1987, pp. 1,110–1,111.
[35] Ikeda, K., S. Horiuchi, T. Tanaka, and W. Susaki, "Design Parameters of Frequency Response of GaAs-(Ga,Al)As Double Heterostructure LED's for Optical Communications," *IEEE Transactions on Electron Devices,* Vol. ED-24, 1977, pp. 1,001–1,005.
[36] Zheng, B., J. Michel, F. Y. G. Ren, L. C. Kimerling, D. C. Jacobson, and J. M. Poate, "Room Temperature Sharp Line Electroluminescence from an Erbium-Doped Silicon Light-Emitting Diode," *Applied Physics Letters,* Vol. 64, 1994, pp. 2,842–2,844.
[37] Nina, T., T. Yamaguchi, T. Yamazawa, T. Ishi, H. Takasu, and K. Inoue, "High Brightness GaP Green LED's," *IEEE Transactions on Electron Devices,* Vol. ED-30, 1983, pp. 264–267.
[38] Kish, F. A., D. A. DeFevere, D. A. Vanderwater, G. R. Trott, R. J. Weiss, and J. S. Major, "High Luminous Flux Semiconductor Wafer-Bonded AlGaInP/GaP Large-Area Emitters," *Electronics Letters,* Vol. 30, 1994, pp. 1,790–1,792.
[39] Fan, X. W. and J. Woods, "Blue Electroluminescence in Forward-Biased ZnSe Diodes," *IEEE Transactions on Electron Devices,* Vol. ED-28, 1981, pp. 428–433.
[40] Ziegler, G., P. Lanig, D. Theis, and C. Weyrich, "Single Crystal Growth of SiC Substrate Material for Blue Light Emitting Diodes," *IEEE Transactions on Electron Devices,* Vol. ED-30, 1983, pp. 277–281.
[41] Burrus, C. A. and B. I. Miller, "Small-Area, Double-Heterostructure Aluminum-Gallium Arsenide Electroluminescent Diode Sources for Optical-Fiber Transmission Lines," *Optics Communications,* Vol. 4, 1971, pp. 307–309.
[42] Ettenberg, M., H. Kressel, and J. P. Wittke, "Very High Radiance Edge-Emitting LED," *IEEE Journal of Quantum Electronics,* Vol. QE-12, 1976, pp. 360–364.
[43] Borsuk, J., "Light Intensity Profiles of Surface-Emitting InGaAsP LED's: Impact on Coupling to Optical Fibers," *IEEE Transactions on Electron Devices,* Vol. ED-30, 1983, pp. 296–303.
[44] Alphonse, G. A., D. B. Gilbert, M. G. Harvey, and M. Ettenberg, "High-Power Superluminescent Diodes," *IEEE Journal of Quantum Electronics,* Vol. 24, 1988, pp. 2,454–2,457.
[45] Patterson, B. D., J. E. Epler, B. Graf, H. W. Lehmann, and H. C. Sigg, "A Superluminescent Diode at 1.3 μm with Very Low Spectral Modulation," *IEEE Journal of Quantum Electronics,* Vol. 30, 1994, pp. 703–712.
[46] Burns, W. K. and C.-L. Chen, "Fiber-Optic Gyroscopes with Broad Band Sources," *IEEE Journal of Lightwave Technology,* Vol. LT-1, 1983, pp. 98–105.
[47] Nakamura, M., A. Yariv, H. W. Yen, S. Somekh, and H. L. Garvin, "Optically Pumped GaAs Surface Laser with Corrugation Feedback," *Applied Physics Letters,* Vol. 22, 1973, pp. 515–516.
[48] Kobayashi, K., and I. Mito, "Single Frequency and Tunable Laser Diodes," *IEEE Journal of Lightwave Technology,* Vol. 6, 1988, pp. 1,623–1,633.
[49] Kogelnik, H., and C. V. Shank, "Coupled-Wave Theory of Distributed Feedback Lasers," *Journal of Applied Physics,* Vol. 43, 1972, pp. 2,327–2,335.
[50] Yariv, A., "Coupled-Mode Theory for Guided-Wave Optics," *IEEE Journal of Quantum Electronics,* Vol. QE-9, 1973, pp. 919–933.

[51] Evans, G. A., N. W. Carlson, J. M. Hammer, and J. K. Butler, "Grating-Outcoupled Surface Emitting Semiconductor Lasers," in *Surface Emitting Semiconductor Lasers and Arrays*, Evans, G., and J. M. Hammer (eds.), Boston: Academic Press, 1993, pp. 119–216.

[52] Hofstetter, D., H. P. Zappe, J. E. Epler, and J. Söchtig, "Single-Growth-Step GaAs/AlGaAs Distributed Bragg Reflector Lasers with Holographically-Defined Recessed Gratings," *Electronics Letters*, Vol. 30, 1994, pp. 1,858–1,859.

[53] Fach, M. A., H. Böttner, K.-H. Schlereth, and M. Tacke, "Embossed-Grating Lead Chalcogenide Buried-Waveguide Distributed Feedback Lasers," *IEEE Journal of Quantum Electronics*, Vol. 30, 1994, pp. 235–240.

[54] Coldren, L. A., and S. W. Corzine, "Continuously-Tunable Single-Frequency Semiconductor Lasers," *IEEE Journal of Quantum Electronics*, Vol. QE-23, 1987, pp. 903–908.

Chapter 10

Photodetectors

. . . detectors in venuvarities, whateither the drugs truth of it, was there an iota of from the faust to the lost.

James Joyce, 1939 [1]

Turn a laser upside down, inject light into it, and measure the current that is generated: this is the essence of a semiconductor photodetector. Although specialized detectors are structurally quite distinct from their light-emitting kin, lasers have indeed been employed as very efficient detectors [2], which suggests that these components may relatively easily be monolithically integrated. The detector structures we will consider in this chapter, namely photodiodes, represent a subset of semiconductor photodetectors, which in general also include photoconductors, phototransistors, and solar cells. Much of the driving force behind detector development has been the need for effective receivers for optical communications systems [3]. We will focus on devices that are useful as discrete components as well as those suitable for monolithic integration with lasers and waveguides, and we'll provide a brief overview of some of the relevant design issues. Treatment of electroabsorption and detector integration with other optoelectronic structures will be deferred to Chapters 11 and 12, respectively.

The detector configuration often most useful for semiconductor integrated optical systems is the photodiode: a pn diode is operated in reverse bias and the resultant electric field pulls the electron/hole pairs generated in the depletion region apart, giving rise to the detector current. Photon absorption in the depletion region is the heart of detector operation, and the absorption wavelength is defined by the bandgap of the material. Three types of devices are particularly interesting and will draw our attention in the last sections of this chapter: the p-i-n diode employs an intrinsic region in the pn junction; the MSM photodiode uses two back-to-back Schottky barriers; and the avalanche photodiode relies on avalanche multiplication for intrinsic amplification of the detector signal.

10.1 GENERAL PRINCIPLES

A semiconductor photodetector creates electron hole pairs by the absorption of optical radiation. The same fundamental transitions we discussed in connection with lasers in Section 9.2 are at work in a detector, and many of the material issues that define the relevant and useful wavelength range translate to the case of a photodetector as well. Detection of optical radiation using semiconductors is a well-established field with a rich research history [4], and we will skim a few of the central aspects. One relevant consideration, namely detector speed, will not be treated here, and we refer the interested reader to the extensive literature.[1]

10.1.1 Photon to Electron Conversion

In a bulk semiconductor, incident optical power, P_{in} [W], at an energy, hν [J], is converted with efficiency, η [dimensionless], into a current, I [A]. Because we can express current as n [cm^{-3}] carriers with a lifetime τ [s], the above relation implies simply

$$P_{in} \cdot \left(\frac{1}{h\nu}\right)\left(\frac{1}{\text{volume}}\right) \cdot \eta = n \cdot \left(\frac{1}{\tau}\right) \tag{10.1}$$

for an excitation in a prescribed volume.

Usually, a photodetector employs an applied electric field, $\mathcal{E}$ [V/m], which generates a current flow and thus a useful detector output signal. Because current is given by

$$I = q\mu n\mathcal{E} \cdot \text{area} \tag{10.2}$$

for electronic charge, q [C], mobility, μ [cm^2/Vs], and a given cross-sectional area, we can relate the incident optical power to the subsequently generated current as

$$I = \frac{q\mu\tau}{h\nu}\eta P_{in} \cdot \left(\frac{\mathcal{E}}{L}\right) \tag{10.3}$$

The length, L [m], is the ratio of the volume in which optical carrier generation takes place to the cross-sectional area of current flow, or the length over which the applied voltage is dropped.

A useful detector is characterized by a high overall efficiency and high sensitivity. The former criterion implies a maximum generated current for a particular P_{in}, which is achieved for large mobility, carrier lifetime, and of course a high quantum efficiency, as we can see directly from Equation 10.3. Sensitivity defines the minimum P_{in} detectable, and is usually limited by noise from internal and external sources, which give rise to a detector current unrelated to the signal of interest.

[1]See for example: Bowers, J.E., and C. A. Burrus, "Ultrawide-Band Long-Wavelength P-I-N Photodetectors," *IEEE Journal of Lightwave Technology*, Vol. LT-5, 1987, pp. 1,339–1,350.

10.1.2 Efficiency

Analogously to the laser, detector efficiency is defined as the number of electrons generated per incident photon. We can define the efficiency, η, by the simple relationship

$$\eta = \frac{h\nu}{q}\left(\frac{I}{P_{in}}\right) \tag{10.4}$$

A related quantity often used to characterize a photodetector is the responsivity, $\mathcal{R}$ [A/W], defined as

$$\mathcal{R} = \frac{I}{P_{in}} = \eta\frac{q}{h\nu} \tag{10.5}$$

Responsivity simply relates the amount of current generated in amperes to the amount of optical power incident in watts. It is usually directly read from the slope of a plot of detector output current versus input optical power. We note that responsivity varies with incident energy, decreasing as photon energy increases, whereas efficiency, η, is nominally independent of wavelength for $h\nu > \mathcal{E}_g$; this is because it is the *number* rather than energy of photons arriving that gives rise to a particular detector signal.

Example

Let's take a detector for which we have measured an output characteristic showing 500 nA generated for an optical field of 1 μW at 820 nm. Such data implies a responsivity of $\mathcal{R} = I/P = 0.5$ A/W. We can see directly from Equation 10.4 that the efficiency is given by $\eta = h\nu/q \cdot (I/P) = hc/\lambda q \cdot (I/P)$ and has a value of 76%. If we assume the same efficiency for a signal at a wavelength of 1.3 μm, we note that the responsivity rises to $\mathcal{R} = \eta \cdot q\lambda/hc = 0.80$ A/W.

10.1.3 Noise

The sensitivity of photodetectors is usually characterized by the minimum optical signal that gives rise to a current that can be distinguished from the background current due to various sources of noise. Contributions to noise currents include the dark current, I_d [A], resulting from thermal generation of carriers, either in the bulk or at the surface. The dark current is equivalent to the reverse saturation current of the pn junction (Equation 2.57), so that we can write

$$I_d = qn_i^2\left(\sqrt{\frac{D_p}{\tau_p N_D^2}} + \sqrt{\frac{D_n}{\tau_n N_A^2}}\right)(1 - e^{-qV_a/k_BT})(LW) \tag{10.6}$$

for doping levels N_D and N_A, diffusion constants D_n and D_p, and lifetimes τ_n and τ_p, for electrons and holes, respectively, plus a junction area (LW) and applied junction voltage V_a. The temperature dependence of the dark current is primarily due to the rapid increase of n_i with increasing temperature (Figure 2.11), which, if we recall that $n_i \propto \exp(-E_g/2k_BT)$ (Equation 2.21), is a particularly strong effect in narrow bandgap materials.

Because the discrete arrival of photons is a stochastic process, there is a time-varying fluctuation in the signal current that gives rise to shot noise. The mean square value of photodiode current due only to the optical signal is given by

$$\langle I_{\text{pd}}^2 \rangle = 2q\mathscr{B} \cdot I_{\text{sig}} \tag{10.7}$$

where I_{sig} is the time average value of the current due to the optical signal of interest and $\mathscr{B}$ [Hz] is the detector bandwidth [5]. We remember that the mean square value of a quantity ζ is given by

$$\langle \Delta\zeta^2 \rangle = \overline{(\zeta - \bar{\zeta})^2} \tag{10.8}$$

If we include the dark current in Equation 10.7, we obtain a total mean square photodiode current

$$\langle I_{\text{pd}}^2 \rangle = 2q\ \mathscr{B} \cdot (I_{\text{sig}} + I_d) \tag{10.9}$$

Realizing that the signal detector current is simply

$$I_{\text{sig}} = \frac{q}{h\nu}\eta \cdot P_{\text{in}} \tag{10.10}$$

and that the other noise contributions can be written similarly, Equation 10.9 can be rewritten as

$$\langle I_{\text{pd}}^2 \rangle = \frac{2q^2\eta\mathscr{B}}{h\nu} \cdot (P_{\text{in}} + P_d) \tag{10.11}$$

where P_d [W] is the power generated by the dark current. This last expression is the total detector current due to signal and noise sources. Combining the last two expressions, we are able to derive a power signal-to-noise ratio, S/N [dimensionless], which relates the signal to background currents, namely

$$\frac{S}{N} = \frac{I_{sig}^2}{\langle I_{pd}^2 \rangle} = \frac{\eta}{2\mathcal{B}h\nu}\left(\frac{P_{in}^2}{P_{in} + P_d}\right) \qquad (10.12)$$

Thus for a high S/N, as is intuitively pretty obvious, the nonsignal sources of detector current should be minimized; a decrease in temperature helps raise the signal-to-noise ratio, explaining why low temperature (liquid nitrogen) ambients are often required for the operation of sensitive infrared detectors.

A further useful parameter often quoted in the characterization of a photodetector is the noise-equivalent-power, NEP [W], which defines the magnitude of signal required so that the S/N becomes unity. Using Equation 10.12 and the assumption that the signal power, $P_{in} \ll P_d$, we obtain

$$\text{NEP} = \sqrt{\frac{2\mathcal{B}h\nu}{\eta}P_d} \qquad (10.13)$$

Because we can think of the NEP as the minimum reliably detectable signal, we can see from the last expression that this minimum input power increases with the root of bandwidth.

For the case of an avalanche photodiode, carrier multiplication affects not only the signal carriers but also any carriers from noise sources. Since avalanche multiplication is essentially a bulk effect, we separate the generated dark currents into bulk, I_{dB}, and surface, I_{dS}, components. The mean square diode current then becomes

$$\langle I_{pd}^2 \rangle = 2q\mathcal{B} \cdot (M^2 I_{sig} + M^2 I_{dB} + I_{dS}) \qquad (10.14)$$

for multiplication factor M. Thus, high dark currents can offset any amplification advantage an avalanche diode may have, limiting the application of these devices in some, particularly narrow-bandgap, materials. There are numerous further noise sources important for a complete analysis of detectors, which we have not considered. These include thermal noise, due to temperature variations of the detector or background; Johnson noise, due to stochastic carrier transport; and 1/f noise, an increase in noise intensity as energy decreases.

Example

We can analyze the noise behavior of the detector we studied in the last example. Take an input power of 1 μW at 820 nm and 300°K and an assumed bandwidth of $\mathcal{B}$ = 50 MHz. Assume a dark current of I_d = 1 μA, which with a diode resistance of 10 Ω yields a dark current power of 10 pW. From Equation 10.12, we can directly calculate the signal to noise ratio as S/N = $3 \cdot 10^4$, which is roughly 45 dB, a thoroughly usable value. If we repeat this calculation for a 0.1 nW input signal, we only reach S/N ≈ 3, about 5 dB. The low power limit for this detector is best given by the NEP for these noise sources, which, from Equation 10.13, is roughly NEP = 18 pW.

10.2 SEMICONDUCTOR DETECTOR MATERIALS

Photodetectors rely on absorption, so materials that have an energy gap, $E_g < h\nu$, for an optical signal of interest with energy $h\nu$, are required. Many materials have values for absorption coefficient in the range $\alpha \sim 10^4$ to 10^5 cm^{-1} for optical excitation above the edge of the bandgap. Since most detectors employ an absorption region of finite width for carrier generation, high values for α imply that this region can be made thin, thereby reducing device dimensions. Although higher input energies likewise lead to increased absorption and a reduced path for incident photons, excessive absorption near the semiconductor surface can reduce efficiency. This is because surface states can seriously reduce the carrier lifetime, τ in Equation 10.3. Thus, a match between material and intended application wavelength is important.

In Figure 10.1 we see the absorption spectra of several semiconductor materials commonly employed for photodetection. It is clear that at energies close to the bandgap, absorption increases rapidly with decreasing wavelength. Silicon is probably the most popular photodetector material for wavelengths below 1.1 μm; despite the indirect bandgap, superior Si technology, material purity, and device performance have proven to be decided advantages for many applications. The useful range of silicon-based photodiodes can be extended to longer wavelengths by the use of Si/Ge [6]. For the midinfrared, germanium has enjoyed a modicum of popularity for wavelengths from 600 to 1900 nm, but has a high leakage current and usually needs to be operated cooled at 77 K.

Photodetectors based on III-Vs have been developed both to provide devices suitable for integration with active optical components [8] and to extend the range of useful wavelengths. GaAs is useful as a detector for near-infrared wavelengths (< 850 nm) and has a sharper increase in absorption than Si due to its direct bandgap. A further widely employed room-temperature infrared material is InGaAs on InP [9]. This ternary material absorbs from 0.92 μm (InP) to 1.65 μm ($In_{0.53}Ga_{0.47}As$), from which it is clear that the InP substrate is transparent for the longer wavelengths. Thus, illumination from the wafer backside is possible; this is a configuration with numerous advantages for discrete detector applications. The InGaAsP and AlGaSb systems are interesting in the same wavelength range, where the quaternary material provides a bit of extra flexibility in that the material can be grown lattice matched to a number of substrate types [10].

Further into the midinfrared wavelengths, PbSe absorbs from about 1 μm to 5 μm, a regime in which InAs and InSb are also useful, albeit only at liquid nitrogen temperatures. The workhorse material for wavelengths into the far infrared is

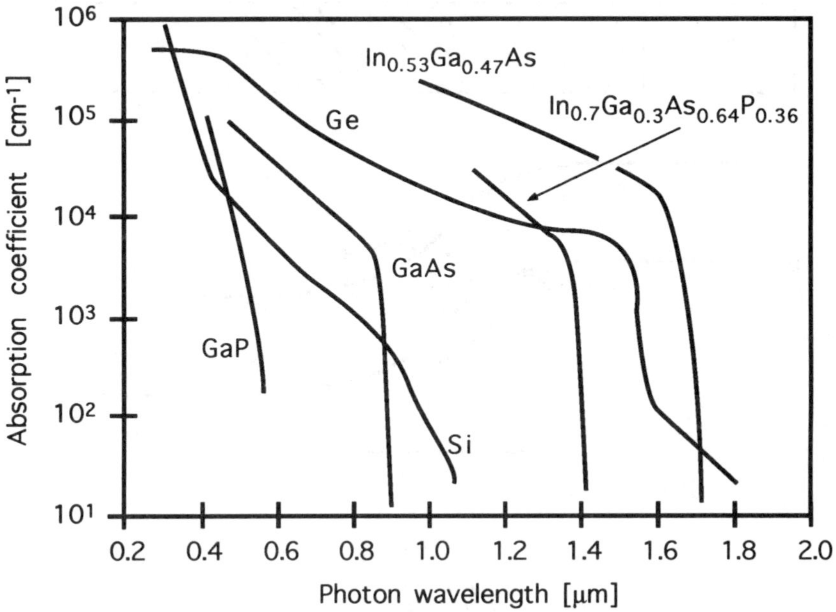

Figure 10.1 Room-temperature absorption spectra for several materials popular for the fabrication of semiconductor photodetectors. Data from [7] used with permission.

HgCdTe, whose bandgap can be reduced to zero. The use of intersubband transitions, absorption of photons which promotes an electron from one quantum-well subband to another, both within the conduction band, has allowed the use of standard III-V materials at considerably longer wavelengths [11]. Since most detectors for long wavelengths must be operated at low temperatures, the high cost usually limits their use to specialized applications.

10.3 THE P-I-N PHOTODIODE

As we saw in section 2.6.6 and have sketched in Figure 10.2, the p-i-n diode consists of an undoped, "intrinsic" region sandwiched between n- and p-doped ends; we write intrinsic in quotes because real structures have low net doping, but still at levels above n_i, and the reduced carrier concentration is usually achieved by compensation. The intrinsic region is relatively free of carriers, so the junction voltage dropped across it results in a relatively constant electric field and a depletion layer extends the entire length of the "i" region. The entire intrinsic layer then acts as a photon absorber and the thickness of the depletion region is enhanced considerably when compared to a normal pn junction. Photons absorbed in the intrinsic layer generate

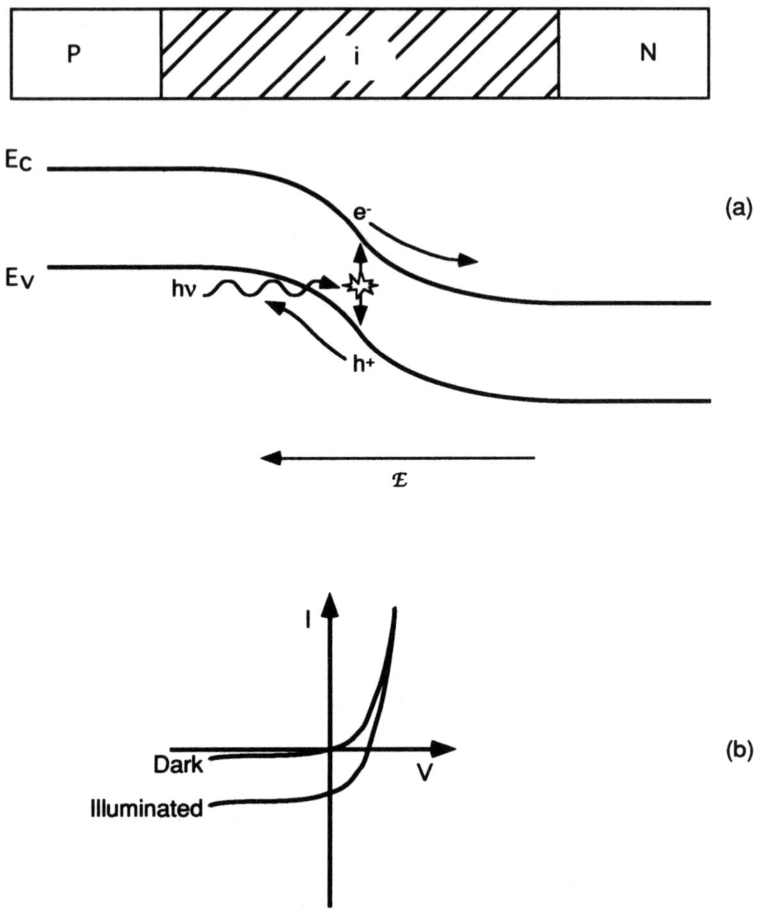

Figure 10.2 The p-i-n diode used as a photodetector. (a) The electric field due to junction bias is dropped across the central "i" region, which is fully depleted. The band diagram shows how generated electron/hole pairs are separated by the field and contribute to diode signal current when they reach the p- and n-doped regions. (b) The diode IV characteristic is shifted in response to an optical signal.

electron/hole pairs that are drawn to opposite sides of the junction by the field, where they lead to the detector current. The shift in the p-i-n diode IV characteristic along the current axis is then the photodiode output signal.

10.3.1 Waveguide Detector

Two basic p-i-n diode configurations commonly used for integrated optics applications, the waveguide and surface detector, are shown in Figure 10.3. The waveguide

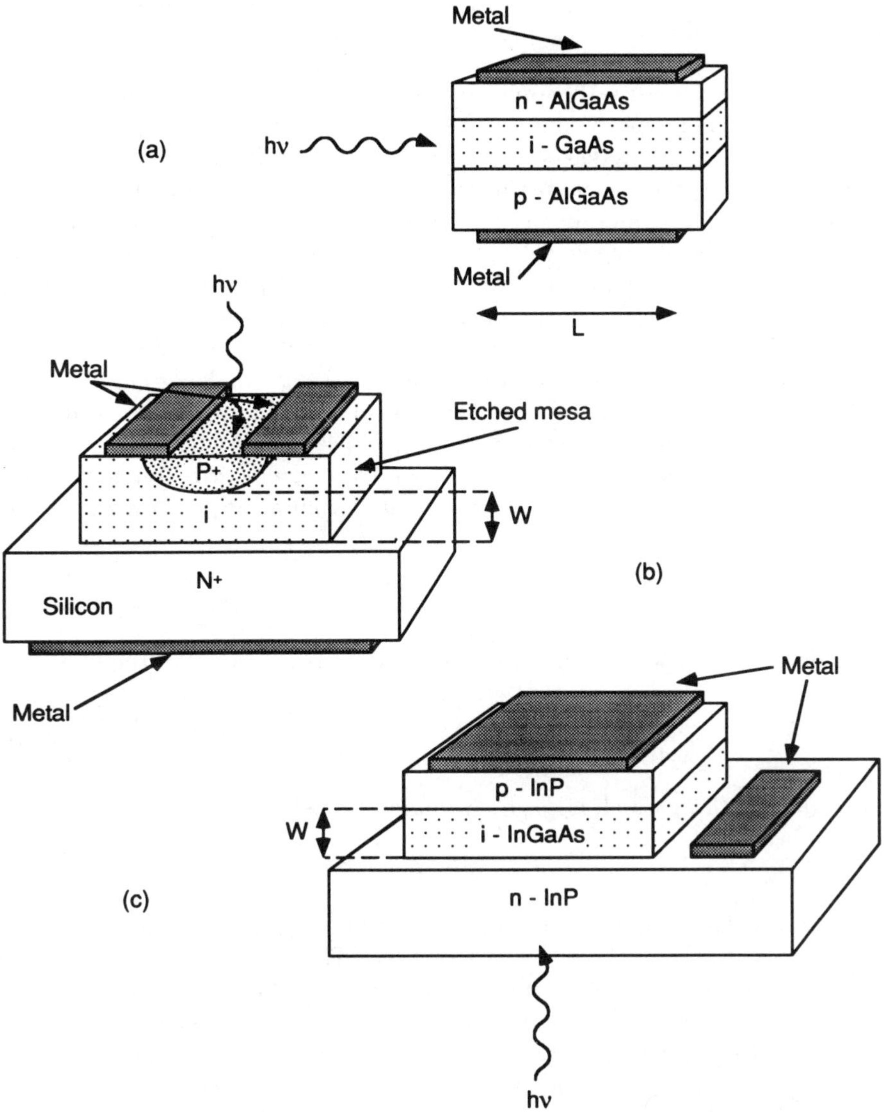

Figure 10.3 Three typical p-i-n diode detector configurations and materials useful for integrated optics applications: (a) waveguide detector using the GaAs/AlGaAs-based heterostructure layers of a waveguide or laser, (b) Si-based surface illuminated detector defined by an etched mesa that requires illumination from the top of the device, and (c) an InGaAs on InP surface mesa p-i-n detector, which can detect an optical field from the backside since InP is transparent at InGaAs absorption wavelengths.

detector, Figure 10.3(a), can employ a material layer sequence similar, or even identical, to that of a waveguide or laser; the techniques for lateral structuring of the waveguide are also directly applicable to waveguide detectors. This configuration is suitable for the high-speed detection of lightwave communications signals [12],and this form of the p-i-n photodetector is a workhorse for optical systems. In the example shown in the figure, the upper and lower cladding layers are p- and n-doped, respectively, and the core becomes the "i" region of the p-i-n diode. Since a heterostructure is often used, the absorption wavelength will be that of the core material. The optical field is incident from the side, the junction is reverse biased, and photon absorption leading to a detector current takes place along a length of waveguide L [m].

We will discuss in Chapter 12 how one goes about attaching an absorbing detector to an otherwise transparent waveguide, but let us take at face value for the moment that the detector intrinsic region has an absorption coefficient of α [cm^{-1}]. For an input **z**-directed photon flux, $\Phi_p = (P_{in}/h\nu)(1/\text{area})$ [cm^{-2}s^{-1}], defining the number of incident photons per unit time and waveguide cross-sectional area, the number of electron/hole pairs generated per unit volume at position z is given by

$$\frac{\text{carriers}}{\text{volume}} = \Phi_p \alpha \cdot e^{-\alpha z} \tag{10.15}$$

We can derive the generated detector current density J_{pd} [A/cm^2] from

$$J_{\text{pd}} = -q\int_0^L \Phi_p \alpha \cdot e^{-\alpha z} dz \tag{10.16}$$

for detector length L, which is easily evaluated to yield

$$J_{\text{pd}} = q\Phi_p(1 - e^{-\alpha L}) \tag{10.17}$$

From this last expression, it seems clear that a high overall detector efficiency or responsivity can be achieved by increasing L. Thus, even a low value of α could be compensated for by merely increasing the length of the waveguide detector. Such optimization is possible but limited by photon losses along the detector due to scattering, α_{sc}, and free carrier absorption, α_{fc}; particularly the latter is relevant, since photodetectors are by their nature doped structures and the free carrier concentration tends to be high. If we take the total loss of the detector to be given by

$$\alpha_T = \alpha + \alpha_{\text{sc}} + \alpha_{\text{fc}} \tag{10.18}$$

the detector current density becomes

$$J_{\text{pd}} = q\Phi_p\left(\frac{\alpha}{\alpha_T}\right)(1 - e^{-\alpha_T L}) \tag{10.19}$$

from which we can directly derive the efficiency

$$\eta = \frac{J_{\text{pd}}}{q\Phi_p} = \left(\frac{\alpha}{\alpha_T}\right)(1 - e^{-\alpha_T L}) \tag{10.20}$$

The maximum efficiency of a waveguide detector is thus often limited by parasitic losses, and the doping of the diode must be particularly carefully chosen so that free carrier losses do not seriously degrade performance. The relationships we have derived are useful for a quick understanding of some of the important factors in detector operation and design, but we have ignored carrier transport effects, such as the diffusion of holes into the depletion region. More extensive treatments [13] can overcome these deficiencies. Despite its simplicity, the waveguide p-i-n photodiode has proven itself to be suitable for high-performance receiver applications [14].

Example

The waveguide detector can even be useful for weakly absorbing materials. To see this, let's take a 100-μm-long waveguide detector consisting of an absorber with $\alpha = 10\ \text{cm}^{-1}$ with unity efficiency, $\eta = 1$. Assume that parasitic losses are limited to about 20 dB/cm or $\alpha_{sc} + \alpha_{fc} = 5\ \text{cm}^{-1}$. If we then have 1 μW of input power at 820 nm in a waveguide of lateral cross section $5 \times 0.2\ \mu\text{m}^2$, we find that the number of photons generated per time and area is given by $\Phi_p = P\lambda/\text{h}c(1/\text{area}) \approx 4 \cdot 10^{24}$ photons/s/m^2. Using Equation 10.19, realizing that $\alpha_T = (10 + 5)\ \text{cm}^{-1}$, we obtain a generated current density of $6.24 \cdot 10^4\ \text{A/m}^2$, leading to a current in the waveguide of 62 nA. This current level corresponds to a responsivity of $\mathcal{R} = 0.06$ A/W or an efficiency of $\eta = 9\%$. Thus, while these values are nothing to write home about, the calculation shows that weakly absorbing material can be employed in a waveguide detector. By increasing the length, efficiency can be improved: $\eta = 50\%$ for $L = 1$ mm.

The same detector made with material that has an absorption coefficient of $\alpha = 10^4\ \text{cm}^{-1}$ leads to the generation of 0.66 μA for this input power corresponding to $\mathcal{R} = 0.66$ A/W and $\eta = 1$; this latter result is no surprise since we assumed unity quantum efficiency at the outset.

10.3.2 Surface Detector

The surface detectors of Figure 10.3(b) and (c) employ essentially the same mechanism as the waveguide detector, albeit with a slightly different configuration. As we can see from the figure, a p-i-n diode is illuminated from the top or bottom; a lateral mesa etch is often used to define the detector laterally, where typical device sizes are several hundreds of microns. Illumination from the backside is of course only possible if the substrate is transparent, such as InP for InGaAs, and this is a useful arrangement for fiber coupling in communications systems [15]. The layer structure is similar to that of the waveguide detector, with photon absorption occurring in the central "i" region. However, the optical field is only absorbed for a length equal to the intrinsic layer width, W [m], which is typically less than 1 μm. In addition, Fresnel reflection at the surface reduces the intensity of the available field by a factor $(1 - R)$ for reflectance, R. Thus, the detector current density, given an input photon flux, Φ_p, for a surface detector becomes

$$J_{\text{pd}} = q\Phi_p(1 - R)(1 - e^{-\alpha W}) \tag{10.21}$$

and its associated efficiency becomes

$$\eta = (1 - R)(1 - e^{-\alpha W}) \tag{10.22}$$

Whereas the loss due to reflection can be minimized by proper antireflection coating of the surface, the thin absorption region limits the attainable efficiency of such a detector. As above, we have ignored the generation and transport of carriers due to absorption in the p- and n-regions, a phenomenon that improves efficiency somewhat while degrading high-speed performance. Surface photodiode detectors are somewhat less useful for real integrated optics than are their waveguide counterparts, but they are quite widely applied as discrete components where the potentially large detector area is an advantage.

Example

By examining a surface photodetector with the same material properties as in the previous example, we can see some of the limitations of this device. Taking the weak absorber (α = 10 cm^{-1}) and a surface reflectance of R = 0.31, we can calculate the efficiency of the device for an absorption width W = 0.5 μm using Equation 10.22. The weak absorber gives rise to η = 0.03%, a virtually unusable value. A stronger absorber (α = 10^4 cm^{-1}) permits an efficiency of η = 27%, despite the short absorption length. The surface device benefits from a large collection area, but requires efficient absorption or techniques to confine the light within the absorption width.

10.4 SCHOTTKY AND MSM DETECTORS

The rectifying behavior of the pn junction, namely the separating of the generated electrons and holes, is key to the generation of an electrical signal from an optical field. A rectifying junction may also be fabricated using metals in direct contact with a doped semiconductor [16], usually by replacing the p region with a metallic layer.

This structure is known as a Schottky barrier and forms the basis for a popular photodetector configuration.

10.4.1 Schottky Barrier

Whereas some metals form an ohmic contact when deposited on a semiconductor surface, meaning the relationship between current and applied voltage is linear for both positive and negative bias, others form a Schottky barrier. A Schottky barrier rectifies in a manner similar to the pn diode. Aluminum or Au-Ti are useful Schottky barrier metals for GaAs, on which a rectifying contact can be made with most metals. Al is also used for Si Schottky diodes, usually on lightly n-doped surfaces, and occasionally employs Pt_5Si_2 intermediary layers. Since the depletion region formation of Schottky barriers is similar to the standard pn junction, they may be used as photodetectors in either a metal-semiconductor (surface) or lateral metal-semiconductor-metal (MSM) configuration. The energy band diagram of a Schottky barrier is schematically sketched in Figure 10.4 for the case of no applied bias and ignoring semiconductor surface states. The requirement that the Fermi levels are constant across the junction results in a bending of the bands in the semiconductor. This bending gives rise to a potential barrier for electrons from the metal of energy height approximately qV_{SB} and to a depletion region that extends into the semiconductor a distance, W. The barrier height is a function of the specific metal and semiconductor

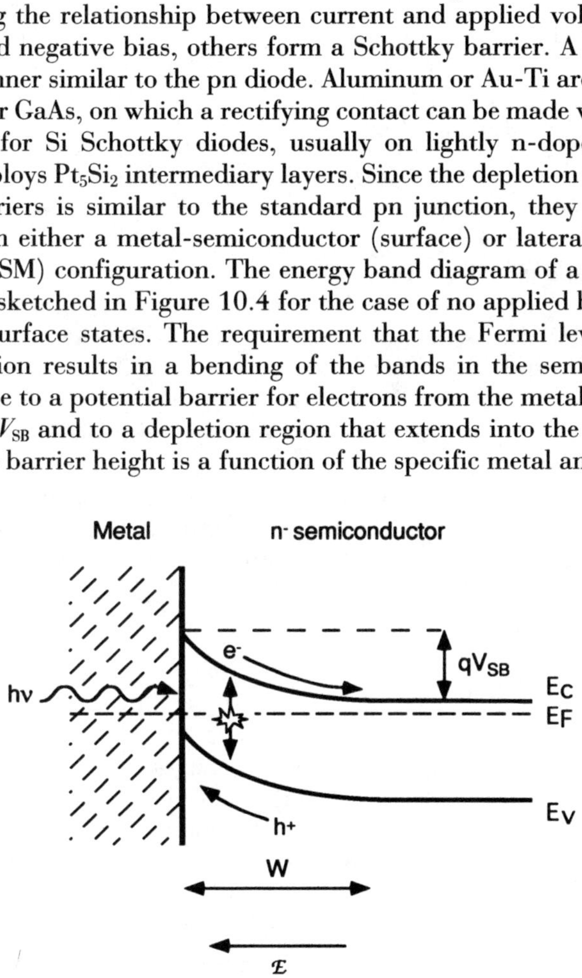

Figure 10.4 The schematic energy band diagram of an unbiased Schottky barrier diode. The particular metal and semiconductor used defines the height of the Schottky barrier, qV_{SB}, and thus the width of the depletion region, W. Carriers generated in the depletion layer are swept either to the metal (holes) or semiconductor bulk (electrons) and thus contribute to detector current.

used and is typically on the order of 0.8 V for most Schottky metals on Si or GaAs. The depletion region, and its characteristic of pulling generated electrons and holes to opposite sides of the junction where they contribute to detector current, performs the same function as that in the p-i-n photodiode, and the electrical behavior is similar.

10.4.2 MSM Detector

Two Schottky barrier photodiode configurations are typically employed, namely waveguide and surface illuminated devices. The waveguide-based structures shown in Figure 10.5 use an interlaced finger pattern of metal-semiconductor contacts, so that the detector is a metal-semiconductor-metal, or MSM, sequence [17]. The fingers of the pattern shown in the figure have alternating bias values, so that a lateral electric field is established between any pair of contacts. This field separates the generated holes and electrons, allowing the generation of a detector current. The waveguide-based MSM structure requires an optical field incident from the sides and can use a long absorption region to increase the total device efficiency; the same limitations as for the p-i-n detector apply.

The essential difference between this and the p-i-n diode is that the electric field is not vertical, as was the case for the previous device since the junction planes are displaced vertically. Instead, the field is lateral, since all the Schottky contacts are on the waveguide surface. The performance of MSM detectors for high-speed applications is usually good because the surface contacts give rise to low capacitance and thus also a reduction in noise [18].

10.4.3 Surface Illumination

The surface illuminated Schottky photodiode behaves much like the vertical p-i-n detector. Absorption takes place in either the thin metal or depletion regions, with a typical total thickness of 0.1 μm or less; the device is schematically sketched in Figure 10.6. With sufficient incident photon energy, electrons generated in the metal can surmount the barrier and transfer into the semiconductor, contributing to detector current. Efficiency of this type of detector is limited by the thin absorption region, but can be enhanced through the use of transparent or semitransparent metals for the contacts. Indium-tin-oxide (ITO) is popular for this application since it couples transparency with good electrical conductivity [19]. Antireflection coatings are also important, where dielectrics such as Si_3N_4 are common. Nevertheless, the dominance of surface absorption implies that this type of device is primarily interesting for visible and ultraviolet wavelengths.

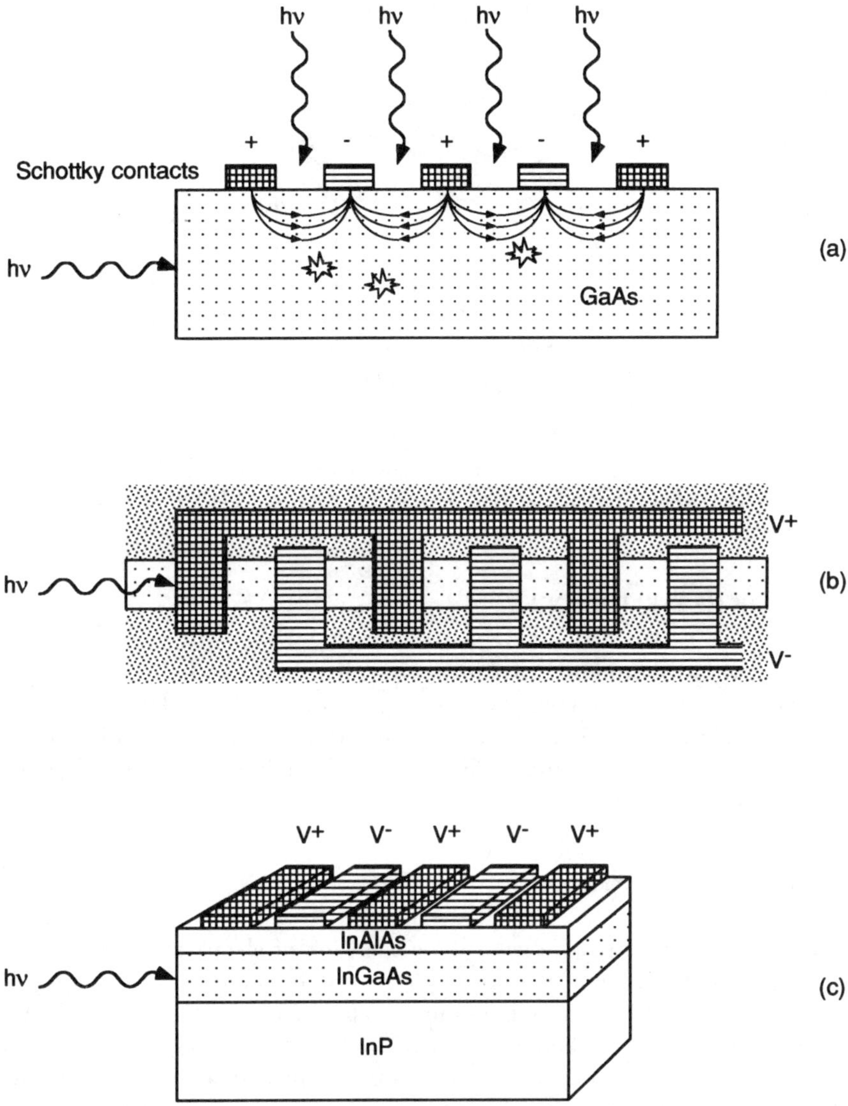

Figure 10.5 MSM Schottky diode based photodetectors: (a) cross section of a generic MSM photodetector configuration using an arrangement of fingers with alternating electrical bias on a GaAs substrate, (b) top view of a waveguide MSM finger pattern, and (c) the MSM pattern for side illumination using InAlAs with an InGaAs absorber.

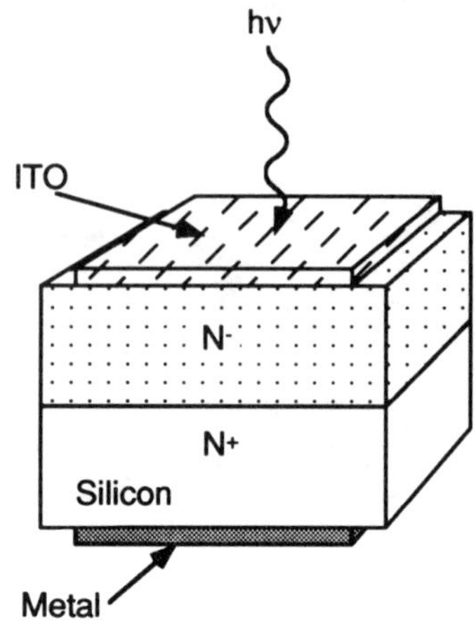

Figure 10.6 A surface Schottky photodiode using a semitransparent metal (ITO) as the electrical contact on a Si substrate.

10.5 AVALANCHE PHOTODIODES

Both of the previous photodiode structures typically require some form of amplification after the detector to reach usable circuit signal levels. The avalanche photodetector makes use of the intrinsic gain achievable through avalanche multiplication of the generated carriers and can thus yield considerably higher signal levels than devices that have a quantum yield of maximally one [20].

10.5.1 Avalanche Multiplication

Avalanche multiplication results from the generation of electron/hole pairs through impact ionization of existing electrons accelerated to high energies by a high electric field [21]. As we see from the schematic sketch of Figure 10.7, a single electron accelerated by a high electric field can smash into the crystal lattice creating a further electron, both of which can generate more electrons, until an "avalanche" of carriers propagates through the material. In the limiting case, this effect leads to junction breakdown at the breakdown voltage, V_{bd} [V]. For applied bias, V_a, just below V_{bd}, avalanche can multiply the generated carrier density, so that for a current due to photon absorption, I_{sig}, the detector current, I_{pd}, becomes

$$I_{pd} = M \cdot I_{sig} \tag{10.23}$$

The multiplication factor M [dimensionless] is given by

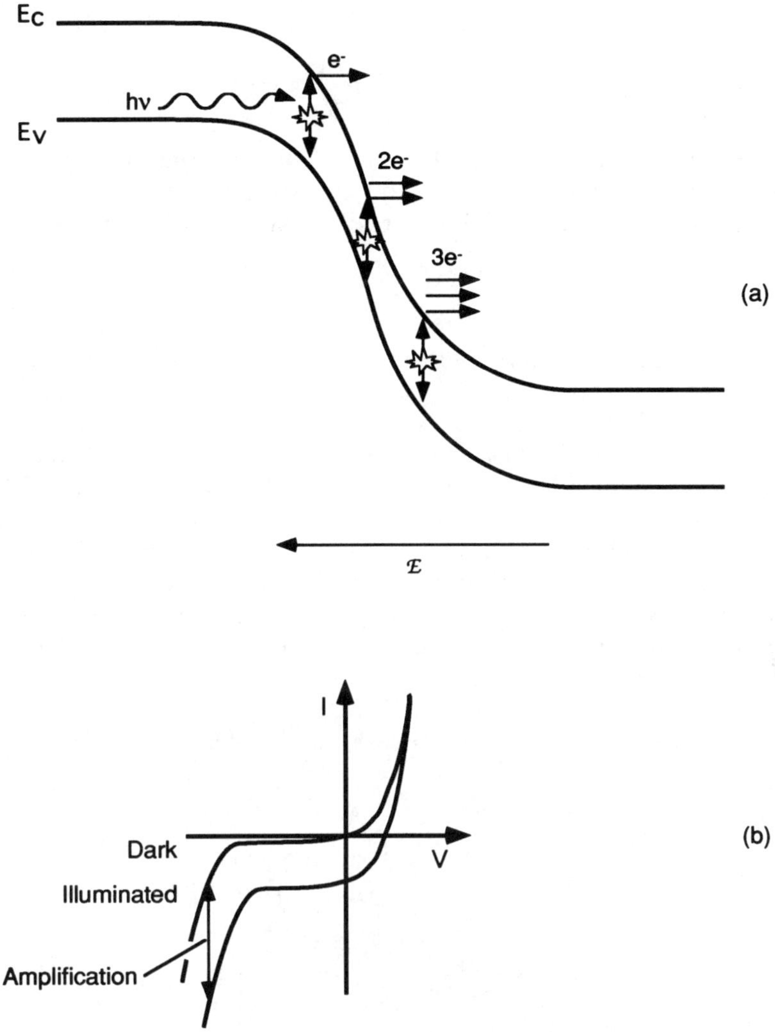

Figure 10.7 (a) The band diagram of a pn junction under high reverse bias so that avalanche multiplication of carriers takes place. An electron generated by an incident optical field will generate further electrons by impact ionization as it traverses the high field in the depletion region. (b) The IV characteristic of an avalanche photodetector showing the signal amplification due to electron multiplication.

$$M = \frac{1}{1 - \left(\frac{V_a}{V_{\text{bd}}}\right)^n} \tag{10.24}$$

where the exponential factor, n, is a function of the material, its doping, and the wavelength of interest. Values for M of 100 are typically achieved with the biases employed, the latter ranging from several tens to hundreds of volts.

10.5.2 Photodetector Structure

Most avalanche detectors are of the surface type shown in Figure 10.8; lightly doped or quasi-intrinsic avalanche regions are often used to generate the fields necessary for avalanche multiplication [22]. These electric fields are in the range $\mathscr{E} \sim 1 - 5 \cdot 10^5$ V/cm for GaAs and Si. Operation is usually at a voltage such that junction breakdown does not quite occur, unless quenching techniques are employed to generate a usable output signal [23]. Avoiding premature breakdown or the formation of microplasmas, localized regions where breakdown can cause light emission, requires high-quality material with a low defect density. Guard rings at the surface are useful to prevent breakdown due to concentration of electric field lines at the curvature of the junctions near the surface. Surface charges can also degrade the maximum usable field and increase the surface leakage.

The same materials as for the p-i-n or Schottky diodes can be used for avalanche detectors. Si has been a pronounced favorite due to the significantly higher ionization

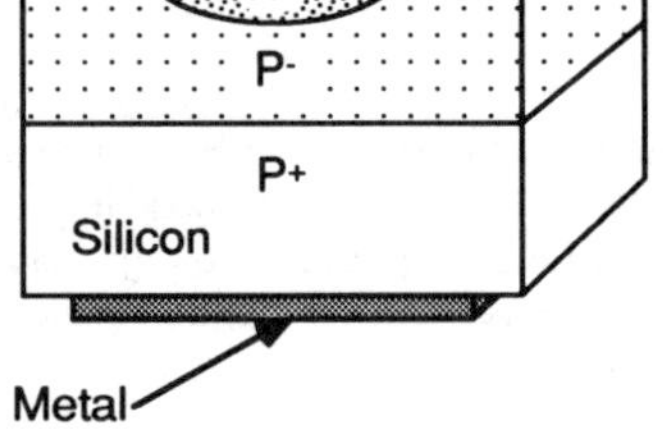

Figure 10.8 An avalanche photodiode for surface illumination; substrate material can be, for example, Si or InP. The high applied voltage results in a field close to the breakdown value in the absorber, which results in multiplication of the generated electrons due to repeated impact ionization.

rate for electrons than for holes, a requirement for low-noise avalanche detectors [24]; Si/Ge avalanche photodiodes have also been used for longer wavelengths than possible with silicon, with comparable performance [25]. Although in GaAs the ionization rates for both carriers are almost identical, the use of superlattice or staircase layer structures has been shown to be effective in artificially enhancing the ratio of electron to hole ionization [26]. Since the avalanche process amplifies both signal electrons and those created by leakage or dark current, the detector material employed must also exhibit low leakage current. Ge and InGaAs both have high leakage, the latter material due to band-to-band tunneling at high fields [27]. For this reason, more complicated structures than the ones we have discussed, such as an InGaAs absorber coupled with a reverse-biased pn junction in InP [28], often need to be used. The indirect bandgap of Si is a positive attribute in this regard, since leakage and interband tunneling is minimal.

Whereas the avalanche photodetector does have an advantage over the previous two photodetectors due to its intrinsic gain, the leverage in performance this might suggest is limited by the multiplication of leakage currents and noise contributions from the avalanche process itself. Thus the use of a low-noise external amplifier with a nonavalanche photodiode may provide comparable sensitivity. In addition, frequency response for avalanche photodiodes can be reduced due to the finite times required for the avalanche process to occur; such a limitation is, of course, critical for high-bandwidth optical systems.

REFERENCES

[1] Joyce, J., *Finnegans Wake*, London: Faber and Faber, 1939, 355.36.

[2] Moss, D., D. Landheer, D. Conn, D. Halliday, S. Charbonneau, G. Aers, R. Barber, and F. Chatenoud, "High-Speed Photodetection in a Reverse Biased GaAs/AlGaAs GRINSCH SQW Laser Structure," *IEEE Photonics Technology Letters*, Vol. 4, 1992, pp. 609–611.

[3] Brain, M., and T.-P. Lee, "Optical Receivers for Lightwave Communication Systems," *IEEE Journal of Lightwave Technology*, Vol. LT-3, 1985, pp. 1,281–1,300.

[4] Smith, R. A., F. E. Jones, and R. P. Chasmer, *The Detection and Measurement of Infra-red Radiation*, Oxford, England: Clarendon Press, 1968.

[5] Kingston, R. H., *Detection of Optical and Infrared Radiation*, Berlin: Springer-Verlag, 1978.

[6] Temkin, H., T. P. Pearsall, J. C. Bean, R. A. Logan, and S. Luryi, "Ge_xSi_{1-x} Strained-Layer Superlattice Waveguide Photodetectors Operating Near 1.3 μm," *Applied Physics Letters*, Vol. 48, 1986, pp. 963–965.

[7] Melchior, H., "Demodulation and Photodetection Techniques," in *Laser Handbook*, Arecchi, F. T. and E. O. Schulz-Dubois (eds.), Vol. 1, Amsterdam: North-Holland, 1972, pp. 725–835.

[8] Deri, R. J., "Monolithic Integration of Optical Waveguide Circuitry with III-V Photodetector for Advanced Lightwave Receivers," *IEEE Journal of Lightwave Technology*, Vol. 11, 1993, pp. 1,296–1,313.

[9] Pearsall, T. P., and M. A. Pollack, "Compound Semiconductor Photodiodes," in *Semiconductors and Semimetals*, Tsang, W. T. (ed.), Vol. 22D, San Diego: Academic Press, 1985, pp. 173–245.

[10] Stillman, G. E., L. W. Cook, N. Tabatabaie, G. E. Bulman, and V. M. Robbins, "InGaAsP Photodiodes," *IEEE Transactions on Electron Devices*, Vol. ED-23, 1983, pp. 364–381.

[11] Schneider, H., F. Fuchs, B. Dischler, J. D. Ralston, and P. Koidl, "Intersubband Absorption and

Infrared Photodetection at 3.5 and 4.2 μm in GaAs Quantum Wells," *Applied Physics Letters*, Vol. 58, 1991, pp. 2,234–2,236.

[12] Williams, A. R., A. L. Kellner, X. S. Jiang, and P. K. L. Yu, "InGaAs/InP Waveguide Photodetector with High Saturation Intensity," *Electronics Letters*, Vol. 28, 1992, pp. 2,258–2,259

[13] Sze, S., *Physics of Semiconductor Devices*, 2nd. Ed., New York: John Wiley & Sons, 1981, Chapter 13.3.

[14] Miyamoto, Y., K. Hagimoto, and K. Kato, "15 Gbit/s Optical Receiver Using Waveguide *PIN* Photodiode," *Electronics Letters*, Vol. 30, 1994, pp. 1,622–1,623.

[15] Lee, T.-P., C. A. Burrus, and A. G. Dentai, "InGaAs/InP Photodiodes for Lightwave Communications at the 0.95–1.65 μm Wavelength," *IEEE Journal of Quantum Electronics*, Vol. QE-17, 1981, pp. 232–238.

[16] Rhoderick, E. H., and R. H. Williams, *Metal-Semiconductor Contacts*, 2nd. Ed., Oxford, England: Clarendon, 1988.

[17] Soole, J. B. D., and H. Schumacher, "InGaAs Metal-Semiconductor-Metal Photodetectors for Long Wavelength Optical Communications," *IEEE Journal of Quantum Electronics*, Vol. 27, 1991, pp. 737–752.

[18] Rogers, D. L., "Integrated Optical Receivers Using MSM Detectors," *IEEE Journal of Lightwave Technology*, Vol. 9, 1991, pp. 1,635–1,638.

[19] Aliyu, Y. H., D. V. Morgan, and R. W. Bunce, "Thermal Stability of Indium-Tin-Oxide/n-GaAs Schottky Diodes," *Electronics Letters*, Vol. 28, 1992, pp. 142–144.

[20] Kaneda, T., "Silicon and Germanium Photodiodes" in *Semiconductors and Semimetals*, Tsang, W. T. (ed.), Vol. 22D, San Diego: Academic Press, 1985, pp. 247–328.

[21] Sze, S., *op.cit.*, Chapter 2.5.3.

[22] Melchior, H., A. R. Hartman, D. P. Schinke, and T. E. Seidel, "Planar Epitaxial Silicon Avalanche Photodiode," *Bell System Technical Journal*, Vol. 57, 1978, pp. 1,791–1,807.

[23] Owens, P. C. M., J. G. Rarity, P. R. Tapster, D. Knight, and P. D. Townsend, "Photon Counting with Passivley Quenched Germanium Avalanche," *Applied Optics*, Vol. 33, 1994, pp. 6,895–6,901.

[24] McIntyre, R. J., "Multiplication Noise in Uniform Avalanche Diodes," *IEEE Transactions on Electron Devices*, Vol. ED-13, 1966, pp. 164–168.

[25] Pearsall, T. P., H. Temkin, J. C. Bean, and S. Luryi, "Avalanche Gain in Ge_xSi_{1-x}/Si Infrared Waveguides," *IEEE Electron Device Letters*, Vol. EDL-7, 1986, pp. 330–332.

[26] Capasso, F., W.-T. Tsang, and G. F. Williams, "Staircase Solid-State Photomultipliers and Avalanche Photodiodes with Enhanced Ionization Rates Ratio," *IEEE Transactions on Electron Devices*, Vol. EDL-30, 1983, pp. 381–390.

[27] Forrest, S. R., M. DiDominico, R. G. Smith, and H. J. Stocker, "Evidence for Tunneling in Reverse-Biased III-V Photodetector Diodes," *Applied Physics Letters*, Vol. 36, 1980, pp. 580–582.

[28] Takanashi, T., and Y. Horikoshi, "Temperature Dependence of Ionization Coefficients for InP and 1.3 μm InGaAsP Avalanche Photodiodes," *Japanese Journal of Applied Physics*, Vol. 20, 1981, pp. 1,907–1,913.

Chapter 11

Modulators

In general, whether in long swords or in the manner of holding a sword, I dislike rigidity. Rigidity means a dead hand and flexibility means a living hand. One must understand this fully.

Miyamoto Musashi, 1645 [1]

Now that generation and detection of light belong to our palette of integrated optical capabilities, we can enhance our competence further by allowing for some controlled means to change the direction, phase, or intensity of a light beam. The amplitude of the emission from a laser, for example, can be modulated by an external device, generating the series of digital optical pulses that forms the basis for optical communications. Alternatively, we may wish to electrically change the direction of a wave propagating through a slab waveguide. The key to both of these functions, and numerous related ones, is the modulator, a device that provides considerable extra flexibility in the design of an integrated optical circuit for digital signal modulation, amplification, or switching of optical signals.

Following an overview of some modulation schemes possible with semiconductor integrated optical devices, we will focus on the electro-optic effect, the change in complex refractive index with applied electric field, since this forms the basis for many semiconductor-based modulator concepts. Quantum-well and interferometric modulators, with which we finish the section, often employ the electro-optic effect as the source of modulator action.

11.1 PHASE AND INTENSITY MODULATION

Modulation implies an induced change in an optical field. With several exceptions, some of which we will encounter below, most practical modulators are based on a shift in the real and/or imaginary refractive index of the material in which an optical field propagates. The induced index changes then often lead to variations in macroscopically measurable quantities, usually intensity or direction. We thus begin with a brief review of the effect of refractive index on propagation. Recall that the refractive index (Equation 3.42) of an optical material has a real and imaginary component,

$n' = n_R + jn_I$, from which we derived a complex propagation constant k [m^{-1}] (Equation 3.43), which is likewise complex. We can write the latter as $k = \beta + j\alpha/2$—in this case, $\beta \equiv k_z = 2\pi n/\lambda_0$ [m^{-1}], the real part of the propagation constant involving the real index refractive index $n \equiv n_R$, and α [m^{-1}] is the loss. Thus the intensity, $\mathscr{I}$, of a propagating wave, proportional to the square of the electric field, has the old familiar form

$$\mathscr{I} = \mathscr{I}_0 e^{j2kz} = \mathscr{I}_0 e^{(j2\beta - \alpha)z} \tag{11.1}$$

for propagation in the z direction. The function of a modulator is to change the index, n, or the loss, α, or both, which we can write as

$$\begin{aligned} n &\rightarrow n_0 + \Delta n \\ \alpha &\rightarrow \alpha_0 + \Delta\alpha \end{aligned} \tag{11.2}$$

for unmodulated index and loss, n_0 and α_0, respectively. Thus, the modulated intensity, $\mathscr{I}'$, after propagation for a distance, L [m], is given by

$$\mathscr{I}' = \mathscr{I}_0 \cdot e^{j\left[\frac{4\pi}{\lambda_0}(n_0 + \Delta n)\right]L} \cdot e^{-(\alpha_0 + \Delta\alpha)L} \tag{11.3}$$

From Equation 11.3, we can see that the modulation can give rise to both a phase change

$$\Delta\phi = \frac{4\pi}{\lambda_0} L \Delta n \tag{11.4}$$

and an intensity change

$$\frac{\mathscr{I}}{\mathscr{I}'} = e^{\Delta\alpha L} \tag{11.5}$$

where the ratio $\mathscr{I}/\mathscr{I}'$ is known variously as the contrast ratio, on/off ratio, or modulation depth. Using decibel (dB) units, as is often done, we can alternatively define the extinction ratio as $10 \cdot \log_{10}(\mathscr{I}/\mathscr{I}')$. In the following sections, we will see how phase and intensity modulation are achieved.

11.2 MODULATOR STRUCTURES: A BRIEF SURVEY

The goal of a modulator is usually to change the phase, intensity, polarization, or direction of light. This modulation may be accomplished by one of a wide variety of structures, depending on the exact function required and the materials and device structures available.[1] We survey a selection of the more popular modulator concepts in this section; electro-optics, interferometers, and quantum-well modulators, which represent perhaps the most advanced and popularly employed of these, will receive a more detailed treatment in subsequent sections.

11.2.1 Electro-Optic Effects

The application of an electric field, $\mathcal{E}$ [V/m], to many semiconductor materials leads directly to a change in the real and imaginary refractive indices. The shift in real refractive index with electric field, Δn, is known as electrorefraction, and the shift in the imaginary component, $\Delta\alpha$, is known as electroabsorption. The required fields are usually generated in semiconductor structures through the use of a surface Schottky barrier or p-i-n diode.

Numerous phenomena may be at work in the generation of field-induced index changes: the Pockels and Kerr effects are otherwise known as the linear and quadratic electro-optical effects, respectively [2]. The Pockels effect causes a shift in real refractive index, $\Delta n \propto \mathcal{E}$, but only exists in semiconductor materials whose crystal structure is noncentrosymmetric (i.e., not in Si). The quadratic dependence on electric field exhibited by the Kerr effect, where $\Delta n \propto \mathcal{E}^2$, is found in many materials but is relatively small in III-Vs and Si. In addition, the Franz-Keldysh effect, to which we were briefly introduced in Section 3.2.5, gives rise to *both* electrorefraction and electroabsorption, $\Delta n(\mathcal{E})$ and $\Delta\alpha(\mathcal{E})$ [3]. We will discuss all of these in greater depth in Section 11.3

11.2.2 Directional Couplers

Electrorefraction can be put to direct use for optical switching in a directional coupler modulator based on the waveguide proximity coupler we saw in Section 8.6.2. In that section, we learned that the optical field in a channel waveguide can couple into a second waveguide in close proximity, the coupling constant, κ, having been determined by a solution of the coupled mode equations of the system. For a particular coupling length, L_π [m], all of the power is transferred from one waveguide to the other; L_π is a function of waveguide spacing, width, and lateral confinement.

In Figure 11.1(a) we see a variation of the directional coupler structure, the uniform-$\Delta\beta$ modulator. In the region of optical field overlap, metal contacts have been placed on the waveguide, allowing the application of an electric field normal to the semiconductor surface. Through electrorefraction, the presence of $\mathcal{E}$ in the

[1] A more extensive treatment may be found in: Nishihara, H., M. Haruna, and T. Suhara, *Optical Integrated Circuits*, New York: McGraw Hill, 1989, Chapter 10.

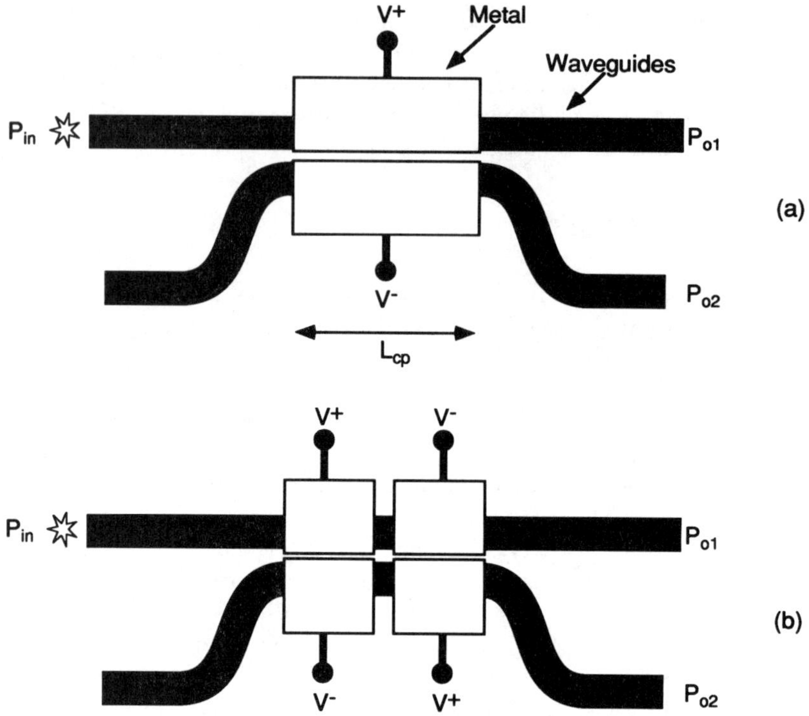

Figure 11.1 Directional couplers employing electrorefraction for waveguide optical switching; both devices have a coupling region of length L_{cp}, where metal contact pads allow application of an electric field normal to the plane of the device. (a) The uniform–$\Delta\beta$ modulator, and (b) the reverse–$\Delta\beta$ modulator.

waveguide leads to a change in the waveguide index, thus the effective index and thus finally the propagation constant, β. Therefore the coupling coefficient, κ, and the coupling length, L_π, can be electrically controlled [4].

The coupler is designed so that, in the absence of bias, all of the input power, P_{in}, exits at P_{o2}, which is known as the cross-over condition. Cross–over occurs when the physical coupler length, L_{cp}, is equal to the coupling length, L_π. By biasing one of the contact pads, the electric field changes κ and thus L_π, so that the output power is split between P_{o1} and P_{o2}. The required change in waveguide index for complete switching to take place (i.e., all power exiting at P_{o1}) can be shown to be [5]

$$\Delta n = \frac{\sqrt{3}\pi}{k_0 L_{cp}} \tag{11.6}$$

for free space propagation constant k_0 and physical coupler length L_{cp}. Switching between the two output ports is periodic in applied voltage. This is because the conditions for complete coupling to P_{o2} may again be fulfilled for a sufficiently high bias.

The uniform-$\Delta\beta$ coupler requires accurate design and fabrication to assure that the coupling length, L_{cp}, is exactly equal to L_π in the absence of bias. This requirement can be relaxed somewhat by the use of the reverse-$\Delta\beta$ coupler of Figure 11.1(b), which uses two pairs of modulator contacts with applied biases of opposite polarity; the restriction on device length is now only that $L_\pi < L_{cp} < 3L_\pi$ [6]. Using this electrode structure, cross–over (all power in P_{o2}) can be achieved for nonzero bias conditions, so that both conditions (all power in P_{o2} or all power in P_{o1}) can be adjusted electrically. The tolerances on physical device length, waveguide etch depth, and spacing to ensure that $L_{cp} = L_\pi$ are thus somewhat less stringent.

A variation on the directional coupler theme uses a waveguide in which the guided waves, in the absence of an applied bias, are close to cutoff; usually very small index steps between cladding and core in both doping-induced and heterostructure waveguides are used [7]. When a field is applied to the waveguide, the reduction in the waveguide core index leads to a cutoff condition, so that any propagating light radiates into the substrate. This configuration forms a switch similar to the directional coupler, except that in the "off" state all light is lost in the bulk of the semiconductor.

11.2.3 Total Internal Reflection Deflectors

The shift in real refractive index generated by an applied electric field may also be used in a slab waveguide configuration. The schematic of Figure 11.2 shows an optical beam propagating in a slab waveguide incident on a region covered by a metal contact to which a bias is applied. Because electrorefraction can give rise to a negative Δn, the region under the electrode can be made to have a slightly lower refractive index than the rest of the slab. An incident wave is thus reflected and, for angles greater than the critical angle, θ_c, total internal reflection takes place [8].

The behavior of waves incident on an interface was the subject of Chapter 6; using the definition of the critical angle as defined in Equation 6.4, we see that for this structure

$$\theta_c = \sin^{-1}\left(\frac{\mathrm{n}_0 - \Delta\mathrm{n}}{\mathrm{n}_0}\right) \tag{11.7}$$

for a slab waveguide of index n_0 and an electrorefractively induced change in index of $-\Delta$n. Thus, for properly chosen geometry and bias voltages, the optical beam incident on the modulator with an angle greater than the critical angle may be electrically switched between two possible output paths. For incident angles below

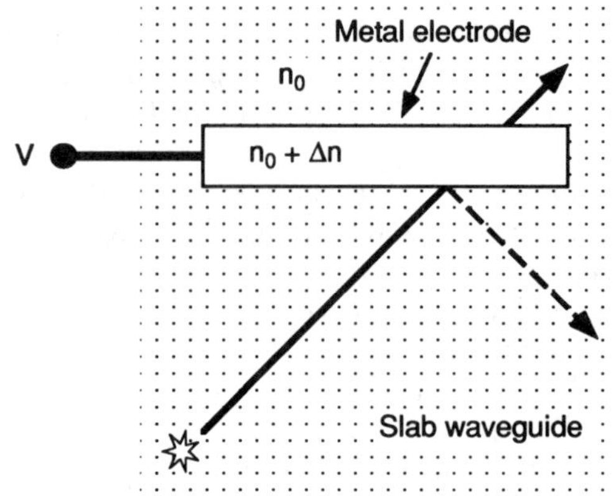

Figure 11.2 A total internal reflection deflector using a slab waveguide, as seen from above. An electro-optically induced reduction in refractive index under the metal electrode can lead to total internal reflection at the interface and thus optical switching.

the critical angle, the structure may be used to scan a transmitted beam over a range of output angles, as the transmission angle is given by Snell's law. While we have drawn a slab waveguide configuration, this scheme may also be realized using channel waveguides.

Example

To visualize the form an internal reflection modulator would take, we can calculate the critical angle for a GaAs slab waveguide with $n_0 = 3.45$ and an induced index change of $\Delta n = 10^{-3}$. Using Equation 11.7, we find $\theta_c = 88.6°$, implying that grazing incidence is required for the device to function and that the difference in propagation angle of the transmitted and reflected beams is at best about 3°.

11.2.4 Bragg Gratings

Optical beams propagating in a slab or channel waveguide may be deflected by an induced refractive index variation, which takes the form of a Bragg grating [9]. Using an interleaved finger electrode structure, as shown in Figure 11.3, we can create a periodic variation of refractive index through electrorefraction. The Bragg angle, θ_B, is dependent on the optical wavelength, λ; waveguide index, n_0; and Bragg grating period, Λ; all related by the Bragg condition

$$\theta_B = \sin^{-1}\left(\frac{\lambda}{2n_0\Lambda}\right) \tag{11.8}$$

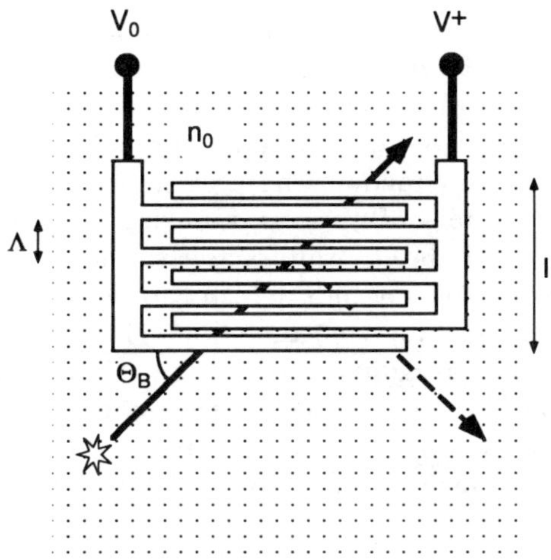

Figure 11.3 Using electrorefraction under the metal finger electrode structure, a region of periodically varying refractive index forms a Bragg grating of period Λ. The reflection of an incident light beam can be electrically switched or scanned due to interaction with the refractive index grating of length l. The Bragg angle θ_B is a function of the optical wavelength, Λ, and n_0.

Note that the magnitude of the induced index change, Δn, does not appear in Equation 11.8. The advantage encountered when using an electro-optic Bragg grating is that even small induced Δn can lead to strong reflection, due to the distributed nature of the Bragg reflection. This latter point does imply that the interaction lengths, l [m], often need to be in the millimeter range.

The induced Bragg grating may be used for beam switching in a manner completely analogous to the total internal reflection deflector. The Bragg structure, however, is wavelength dependent and can thus be used as an effective electrically controlled optical filter as well.

11.2.5 Surface Acoustic Wave Devices

Rather than use electrorefraction to induce a periodic refractive index variation in a waveguide, we can alternatively take advantage of the photoelastic effect, in which induced strain in the crystal leads to a shift in the real refractive index. By employing a piezoelectric substrate or transducer, we can electrically launch acoustic waves through the material. Because these are compressional waves, they create a region of periodically varying refractive index. This effect is the basis for acousto-optic devices, which include surface acoustic wave (SAW) and bulk acousto-optic structures [10].

A typical SAW structure is that shown in Figure 11.4. A piezoelectric transducer, using a finger electrode configuration, launches an acoustic wave along the semiconductor surface, usually to a depth of an acoustic wavelength. The sound wave, and the resultant periodic refractive index variation, propagates along a slab waveguide, again allowing an optical interaction length of several millimeters. Typical Bragg periods are one to several tens of microns, implying an acoustic frequency of 100 MHz to several GHz. As with the electrorefractively induced Bragg deflector, the long interaction length implies that the induced Δn can be small; indeed, values on the order of $\Delta n \sim 10^{-4}$ are typical for acoustic powers of 0.1 kW/cm^2.

Using a GaAs-based waveguide substrate, the transducer can be placed on the semiconductor directly [11]. For Si-based optics, piezoelectric films such as ZnO on SiO_2 or Si are commonly employed. Variations in the transducer physical design can considerably enhance the flexibility of a SAW device: using multiwavelength, curved,

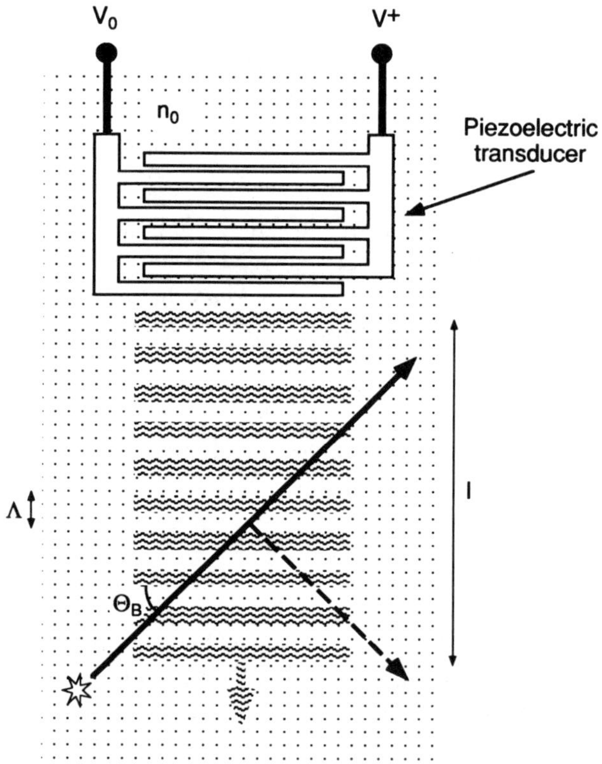

Figure 11.4 A surface acoustic wave modulator. The piezoelectric transducer launches an acoustic wave downward along the slab waveguide, where Bragg reflection takes place over an interaction length, *l*. The device can be used to switch, scan, or filter an optical input beam.

or chirped (period spatially varying) structures, this type of modulator is employed to scan or sweep an optical beam, or continuously filter the optical wavelength.

11.2.6 Interferometric Modulators

The use of an interferometer greatly enhances the performance of electro-optics for modulator applications. Since electrorefraction causes a change in real refractive index, which for a wave propagating in a waveguide implies a shift in phase, the interferometer is a useful means to translate this phase change into a modulation of intensity [12]. A workhorse for interferometric modulation is the Mach-Zehnder interferometer (see Figure 11.10), whereby the input optical field is split into two halves, one of which is subject to a change in phase. When the two halves are fed together again, interference leads to a large change in intensity as the output moves from constructive to destructive interference. Interferometers are the subject of more detailed scrutiny in Section 11.5.

11.2.7 Traveling Wave Modulators

The traveling wave configuration is not really a distinct form of modulator, but rather a refinement of some of the channel waveguide-based devices we considered above. In contrast to the modulator electrodes we considered previously, termed lumped electrodes, the traveling wave structure employs a transmission line electrode for application of the electric field [13]. Thus, the modulation electric field moves at the same velocity along the waveguide as the optical field, and the impedance of the modulator electrode is matched to the driving electronics, usually 50 Ω. The traveling wave approach is useful for microwave optical signal modulation, encountered in telecommunications, where the lumped electrode approach would limit system bandwidth.

11.2.8 Mechanical Beam Steering

The perhaps conceptually simplest approach for deflecting an optical beam, reflecting it off of a physically moving mirror, has had to await the refinements in micromachining to become a viable alternative for integrated optics applications. The use of Si machining techniques—primarily the selective etching of various layers, which allows the fabrication of micromechanical structures—has permitted the fabrication of micromirrors, made from Si or Si_3N_4. Deflection of the reflecting surface can take place electrostatically, where bandwidths of 1 MHz have been demonstrated [14]. The utility of such structures can be found both in integrated optical circuits, where deflection may be in the plane of a waveguide, and as arrays that deflect beams incident normally for free-space optical applications.

11.3 ELECTRO-OPTICS

As we found in our survey above, the changes in refractive index resulting from the application of an electric field to a semiconductor are central to the operation of numerous types of optical modulators. Electrically induced changes in material optical properties come under the general heading of electro-optics and consist of several distinct phenomena.[2] The Pockels, Kerr, and Franz-Keldysh effects combine to define the electro-optical behavior; in addition, plasma effects due to free carriers and thermo-optic effects can also play a role, particularly in silicon. In this section, we discuss these phenomena in greater detail for various material systems.

11.3.1 Pockels Effect

The linear variation of real refractive index with electric field is known as the linear electro-optic, or Pockels, effect. We saw in Section 3.6 that the dielectric constant and, by extension, the refractive index is in general crystalline-direction dependent. The same is true for the electro-optic effects [15], so that the electro-optic coefficients are usually tensor quantities.

In the absence of an electric field, the index ellipsoid defining the refractive indices in the three crystalline directions **x, y, z**, is given by

$$\frac{x^2}{n_x^2} + \frac{y^2}{n_y^2} + \frac{z^2}{n_z^2} = 1 \tag{11.9}$$

for diamond and zincblende semiconductors such as Si or GaAs. In these materials, the index values are isotropic so that $n_x = n_y = n_z = n_0$. When an electric field, $\mathscr{E}$, is applied, this relationship becomes slightly more complicated, namely [16]

$$\frac{x^2 + y^2 + z^2}{n_0^2} + 2r_{41}(\mathscr{E}_x yz + \mathscr{E}_y zx + \mathscr{E}_z xy) = 1 \tag{11.10}$$

where r_{41} is one of the three equivalent nonzero terms of the 6×3 electro-optic tensor for zincblende whose scalar value is usually employed for the definition of the Pockels effect.

By undertaking a transformation of coordinates to eliminate all the cross terms for the GaAs system, we can derive a simpler relationship.[3] If we take an electric

[2]For a thorough discussion of the Pockels and Kerr effects, see, for example: Narasimhamurty, T. S., *Photoelastic and Electro-Optic Properties of Crystals*, New York: Plenum, 1981, Chapter 8.

[3]The derivation of Equation 11.12 is not particularly obvious; Yariv has done it nicely. See, for example, Yariv, A., *Quantum Electronics*, 2nd Ed., New York: John Wiley & Sons, 1975, Chapter 14, or *Optical Electronics*, 4th. Ed., Philadelphia: Saunders, 1991, Appendix C.

field, $\mathcal{E}$, along the [100] direction (i.e., normal to the (100) surface, which is usually the case for GaAs or Si), then the index values in the three relevant orthogonal directions are [17]

$$
\begin{aligned}
n' &= n_0 - \Delta n \quad && \text{for } [110] \\
n' &= n_0 + \Delta n \quad && \text{for } [\bar{1}10] \\
n' &= n_0 \quad && \text{for } [001]
\end{aligned}
\tag{11.11}
$$

where the electrorefraction-induced change in index is given by

$$
\Delta n = \frac{1}{2} n_0^3 r_{41} \mathcal{E}
\tag{11.12}
$$

Note that, although the $[\bar{1}10]$ and [110] directions are both members of the <110> family of crystalline directions, the two are not identical electro-optically. The reader with elephantine memory will recall that in Section 5.3.4 we saw that chemically etched waveguides yield different profiles if cleaved along the (110) or (110) planes (Figure 5.9), a similar asymmetry. From the calculations above, it is clear that the electro-optical properties are also direction dependent, reflected in the sign of Δn.

The magnitude of the linear electro-optic coefficient, r_{41}, is a function of material. For GaAs, at a wavelength of 1 μm, $r_{41} \approx -1.44 \cdot 10^{-12}$ m/V with $n_0 = 3.45$. InP and InGaAsP are similar, $r_{41} \approx -1.45 \cdot 10^{-12}$ m/V with $n_0 = 3.29$ for the former, and for GaP, $r_{41} \approx -1.1 \cdot 10^{-12}$ m/V and $n_0 = 3.11$ [18]. These values are only weakly dependent on wavelength and suggest that index shifts of $\Delta n \sim 10^{-3}$ can be expected for applied fields of $\mathcal{E} = 4 \cdot 10^5$ V/cm in GaAs. A selection of coefficients for numerous materials may be found in [19]. Since it has a centrosymmetric diamond crystal lattice, silicon exhibits no linear electro-optic effect.

The Pockels effect is attractive due to the linear variation of n with applied field (and thus applied bias), and its relative insensitivity to wavelength. One can operate a modulator based on the linear electro-optic effect far from the bandgap, so that residual interband absorption is minimal.

11.3.2 Kerr Effect

The quadratic dependence of refractive index on applied electric field, $\Delta n \propto \mathcal{E}^2$, known as the Kerr effect, is of small magnitude in III-V semiconductors. The Kerr effect does exist in Si, but is similarly very weak and only measurable for very high applied electric fields. The real refractive index in this case varies quadratically with the field as

$$\Delta n = \frac{1}{2} n_0^3 s_{33} \mathcal{E}^2 \tag{11.13}$$

and the sign of the change does *not* vary with crystalline direction. In bulk GaAs, the Kerr coefficient s_{33} [m^2/V^2] varies from $s_{33} = -5.4 \cdot 10^{-20}$ m^2/V^2 at $\lambda = 1.09$ μm to $s_{33} = -4.0 \cdot 10^{-20}$ m^2/V^2 at $\lambda = 1.15$ μm [20]. We see that although there is no crystalline orientation dependence, there is a strong wavelength dependence of s_{33}; typical index changes in the wavelength range near 1.1 μm are again about $\Delta n = 10^{-3}$ for $\mathcal{E} = 4 \cdot 10^5$ V/cm, but decrease rapidly for longer wavelengths.

The Kerr coefficient in InGaAsP has been empirically shown [21] to depend on the difference between the energy of the optical field ($h\nu$ [eV]) and the bandgap E_g [eV], $\Delta E = E_g - h\nu$ [eV], as

$$|s_{33}| = 1.5 \cdot 10^{-19} e^{(-8.85 \cdot \Delta E)} [m^2/V^2] \tag{11.14}$$

illustrating the strong wavelength dependence of the effect.

11.3.3 Franz-Keldysh Effect

Whereas the Pockels and Kerr effects represent electrorefraction, the Franz-Keldysh effect, to which we were already briefly introduced in Section 3.2.5, manifests itself primarily as electroabsorption [22]. The tilting or bending of the energy bands in a semiconductor enhances the probability of tunneling for an electron in the gap, as we saw in Figure 3.5. Thus, the absorption edge is shifted to lower energies by the application of an electric field, an effect that we can represent by a reduction in the bandgap energy. In Equation 3.14, we found that the effective bandgap shift ΔE scales as $\mathcal{E}^{2/3}$, namely

$$\Delta E = -\frac{3}{2} (m^*)^{-\frac{1}{3}} (q\hbar \mathcal{E})^{\frac{2}{3}} \tag{11.15}$$

Use of the Franz-Keldysh effect for modulation is straightforward, where the usual Schottky barrier or p-i-n diode on a heterostructure waveguide for application of the electric field is often employed [23]. For operation close to the band edge, absorption can be electrically augmented by several orders of magnitude, typically from background values of several 10 cm^{-1} to several 1,000 cm^{-1}. By monitoring the current of the diode when biased into absorption, the device may also be operated as a detector.

We expect from application of the Kramers-Kronig relations (Section 3.7) that an electrically induced shift in imaginary refractive index, or absorption, will lead to a concomitant shift in real refractive index. This is indeed the case: the change in absorption, $\Delta\alpha$, due to the Franz-Keldysh effect is accompanied by a significant change in real refractive index [24]. Far below the gap energy, the variation in Δn with applied field $\mathcal{E}$ is again quadratic, taking the form

$$\Delta n = F_{fk}\,\mathcal{E}^2 \tag{11.16}$$

where F_{fk} [m^2/V^2] is a proportionality constant. For GaAs at $\lambda = 1.06\ \mu m$, $F_{fk} = 4.5 \cdot 10^{-19}\ m^2/V^2$ and at $\lambda = 1.55\ \mu m$, $F_{fk} = 7.9 \cdot 10^{-20}\ m^2/V^2$. For the quaternary $In_{0.76}Ga_{0.24}As_{0.52}P_{0.48}$ at $\lambda = 1.55\ \mu m$, we have $F_{fk} = 5.8 \cdot 10^{-19}\ m^2/V^2$ [25]. Typical values for refractive index shift at $\lambda = 1\ \mu m$ are $\Delta n = 8 \cdot 10^{-4}$ for a field of $4 \cdot 10^5$ V/cm in GaAs, and a magnitude of $\Delta n = 2.4 \cdot 10^{-3}$ is expected for the same fields at an optical wavelength of 900 nm. The magnitude of Δn can be equal to or larger than that achievable by the Pockels effect at similar electric fields in III-V materials [26]. Table 11.1 summarizes the magnitudes of the various electro-optical effects for GaAs at several wavelengths.

Table 11.1
Magnitude of the Various Electro-Optical Coefficients for GaAs at Different Wavelengths[4]

Effect	*Constant*	*λ [μm]*	*Magnitude*	*f(λ) ?*
Pockels	r_{41}	1.0	$-1.44 \cdot 10^{-12}$ m/V	Weak
Kerr	s_{33}	1.09	$-5.4 \cdot 10^{-20}\ m^2/V^2$	Strong
Kerr	s_{33}	1.15	$-4.0 \cdot 10^{-20}\ m^2/V^2$	Strong
Franz-Keldysh	F_{fk}	1.06	$4.5 \cdot 10^{-19}\ m^2/V^2$	Strong
Franz-Keldysh	F_{fk}	1.55	$7.9 \cdot 10^{-20}\ m^2/V^2$	Strong
QCSE linear TE	r_{41qw}	0.9	$-1.75 \cdot 10^{-12}$ m/V	Weak
QCSE linear TM	r_{41qw}	0.9	~ 0 m/V	Weak
QCSE quadratic	s_{33qw}	$\lambda_g - 10$ nm	$1.4 \cdot 10^{-18}\ m^2/V^2$	Strong

As we have seen, both $\Delta\alpha$ and Δn are strongly wavelength dependent, and operation of a modulator based on the Franz-Keldysh effect is usually at wavelengths quite close to, but under, the bandgap. For wavelengths much longer than the bandgap, the magnitude of absorption and real refractive index change for a particular bias decreases. Finally, the effect is dichroic in GaAs and InGaAsP heterostructures, meaning that for $\mathcal{E}$ applied in the [100] direction, a TM optical field (polarized along $\mathcal{E}$) is more strongly affected than a TE field (polarized perpendicularly to $\mathcal{E}$) [27]. This difference may be used to fabricate a polarization mode filter.

Example

Let's use the Pockels, Kerr, and Franz-Keldysh effects to design a GaAs-based directional coupler modulator, the $\Delta\beta$ modulator of Figure 11.1(a). We will apply a field

[4] λ_g is the absorption edge of the quantum well. The last column qualitatively expresses the wavelength dependence of the effect. Numerical values are from literature references in the text.

of $\mathcal{E} = 4 \cdot 10^5$ V/cm, use a wavelength of $\lambda = 1.55\ \mu$m, and assume $n_0 = 3.45$. We want to know the necessary device length, L_π, for complete switching from one output to the other. Using Equation 11.6, we can apply

$$L_\pi = \frac{\sqrt{3}\pi}{k_0 \Delta n}$$

after we determine Δn for each effect.

Using $|r_{41}| = 1.44 \cdot 10^{-12}$ m/V for the Pockels effect, Equation 11.12 gives us an expected $\Delta n = 1.2 \cdot 10^{-3}$ leading to a coupling length $L_\pi = 1.1$ mm, a reasonable size. If we assume the Kerr coefficient in GaAs to vary with wavelength in roughly the same way as InGaAsP (Equation 11.14), we can use an approximate value $|s_{33}| = 8 \cdot 10^{-22}$ m²/V² leading to, using Equation 11.13, $\Delta n = 2.6 \cdot 10^{-5}$. This would require a coupler size of $L_\pi = 51$ mm, definitely not reasonable (recall that much GaAs processing is still done on 2″ = 50-mm wafers). Finally, taking advantage of the Franz-Keldysh effect at $\lambda = 1.55\ \mu$m, for which $|F_{fk}| = 7.9 \cdot 10^{-20}$ m²/V² and through Equation 11.16, $\Delta n = 1.3 \cdot 10^{-4}$, we obtain $L_\pi = 10$ mm, a tad large but still usable if absolutely necessary.

Example

If we use the Franz-Keldysh effect to make a very efficient GaAs modulator close to the bandgap, with $\lambda = 900$ nm, we find an index shift of $\Delta n = 2.4 \cdot 10^{-3}$ at an applied field of $\mathcal{E} = 4 \cdot 10^5$ V/cm. We expect to be able to make a directional coupler modulator with $L_\pi = 57\ \mu$m, a compact device. If, however, we use Equation 11.15 to determine the expected shift of the bandgap with applied electric field, we find that $\Delta E = 0.18$ eV, implying that the absorption edge has moved to 996 nm. Thus, our operating wavelength is well within the absorption band of the semiconductor, and it is unlikely that anything will emerge from either waveguide of the directional coupler. If, on the other hand, we are able to settle for intensity modulation for which only a single waveguide with modulator is required, the induced electroabsorption would turn out to be quite useful.

11.3.4 Plasma Effects

We saw previously, in Sections 3.4.2 and 8.1.3, that the refractive index may be altered by changing the carrier concentration; indeed, we were able to design Si-based waveguides that employ the higher refractive index of a more lightly doped material to achieve waveguiding. By electrically or optically inducing changes in carrier concentration, the changes in real and imaginary refractive index due to these plasma effects are useful for optical modulation. Extending Equation 8.3, we may calculate the carrier concentration dependence of Δn and Δα from [28]

$$\Delta \mathrm{n} = -\frac{q^2\lambda_0^2}{8\pi^2c^2\mathrm{n}_0\epsilon_0}\left[\frac{\Delta n}{m_e^*} + \frac{\Delta p}{m_h^*}\right] \tag{11.17}$$

and

$$\Delta \alpha = \frac{q^3\lambda_0^2}{4\pi^2c^3\mathrm{n}_0\epsilon_0}\left[\frac{\Delta n}{m_e^{*2}\mu_e} + \frac{\Delta p}{m_h^{*2}\mu_h}\right] \tag{11.18}$$

In these expressions, n and p [cm^{-3}] are the electron and hole concentrations, respectively, and μ_e and μ_h [$\mathrm{cm}^2/\mathrm{Vs}$] are the corresponding electron and hole mobilities. Equation 11.17 and 11.18 show that the index shifts vary linearly with carrier concentration, although we have ignored band filling and emptying, two phenomena that can account for a significant fraction of induced index change [29].

In InP at $\lambda = 1.15\ \mu\mathrm{m}$, a shift in real refractive index of $\Delta \mathrm{n} \sim 10^{-3}$ for an injected carrier concentration of $n = 10^{18}\ \mathrm{cm}^{-3}$ has been determined [30], and at a slightly longer wavelength, $\lambda = 1.52\ \mu\mathrm{m}$, a proportionality of $\Delta \mathrm{n} = 3.63 \cdot 10^{-21} \cdot n$ (for n in cm^{-3}) is found for n-type doping [31]. The variation for holes is slightly less strong, due to their higher effective mass. Since the concomitant losses are essentially due to free carrier absorption, plasma-based modulators provide a means for significant real index modulation with minimal additional absorption [32]. A primary drawback is speed, limited by recombination lifetime, although the use of a depletion region by reverse biasing a heterostructure pn diode somewhat overcomes this restriction [33].

11.3.5 Silicon Electro-Optics

We hinted repeatedly in the previous three sections that the electro-optic effects being considered were either nonexistent or very weak in silicon; indeed, the Pockels, Kerr, and Franz-Keldysh phenomena are generally used with III-V materials. However, carrier concentration–induced effects, in addition to thermo-optic effects, allow the design of some Si-based optical modulators, albeit with some limitations in performance. As these are not strictly electro-optical effects, we will group them under "electro-optics," since they are often electrically activated.

In Si, the linear electro-optic effect is zero due to crystal symmetry. Some measurements have shown that the Pockels effect may be induced in Si through high temperature poling, yielding coefficients on the order of 10^{-12} m/V, but the results are preliminary [34]. The quadratic effect is quite weak, with index changes of Δn $\sim 10^{-4}$ seen for fields of $\mathcal{E} \sim 10^6$ V/cm, although the sign of the index change is actually uncertain. Calculations by Kramers-Kronig have shown that the Franz-Keldysh effect, for which Si/Ge can be used to tune the bandgap, gives index shifts of $\Delta n \sim 1.3 \cdot 10^{-5}$ for $\mathcal{E} \sim 10^5$ V/cm at $\lambda = 1.06$ μm, with a roughly quadratic dependence on field [35].

The small magnitude of these phenomena is offset by the ability to achieve relatively large index changes through plasma effects in silicon. For an injected electron density of 10^{18} cm^{-3}, the shift in real refractive index is seen to be $\Delta n = 0.9 \cdot 10^{-3}$, with only about 1 cm^{-1} of additional loss. Rather high current densities, on the order of kA/cm^2, are required for usable modulation, and the device speed is limited by the carrier recombination lifetime; GaAs actually has an advantage in this respect, due to its direct bandgap. Carrier generation by optical means can also be employed [36], and the thermal contributions to the induced index changes must usually be taken into account as well [37].

One can take advantage of purely thermal effects and use the shift in refractive index due to changes in temperature to produce optical modulation. The change in index is roughly $\Delta n/\Delta T \approx 10^{-5}$ K^{-1} for SiO_2 [38], so that a rather large temperature modulation is required for a usable effect; in silicon, $\Delta n/\Delta T$ is about $2 \cdot 10^{-4}$ K^{-1}. Metal thin film heaters, usually Cr, have been used for thermo-optic modulation in silicon-based systems, although the input powers required are typically in the range of Watts and the response times are slow, in the 100 μs regime.

11.4 QUANTUM-WELL MODULATORS

Quantum wells have become standard fixtures in many III-V-based optoelectronic layer structures; we have seen that they are useful for improving laser performance, reducing the threshold current, and allowing accurate tuning of the emission wavelength. For electro-optical modulation, the efficiency of Δn and $\Delta\alpha$ variation with applied field are significantly enhanced by the presence of a quantum well, so that multiple quantum-well modulators are of considerable interest for research and applications [39].

11.4.1 Quantum-Confined Stark Effect

The quantum well, to which we were first introduced in Section 2.7.2, has a thickness on the order of the electron wavelength and thus forms discrete electron and hole subbands, as shown in Figure 11.5. Interband transitions, such as photon absorption, take place between the subbands, at an energy, $\mathcal{E}_\alpha$. When an electric field is applied, the bands are tilted, as shown in Figure 11.5(b), whereupon the electrons and holes separate in real space and approach each other in energy, so that $\mathcal{E}'_\alpha < \mathcal{E}_\alpha$. In addition, an exciton, essentially an electron and hole bound together by their coulombic

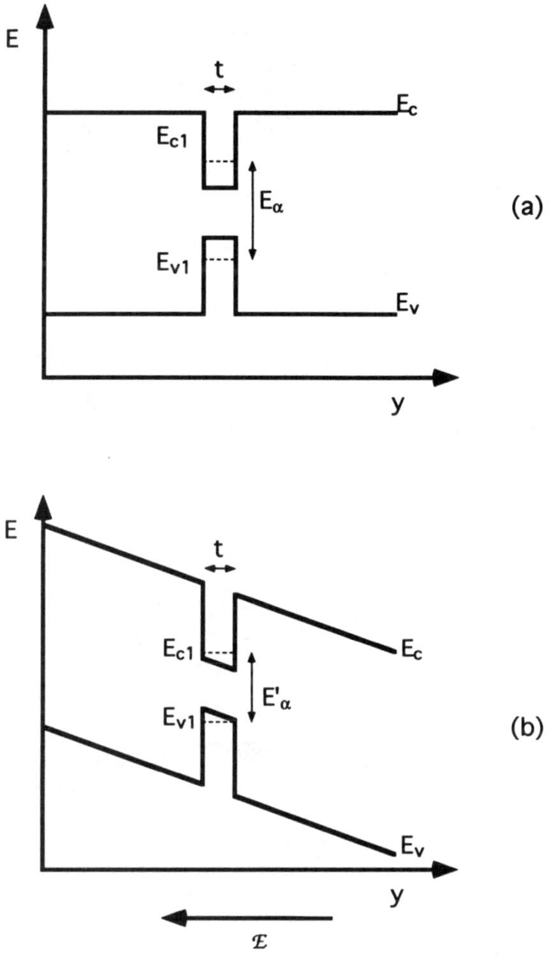

Figure 11.5 A quantum well of thickness *t* and with discrete energy subbands for electrons and holes (a) separated by $\mathscr{E}_\alpha$ without applied bias, and (b) separated by $\mathscr{E}'_\alpha < \mathscr{E}_\alpha$ for an applied electric field.

attraction, can form and exist stably at room temperature and with a high electric field; the presence of excitons leads to numerous optical effects [40].

Of particular interest for quantum-well modulator applications is the enhanced change in absorption with applied electric field for wavelengths close to the bandgap; this phenomenon is referred to as the quantum-confined Stark effect [41] and leads to strong, polarization-dependent electroabsorption [42]. The change in position of the absorption edge to lower energies with applied electric field is shown schematically in Figure 11.6(a). From this quantum-confined electroabsorption, Kramers-Kronig again implies an electrorefractive effect operating in parallel, and the real refractive index change is indeed seen to vary quadratically with applied field [43], as shown schematically in Figure 11.6(b). Although the quantum well gives rise to a variety of additional electro-optical properties [44], we are primarily interested in $\Delta\alpha$ and Δn for optical modulation.

11.4.2 Electrorefraction in Quantum Wells

The shift in real refractive index in a quantum well, Δn, due to the quantum-confined Stark effect and an applied electric field, is strongly dependent on wavelength and polarization. We may write the change in index as

$$\Delta n = \frac{1}{2} n_0^3 (\pm \Gamma_{qw} r_{41qw} \mathscr{E} + \Gamma_{qw} s_{33qw} \mathscr{E}^2) \tag{11.19}$$

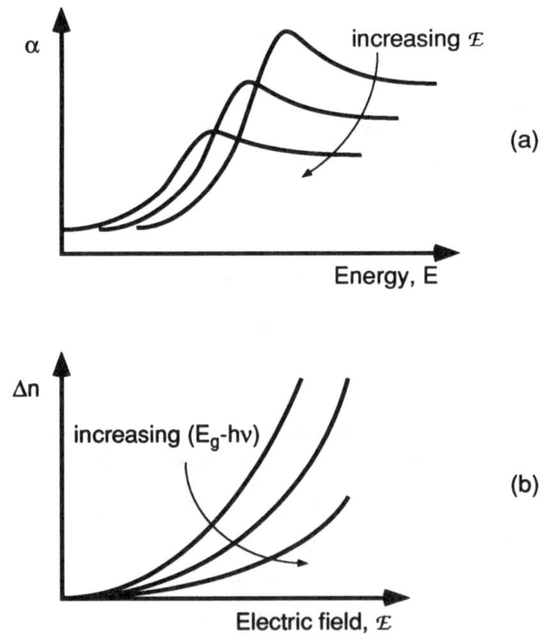

Figure 11.6 Schematic plots of the variation of the (a) absorption edge, and (b) the real refractive index shift with applied electric field for the quantum-confined Stark effect. The peak in the absorption spectrum is due to exciton absorption.

for TE polarization, and

$$\Delta n = \frac{1}{2} n_0^3 (\Gamma_{qw} s_{33qw} \mathcal{E}^2) \tag{11.20}$$

for TM polarization, where r_{41qw} and s_{33qw} are the quantum-confined Stark effect electro-optic coefficients and Γ_{qw} is the confinement factor of the quantum wells. In the first term of Equation 11.19, + is for propagation in the $[\bar{1}10]$ direction and − for the [110] direction. The polarization dependence, reflected in the absence of a linear component for TM polarized waves, is due to the allowed heavy-hole and light-hole transitions for TE polarization, but only the light-hole transitions for TM [45].

The linear coefficient, which is relatively independent of wavelength, is determined [46] to be $r_{41qw} = -1.75 \cdot 10^{-12}$ m/V for the quantum-confined case, compared with the values $r_{41} = -1.44 \cdot 10^{-12}$ m/V for bulk material, both in GaAs/AlGaAs; most other materials also show virtually no enhancement over bulk for the linear term. The quadratic term is very wavelength dependent, peaking for wavelengths close to the quantum-well absorption edge. Quantum-confined values in GaAs of $s_{33qw} = 1.4 \cdot 10^{-18}$ m^2/V^2 at wavelengths 10 nm below the absorption edge are considerably larger than the magnitudes found in bulk GaAs, $|s_{33}| = 5.4 \cdot 10^{-20}$ m^2/V^2, measured at $\lambda = 1.09$ μm. In Figure 11.7, we see the variation in s_{33qw} with distance from the band-edge for 8.5-nm GaAs quantum wells with $Al_{0.3}Ga_{0.7}As$ barriers. In the InGaAs/InAlAs system, $s_{33qw} = 3 \cdot 10^{-18}$ m^2/V^2 for TE polarization at $\lambda = 1.55$ μm [47], for InGaAsP/InP, $s_{33qw} \approx 3 \cdot 10^{-18}$ m^2/V^2 at $\lambda = 1.537$ μm [48], and for strained layer InAs/GaAs measured at the exciton resonance, $s_{33qw} = 3 \cdot 10^{-17}$ m^2/V^2 results [49]. The exact values for the quadratic coefficient s_{33qw} are not only dependent on wave-

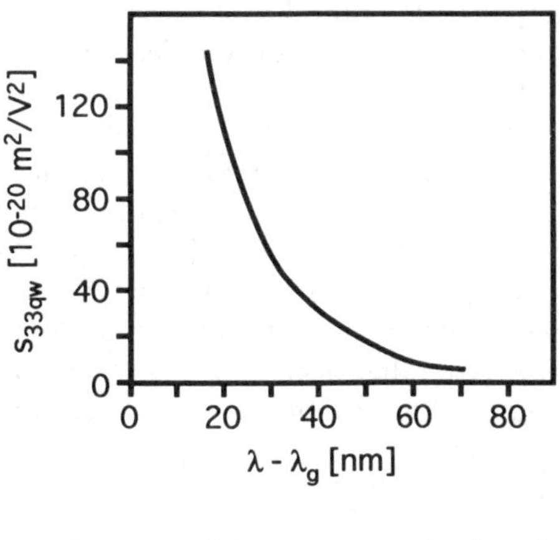

Figure 11.7 Variation of the quadratic term of the quantum-confined Stark effect, s_{33qw}, with wavelength distance from the absorption edge. These values were measured for 8.5-nm GaAs quantum wells separated by 8.5-nm $Al_{0.3}Ga_{0.7}As$ barriers. Data from [46] used with permission.

length, but also to some extent on the quantum-well width and shape [50], allowing some design flexibility. We see thus that, for wavelengths close to the bandgap, the quadratic term dominates the refractive index change, and is considerably stronger than in no-quantum well material.

An important factor, however, is the confinement factor, Γ_{qw}, which appears in Equation 11.19 and 11.20. Since the relevant optical processes for the quantum-confined Stark effect take place in the quantum wells, the overlap of the optical field with the quantum-well width defines the strength of the interaction [51]. We thus refine our definition of confinement factor (Equation 7.72) to include only the quantum wells, namely

$$\Gamma_{qw} = \frac{\int_{qw} \mathscr{E}_x^2(y)\,dy}{\int_{-\infty}^{\infty} \mathscr{E}_x^2(y)\,dy} \tag{11.21}$$

where qw implies an integration only over the quantum-well thickness, for an optical field $\mathscr{E}_x$. Since Γ_{qw} in this case can easily be $\Gamma < 0.1$, the factor is important for accurate calculation of total index shift.

11.4.3 Chirp

From the earlier discussion, we saw that both electroabsorption and electrorefraction are components of the quantum-confined Stark effect. Because most applications require either pure intensity or phase modulation, the chirp, defined as the ratio $\Delta n/\Delta k$, where $\Delta k = \lambda_0 \Delta\alpha / 4\pi$ is the extinction coefficient, will often need to be determined. Chirp in lasers is often referred to as the linewidth enhancement factor, as we saw in Section 9.4.2, with typical magnitudes of about five. Chirp in quantum-well modulators increases roughly linearly with wavelength above the absorption edge, varying from $\Delta n/\Delta k \sim 1$ at $\Delta\lambda = 10$ nm to $\Delta n/\Delta k \sim 3$ at $\Delta\lambda = 30$ nm for TE polarization in 9.4-nm GaAs quantum wells [52]; values for TM are lower.

Acceptable values for chirp are a function of application. Electroabsorption modulators for optical communications, in which the shift in wavelength due to a change in real refractive index is undesirable [53], require low chirp values. If, on the other hand, we are using the quantum-confined Stark effect for purely phase modulation, a field-induced increase in loss will tend to degrade modulator contrast, so that a large chirp is required. It has been found that maximal chirp can be achieved for an operating wavelength 30 to 50 nm below the bandgap [54], a point at which electroabsorption becomes small.

Example

We can examine the performance expected for a quantum-confined Stark effect modulator at a wavelength of 830 nm with TE polarization. Let's use five 4-nm

GaAs quantum wells with 10-nm spacers in a 125-nm-thick $Al_{0.3}Ga_{0.7}As$ core region; for this structure, $\Gamma_{qw} = 0.067$. This well thickness gives an absorption edge at 800 nm, so Figure 11.7 suggests that $s_{33_{qw}} \sim 5 \cdot 10^{-19}$ m^2/V^2 for $\Delta\lambda = 30$ nm. If we apply a bias of 5 V, the resultant field is $\mathscr{E} = 4 \cdot 10^5$ V/cm, and we find from Equation 11.19, assuming the linear effect is negligible, that $|\Delta n| = 1.14 \cdot 10^{-3}$; this is a nice large value.

At $\Delta\lambda = 30$ nm, however, we expect a chirp value of about $\Delta n/\Delta k = 3$. From this we can calculate $\Delta\alpha = 4\pi\Delta k/\lambda_0 = 58$ cm^{-1}, which leads to $58 \cdot (10/\ln 10) \approx 2.5$ dB/100 μm. Thus we expect to lose about a factor of two in intensity for every 100 μm of modulator length, which places limits on our device size.

11.4.4 Quantum-Well Modulator Structures

A multi-quantum-well waveguide modulator relying on the quantum-confined Stark effect employs the standard channel waveguide configuration shown in Figure 11.8; without the quantum wells, the structure would, of course, operate as a bulk electro-optic modulator. The relative amounts of amplitude and phase modulation, as should be clear from the earlier discussion, is a function of the quantum-well width and shape, as well as barrier height and operating wavelength. Optimization of layer structures has been done for the GaAs/AlGaAs [55] as well as InGaAsP/InGaAs/InP systems [56]. In addition, parabolically graded wells [57], or strain, in the InGaAs/InGaAlAs [58] or InGaAsP/InP [59] systems have been shown to provide polarization independence, removing the modulation difference between TE and TM input fields.

The output characteristic of a quantum-confined Stark effect phase modulator is shown in Figure 11.8. This ridge-waveguide-based device employed five GaAs

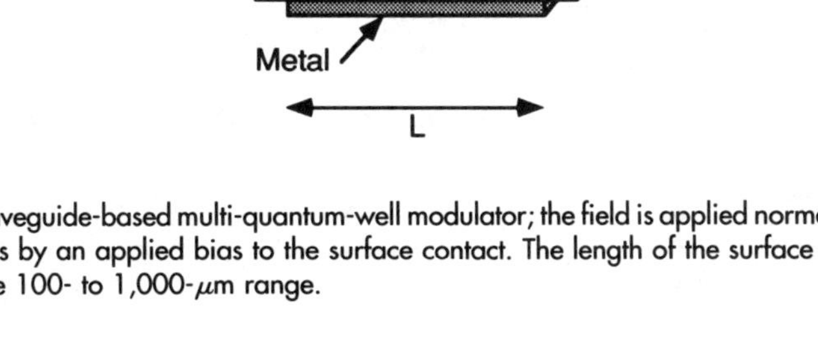

Figure 11.8 A waveguide-based multi-quantum-well modulator; the field is applied normal to the waveguide layers by an applied bias to the surface contact. The length of the surface contact is typically in the 100- to 1,000-μm range.

quantum wells of 4-nm thickness in a 260-nm-thick $Al_{0.3}Ga_{0.7}As$ core; the absorption edge for this quantum-well width is at 800 nm and excitation was at 840 nm. We see from the figure that a modulation depth of about 25 dB is reached at an applied reverse bias of 10 V.

11.5 INTERFEROMETRIC MODULATORS

A large fraction of modulator applications require intensity modulation—the use of lasers for telecommunications, for example, requires that we digitally modulate the emitted light at GHz frequencies to send information. The various types of phase modulation we have seen in the past few sections can be translated into efficient intensity modulation through the use of an interferometer. The Mach-Zehnder interferometer is popular for switching applications [60] and has been operated with modulation frequencies exceeding 10 GHz in InGaAsP on InP [61].

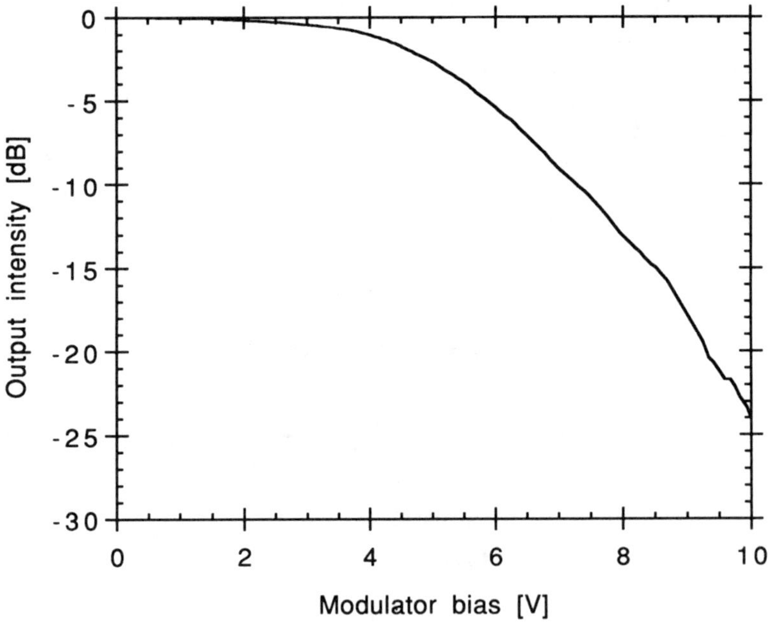

Figure 11.9 A quantum-confined Stark effect intensity modulator in GaAs/AlGaAs using five quantum wells with a width of 4 nm and an excitation wavelength of 840 nm, which is 40 nm longer than the zero-bias absorption edge.

11.5.1 Output Characteristics

In a Mach Zehnder interferometer, as shown in Figure 11.10(a), an input optical field is equally split into two waveguide arms, one of which contains a phase modulator. An applied bias on the modulator leads to an index shift, Δn, leading to a subsequent shift in propagation constant, $\Delta\beta$, and thus phase, $\Delta\phi$. At the output, the two optical fields combine and interfere with one another; for no relative phase shift, $\Delta\phi = 0$, constructive interference leads to a maximum output intensity, and for $\Delta\phi = \pi$, destructive interference implies minimum output power. As we derived in Section

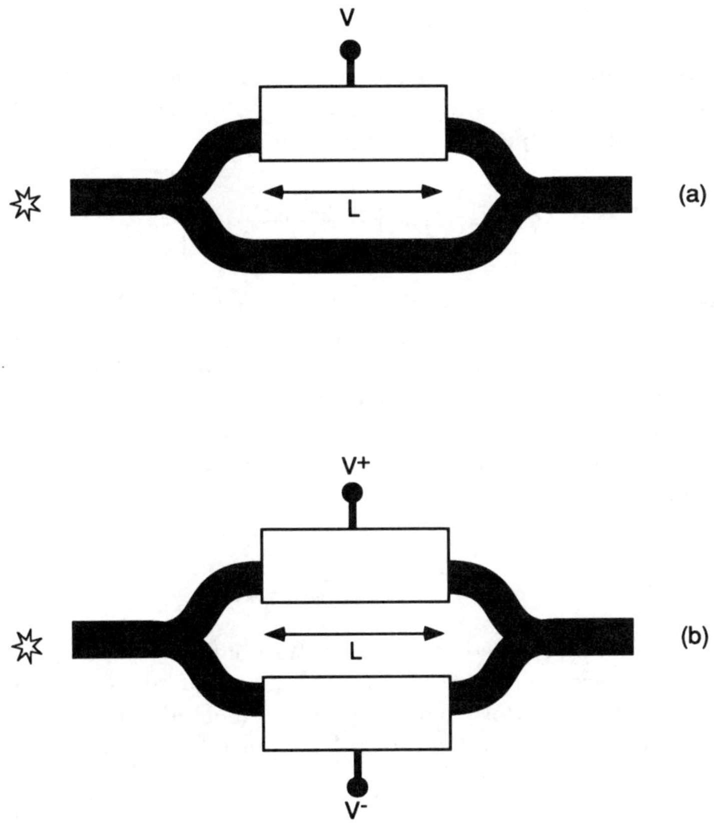

Figure 11.10 The Mach-Zehnder interferometer translates the phase change induced by an electrorefractive modulator into an optical intensity change at the output; (a) a single phase modulator gives rise to a phase difference between the two arms, and (b) the push-pull configuration uses two modulators driven with opposite polarity.

1.4, the time average of output power density, S_T [W/m^2], varies with phase difference between the arms $\Delta\phi$ as

$$\langle S_T \rangle = S_0 \cos^2\left[\frac{\Delta\phi}{2}\right] \tag{11.22}$$

By including the phase shift as a function of induced Δn over a modulator length, L, as given by Equation 11.4, in Equation 11.22 we can rewrite this simply as

$$\langle S_T \rangle = S_0 \cos^2\left[\frac{\pi L \Delta n}{\lambda_0}\right] \tag{11.23}$$

If the real refractive index change is generated by, for example, the quadratic component of the quantum-confined Stark effect as given by Equations 11.19 or 11.20, then the interferometer output power varies with applied modulator field $\mathcal{E}$ as

$$\langle S_T \rangle = S_0 \cos^2\left[\frac{\pi L n_0^3 \Gamma_{qw} s_{33qw}}{2\lambda_0}\mathcal{E}^2\right] \tag{11.24}$$

The required electrical bias to achieve a phase shift of $\Delta\phi = \pi$ is denoted V_π [V] and is one measure of the modulator efficiency; a small V_π implies that the drive voltage required to switch the modulator is reduced. This parameter is also occasionally normalized with modulator length, so that we often quote a voltage · length product, $V \cdot$ mm. If we use the above expressions and assume that the electric field, $\mathcal{E}$, due to an applied voltage, V, is dropped uniformly across a waveguide core region of thickness, t, $\mathcal{E} = V/t$, an expression for V_π,

$$V_\pi = \sqrt{\frac{2\lambda_0 t^2}{L n_0^3 \Gamma_{qw} s_{33qw}}} \tag{11.25}$$

results for the quantum-confined Stark effect. We see from this relation that a small core thickness, t, a large confinement factor, Γ_{qw}, and of course large electro-optic coefficient, s_{33qw}, and modulator length, L, will lead to a reduction in V_π. If we repeat the same exercise with a modulator based on the linear electro-optic effect, the expression for V_π becomes simply

$$V_\pi = \frac{2\lambda_0 t^2}{L n_0^3 \Gamma r_{41}} \tag{11.26}$$

An increase in efficiency also results if the Mach-Zehnder interferometer is operated in a push-pull configuration, as shown in Figure 11.10(b): two electrodes are used, one on either arm, each driven by a voltage of the same magnitude but opposite polarity. Thus, the total induced phase difference for a given applied voltage is doubled with respect to the configuration with only a single modulator.

11.5.2 Interferometer Modulator Performance

Both the linear and quadratic electro-optic effects, either in bulk or as the quantum-confined Stark effect, may be used to induce phase modulation in an interferometer modulator. A Mach-Zehnder interferometer based on the linear electro-optic effect has led to typical values of $V_\pi = 22$ V for $L = 3.3$ mm and a modulation depth of 12.7 dB [62]; a voltage · length product of 72.6 V · mm results. The use of quantum wells and the quantum-confined Stark effect enhances the efficiency, mostly by reducing V_π. Modulation depths of 23.8 dB have been reached in multiple quantum-well structures with $V_\pi \cdot L = 5.5$ V · mm, where one often finds a tradeoff between achievable modulation depth and voltage · length product [63].

Performance limitations include the polarization dependence of the phase modulation, recalling that TE has both a linear and orientation-dependent quadratic term in field, whereas TM has only the quadratic component. The polarization sensitivity can be reduced by orienting the waveguides with modulators at 45° to the {110} cleave planes, leading to a purely quadratic electro-optic effect [64]. Insertion loss must also be considered: optical losses due to scattering, reflection, and coupling can reduce the usable output signal. In addition, the electroabsorption which always exists in tandem with the refractive index effects, reduces the maximum modulation depth as the applied bias increases [65]. Finally, radiation modes from the two waveguide arms can couple and give rise to parasitic interference [66], so that the spacing between arms and the optical isolation between them must be sufficient to minimize this effect.

11.5.3 Other Origins for Phase Shift

We have focused on the use of the various electro-optic effects to achieve phase modulation and thus intensity change in an interferometer. Some of the alternative mechanisms to generate a phase shift, $\Delta\phi$, we discussed previously may also be employed; these are particularly useful for Si and other materials with small or no electro-optic effect. The generation of carriers in one arm of a Mach-Zehnder interferometer by external optical means gives rise to a change in real refractive index [67], although the need for an external light source limits integrated optical applications of this approach. Carriers generated by current injection into Si achieve the same ends [68]: for a 500-μm-long modulator formed by an Al contact, a modulation depth of 4.9 dB was demonstrated with an injection carrier density of

$n = 6.5 \cdot 10^{17}$ cm^{-3} requiring an injection current density of $J = 1.6 \cdot 10^3$ A/cm^2. Switching speeds are on the order of 50 ns.

Finally, we saw that the thermo-optic effect is useful in silicon. Using NiCr heaters over a length of 500 μm along one arm of a 3-μm-wide Si on SiO_2 waveguide interferometer, an input power of 30 mW has been shown to lead to a π phase shift [69]. The rise in waveguide temperature is estimated to be about 6.5 K, using the expected change in index with temperature for Si of $\Delta n/\Delta T = 2 \cdot 10^{-4}$ K^{-1}. Response times, at 50 μs, are relatively slow.

REFERENCES

[1] Musashi, M., *The Book of Five Rings*, New York: Bantam, 1982.

[2] Walker, R. G., "High-Speed Electro-optic Modulation in GaAs/GaAlAs Waveguide Devices," *IEEE Journal of Lightwave Technology*, Vol. LT-5, 1987, pp. 1,444–1,453.

[3] Noda, Y., M. Suzuki, Y. Kushiro, and S. Akiba, "High-sPeed Electroabsorption Modulator with Strip-Loaded GaInAsP Planar Waveguide," *IEEE Journal of Lightwave Technology*, Vol. LT-4, 1986, pp. 1,445–1,986

[4] Kahn, M. N., W. Yang, and A. Gopinath, "Directional Coupler Electro-Optic Modulator in AlGaAs/GaAs with Low Voltage-Length Product," *Applied Physics Letters*, Vol. 62, 1993, pp. 2,033–2,035.

[5] Hunsperger, R. G., *Integrated Optics: Theory and Technology*, 3rd Ed., Berlin: Springer-Verlag, 1991, Chapter 8.4.

[6] Alferness, R. C., "Titanium-Diffused Lithium Niobate Waveguide Devices," in *Guided-Wave Optoelectronics*, Tamir, T. (ed.), Berlin: Springer-Verlag, 1988, pp. 179–180.

[7] Chen, R., and C. S. Tsai, "GaAs-GaAlAs Heterostructure Single-Mode Channel-Waveguide Cutoff Modulator and Modulator Array," *IEEE Journal of Quantum Electronics*, Vol. QE-23, 1987, pp. 2,205–2,208.

[8] Tsai, C. S., B. Kim, and F. R. El-Akkari, "Optical Channel Waveguide Switch and Coupler Using Total Internal Reflection," *IEEE Journal of Quantum Electronics*, Vol. QE-14, 1978, pp. 513–517.

[9] Chen, B., and C. M. Meijer, "Bragg Switch for Optical Channel Waveguides," *Applied Physics Letters*, Vol. 33, 1978, pp. 33–35.

[10] Tsai, C. S. (ed.), *Guided-Wave Acousto-Optics*, Berlin: Springer-Verlag, 1990.

[11] Loh, K. W., W. S. C. Chang, W. R. Smith, and T. Grudkowski, "Bragg Coupling Efficiency for Guided Acoustooptic Interaction in GaAs," *Applied Optics*, Vol. 15, 1976, pp. 156–166.

[12] Soref, R. A., D. L. McDaniel, and B. R. Bennett, "Guided-Wave Intensity Modulators Using Amplitude-and-Phase Perturbations," *IEEE Journal of Lightwave Technology*, Vol. 6, 1988, pp. 437–444.

[13] Wang, S. Y., and S. H. Lin, "High Speed III-V Electro-optic Waveguide Modulators at $\lambda = 1.3$ μm," *IEEE Journal of Lightwave Technology*, Vol. 6, 1988, pp. 758–771.

[14] Goossen, K. W., J. A. Walker, and S. C. Arney, "Silicon Modulator Based on Mechanically-Active Anti-Reflection Layer with 1 Mbit/sec Capability for Fiber-in-the-Loop Applications," *IEEE Photonics Technology Letters*, Vol. 6, 1994, pp. 1,119–1,121.

[15] Eimerl, D., "Crystal Symmetry and the Electro-optic Effect," *IEEE Journal of Quantum Electronics*, Vol. QE-23, 1987, pp. 2,104–2,115.

[16] Namba, S., "Electro-Optical Effect of Zincblende," *Journal of the Optical Society of America*, Vol. 51, 1961, pp. 76–79.

[17] Leonberger, F. J., and J. F. Donnelly, "Semiconductor Integrated Optic Devices," in *Guided-Wave Optoelectronics*, Tamir, T. (ed.), Berlin: Springer-Verlag, 1988, p. 363.

[18] Tada, K., and N. Suzuki, "Linear Electro-optic Properties of InP," *Japanese Journal of Applied Physics*, Vol. 19, 1980, pp. 2,295–2,296.

[19] Yariv, A., *Optical Electronics*, 4th. Ed., Philadelphia: Saunders, 1991, pp. 317–320.

[20] Faist, J., F.-K. Reinhart, D. Martin, and E. Tuncel, "Orientation Dependence of the Phase Modulation in a P-N Junction GaAs/$Al_xGa_{1-x}As$ Waveguide," *Applied Physics Letters*, Vol. 50, 1987, pp. 68–70.
[21] Vinchant, J.-F., J. A. Cavaillès, M. Erman, P. Jarry, and M. Renaud, "InP/GaInAsP Guided-Wave Phase Modulators Based on Carrier-Induced Effects: Theory and Experiment," *IEEE Journal of Lightwave Technology*, Vol. 10, 1992, pp. 63–69.
[22] Bennett, B. R., and R. A. Soref, "Electrorefraction and Electroabsorption in InP, GaAs, GaSb, InAs and InSb," *IEEE Journal of Quantum Electronics*, Vol. QE-23, 1987, pp. 2,159–2,166.
[23] Stillman, G. E., C. M. Wolfe, C. O. Bozler, and J. A. Rossi, "Electroabsorption in GaAs and Its Application to Waveguide Detectors and Modulators," *Applied Physics Letters*, Vol. 28, 1976, pp. 544–546.
[24] Seraphin, B. O., and N. Bottka, "Franz-Keldysh Effect of the Refractive Index in Semiconductors," *Physical Review*, Vol. 139, 1965, pp. A560–A565.
[25] Alping, A., and L. A. Coldren, "Electrorefraction in GaAs and InGaAsP and Its Application to Phase Modulators," *Journal of Applied Physics*, Vol. 61, 1987, pp. 2,430–2,433.
[26] Van Eck, T. E., L. M. Walpita, W. S. C. Chang, and H. H. Wieder, "Franz-Keldysh Electrorefraction and Electroabsorption in Bulk InP and GaAs," *Applied Physics Letters*, Vol. 48, 1986, pp. 451–453.
[27] Reinhart, F. K., "Electroabsorption in $Al_yGa_{1-y}As$-$Al_xGa_{1-x}As$ Double Heterostructures," *Applied Physics Letters*, Vol. 22, 1973, pp. 372–374.
[28] Soref, R. A., and J. P. Lorenzo, "All-Silicon Active and Passive Guided-Wave Components for λ = 1.3 and 1.6 μm," *IEEE Journal of Quantum Electronics*, Vol. QE-22, 1986, pp. 873–879.
[29] Mendoza-Alvarez, J. G., R. H. Yan, and L. A. Coldren, "Contribution of the Band-Filling Effect to the Effective Refractive-Index Change in Double- Heterostructure GaAs/AlGaAs Phase Modulators," *Journal of Applied Physics*, Vol. 62, 1987, pp. 4,548–4,553.
[30] Fiedler, F., and A. Schlachetzki, "Optical Parameters of InP-Based Waveguides," *Solid State Electronics*, Vol. 30, 1987, pp. 73–83.
[31] Vinchant, op.cit.
[32] Chuang, C. L., T. Yoon, R. Jin, G. Khitrova, H. M. Gibbs, R. Fu, and C. S. Hong, "Current Controlled GaAs/AlGaAs Multiple Quantum Well Phase Shifter Operating at 0.9 μm," *Applied Physics Letters*, Vol. 59, 1991, pp. 2,363–2,365.
[33] Alping, A., X. S. Wu, T. R Hausken, and L. A. Coldren, "Highly Efficient Waveguide Phase Modulator for Integrated Optics," *Applied Physics Letters*, Vol. 48, 1986, pp. 1,243–1,245.
[34] Liu, A. C., M. J. F. Digonnet, and G. S. Kino, "Electro-Optic Phase Modulation in a Silica Channel Waveguide," *Optics Letters*, Vol. 19, 1994, pp. 466–468.
[35] Soref, R. A., and B. R. Bennett, "Electro-optical Effects in Silicon," *IEEE Journal of Quantum Electronics*, Vol. QE-23, 1987, pp. 123–129.
[36] Liu, Y. M., X. Xiao, P. R. Prucnal, and J. C. Sturm, "All-Optical Switching in an Asymmetric Silicon Fabry-Perot Etalon Based on the Free-Carrier Plasma Effect," *Applied Optics*, Vol. 33, 1994, pp. 3,871–3,874.
[37] Eichler, H. J., "Optical Multistability in Silicon Observed with a CW Laser at 1.06 μm," *Optics Communications*, Vol. 45, 1983, pp. 62–66.
[38] Takato, N., K. Junguji, M. Yasu, H. Toba, and M. Kawachi, "Silica-Based Single-Mode Waveguides on Silicon and Their Application to Guided-Wave Optical Interferometers," *IEEE Journal of Lightwave Technology*, Vol. 6, 1988, pp. 1,003–1,010.
[39] Wood, T. H., "Multiple Quantum Well (MQW) Waveguide Modulators," *IEEE Journal of Lightwave Technology*, Vol. 6, 1988, pp. 743–757.
[40] Chemla, D. S., and D. A. B. Miller, "Room-Temperature Excitonic Nonlinear-Optical Effects in Semiconductor Quantum-Well Structures," *Journal of the Optical Society of America B*, Vol. 2, 1985, pp. 1,155–1,173.
[41] Miller, D. A. B., D. S. Chemla, T. C. Damen, A. C. Gossard, W. Wiegmann, T. H. Wood, and C.

A. Burrus, "Band-Edge Electroabsorption in Quantum Well Structures: the Quantum Confined Stark Effect," *Physical Review Letters*, Vol. 53, 1984, pp. 2,173–2,176.
[42] Weiner, J. S., D. A. B. Miller, D. S. Chemla, T. C. Damen, C. A. Burrus, T. H. Wood, A. C. Gossard, and W. Wiegmann, "Strong Polarization-Dependent Electroabsorption in GaAs/AlGaAs Quantum Well Waveguides, " *Applied Physics Letters*, Vol. 47, 1985, pp. 1,148–1,150.
[43] Weiner, J. S., D. A. B. Miller, and D. S. Chemla, "Quadratic Electro-Optic Effect Due to the Quantum Confined Stark Effect in Quantum Wells," *Applied Physics Letters*, Vol. 50, 1987, pp. 842–844.
[44] Miller, D. A. B., J. S. Weiner, and D. S. Chemla, "Electric Field Dependence of Linear Optical Properties in Quantum Well Structures: Waveguide Electroabsorption and Sum Rules," *IEEE Journal of Quantum Electronics*, Vol. QE-22, 1986, pp. 1,816–1,830.
[45] Shimizu, J., T. Hiroshima, A. Ajisawa, M. Sugimoto, and Y. Ohta, "Measurements of the Polarization Dependence of Field-Induced Refractive Index Changes in GaAs/AlGaAs Multiple Quantum Well Structures," *Applied Physics Letters*, Vol. 53, 1988, pp. 86–88.
[46] Jennings, A., C. D. W. Wilkinson, and J. S. Roberts, "Spectral Characterization of the Linear and Quadratic Electro-Optic Effects in GaAs/AlGaAs Multiple Quantum Well Waveguides," *Semiconductor Science and Technology*, Vol. 7, 1992, pp. 60–65.
[47] Nishimura, S., H. Inoue, H. Sano, and K. Ishida, "Electro-optic Effects in an InGaAs/InAlAs Multiquantum Well Structure," *IEEE Photonics Technology Letters*, Vol. 4, 1992, pp. 1,123–1,126.
[48] Zucker, J. E., I. Bar-Joseph, B. I. Miller, U. Koren, and D. S. Chemla, "Quaternary Quantum Wells for Electro-Optic Intensity and Phase Modulation at 1.3 and 1.55 μm," *Applied Physics Letters*, Vol. 54, 1989, pp. 10–12.
[49] Koehler, S. D., E. M. Garmire, M. H. Jupina, T. C. Hasenberg, and A. R. Kost, "Direct Interferometric Measurements of Quantum-Confined Stark Effect in InAs/GaAs Short Period Strained-Layer Superlattice MQWs," *Electronics Letters*, Vol. 27, 1991, pp. 2,303–2,304.
[50] Chen, W., and T. G. Andersson, "Quantum-Confined Stark Shift for Differently Shaped Quantum Wells," *Semiconductor Science and Technology*, Vol. 7, 1992, pp. 828–836.
[51] Wood, T. H., R. W. Tkach, and A. R. Chraplyvy, "Observation of Large Quadratic Electro-Optic Effect in GaAs/AlGaAs Multiple Quantum Wells," *Applied Physics Letters*, Vol. 50, 1987, pp. 798–800.
[52] Zucker, J. E., T. L. Hendrickson, and C. A. Burrus, "Electro-Optic Phase Modulation in GaAs/AlGaAs Quantum Well Waveguides," *Applied Physics Letters*, Vol. 52, 1988, pp. 945–947.
[53] Dorgeuille, F., and F. Devaux, "On the Transmission Performances and the Chirp Parameter of a Multiple-Quantum-Well Electroabsorption Modulator," *IEEE Journal of Quantum Electronics*, Vol. 30, 1994, pp. 2,565–2,572.
[54] Bloemer, M. J., and K. Myneni, "Electro-Optic Properties Near the Absorption Edge of GaAs/AlGaAs Multiple-Quantum-Well Waveguides," *Journal of Applied Physics*, Vol. 74, 1993, pp. 4,849–4,859.
[55] Susa, N., and T. Nakahara, "Design of AlGaAs/GaAs Quantum Wells for Electroabsorption Modulators," *Solid State Electronics*, Vol. 36, 1993, pp. 1,277–1,287.
[56] Martin, R. W., S. L. Wong, R. J. Nicholas, K. Styke, M. Gibbon, and E. J. Thrush, "The Design of Quantum-Confined Stark Effect Modulators for Integration with 1.55 μm Lasers," *Semiconductor Science and Technology*, Vol. 8, 1993, pp. 1,173–1,178.
[57] Tada, K., S. Nishimura, and T. Ishikawa, "Polarization-Independent Optical Waveguide Intensity Switch with Parabolic Quantum Well," *Applied Physics Letters*, Vol. 59, 1991, pp. 2,778–2,780.
[58] Chen, Y., J. E. Zucker, N. J. Sauer, and T. Y. Chang, "Polarization-Independent Strained InGaAs/InGaAlAs Quantum-Well Phase Modulators," *IEEE Photonics Technology Letters*, Vol. 4, 1992, pp. 1,120–1,123.
[59] Devaux, F., S. Chelles, A. Ougazzaden, A. Mircea, M. Carré, F. Huet, A. Carenco, Y. Sorel, J. F. Kerdiles, and M. Henry, "Full Polarization Insensitivity of a 20 Gb/s Strained-MQW Electroabsorption Modulator," *IEEE Photonics Technology Letters*, Vol. 6, 1994, pp. 1,203–1,206.
[60] Erman, M., P. Jarry, R. Gamonal, P. Autier, J.-P. Chanè, and P. Frijlink, "Mach-Zehnder Modulators

and Optical Switches on III-V Semiconductors," *IEEE Journal of Lightwave Technology*, Vol. 6, 1988, pp. 837–846.

[61] Rolland, C., R. S. Moore, F. Shepherd, and G. Hillier, "10 Gbit/s, 1.56 μm Multiquantum Well InP/InGaAsP Mach-Zehnder Optical Modulator," Electronics Letters, Vol. 29, 1993, pp. 471–472.

[62] Buchmann, P., H. Kaufmann, H. Melchior, and G. Guekos, "Broadband Y-Branch Electro-Optic GaAs Waveguide Interferometer for 1.3 μm," *Applied Physics Letters*, Vol. 46, 1985, pp. 462–464.

[63] Cites, J. S., and P. R. Ashley, "High-Performance Mach-Zehnder Modulators in Multiple Quantum Well GaAs/AlGaAs," *IEEE Journal of Lightwave Technology*, Vol. 12, 1992, pp. 1,167–1,173.

[64] Bachmann, M., E. Gini, and H. Melchior, "Polarization Insensitive Waveguide Modulator Using InGaAsP/InP Mach Zehnder Interferometer," *European Conference on Optical Communications*, Berlin, 1992, pp. 345–348.

[65] Chin, M. K., "Effect of Electroabsorption on Electrorefractive Intensity Modulators," *IEEE Photonics Technology Letters*, Vol. 4, 1992, pp. 583–585.

[66] Khalil, D. A. M., and S. Tedjini, "Coherent Coupling of Radiation Modes in Mach-Zehnder Electro-optic Modulators," *IEEE Journal of Quantum Electronics*, Vol. 28, 1992, pp. 1,236–1,238.

[67] Cheng, Z. Y., and C. S. Tsai, "Optically Activated Integrated Optic Mach-Zehnder Interferometer on GaAs," *Applied Physics Letters*, Vol. 59, 1991, pp. 2,222–2,224.

[68] Treyz, G. V., P. G. May, and J.-M. Halbout, "Silicon Mach-Zehnder Waveguide Interferometers Based on the Plasma Dispersion Effect," *Applied Physics Letters*, Vol. 59, 1991, pp. 771–773.

[69] Treyz, G. V., "Silicon Mach-Zehnder Waveguide Interferometers Operating at 1.3 μm," *Electronics Letters*, Vol. 27, 1991, pp. 118–119.

Chapter 12

Hybridization and Monolithic Integration

Incidents! All we get is incidents! Dear God, is it too much to expect a little sustained action?!

Tom Stoppard, 1967 [1]

By this point we have learned how to take a hunk of semiconductor and mold from it a few types of optical devices to suit our fancy: light emitters, light absorbers, light modulators, and waveguides to interconnect them all. Bringing some or all of these components together into an optical integrated circuit may, in a manner analogous to electrical integrated circuits, result in a reduction in size, an increase in density, a simplification of assembly and packaging, an enhancement of robustness, a removal of parasitics, an improvement in speed and, possibly, a decrease in cost. True integrated semiconductor optics take advantage of the batch fabrication inherent to semiconductor processing to yield an integrated optical microsystem on a chip.

The capabilities of this "laser beam circuitry," as integrated optics was originally conceived [2], have grown from the clever use of guided wave structures [3] and discrete devices [4] to the point where complete optical circuits on a single piece of semiconductor may now be more or less routinely fabricated [5]. Such structures are generally referred to as photonic integrated circuits (PICs) [6] if they consist only of optical, or so-called photonic, components, and optoelectronic integrated circuits (OEICs) [7] if electronic drive circuitry is also included; the distinction between the two is often blurred and not really relevant.

In this final chapter, we discuss some of the techniques employed for the fabrication of these integrated optical systems. Both monolithic integration, where all components are fabricated from a single piece of semiconductor, and hybridization, where parts of various material origins are assembled, will be considered. We will focus on *integrated* optical circuits, where the optical functions are performed on-chip, and which are for that reason frequently waveguide based. A closely related, but conceptually distinct, semiconductor optics field uses vertical light emitters, often in association with integrated electronics, for fiber communications [8] or free-space optical interconnect; this latter area is particularly interesting and undergoing rapid development at present.[1] This chapter will restrict itself, mostly, to on-chip optical systems; this field is exciting and expanding, so our survey can provide only a tantalizing overview.

12.1 INTEGRATION VERSUS HYBRIDIZATION

We have basically two options when trying to fabricate a semiconductor-based integrated optical circuit. The first, monolithic integration takes the approach that electrical integrated circuits take and fabricate all of the necessary components on a single piece of semiconductor. This technique allows us to take full advantage of semiconductor batch processing, as all components are fabricated simultaneously. Monolithic integration yields structures that are aligned photolithographically and are thus robust, rather insensitive to vibration or aging and relatively easy to package. On the other hand, the technology required to fabricate different components on a single wafer can be complex, isolation issues can become bothersome, and the materials employed often represent a compromise in performance for at least some of the devices.

Hybridization, the second alternative, permits us to select the optimal material for any given function, such as GaAs for a laser, Si for circuitry, and perhaps a polymer for a mechanical structure; this inherent flexibility can represent an important advantage. The fabrication technology for each device or subsystem can be simplified, but the subsequent assembly and packaging of the components to form the complete integrated optical system may be complicated and expensive. In addition, issues such as mechanical reliability and durability are more prominent than for monolithic integration.

Although it is often argued that one or the other of these alternatives is the one true answer to all of the world's problems, we can see by an appraisal of the relative merits of each that the choice of monolithic integration or hybridization will depend upon the demands of the application. The questions an integrated optician will need to ask include: Which of the above considerations are relevant? Of what is our technology capable? And, most often the bottom line, what can it cost? Based on such an analysis, one application may require the robustness of a monolithic structure, whereas another may benefit more from the flexibility of a hybrid construction.

By having a look at some examples of integrated and hybridized semiconductor optical systems, we may obtain a bit of a feeling for when it makes sense to choose one over the other. To this end, the rest of this section provides a brief overview of several types of integrated optical circuits of current interest. Many of the developments in this area have been and continue to be driven by needs of telecommunica-

[1]See, for example, the special issue on optical interconnect and packaging, *Optical Engineering*, Vol. 33, May 1994, pp. 1,511–1,603.

tions, and are thus designed with optical fiber applications in mind. In this field, performance[2] is often, but not always, a more cardinal consideration than price.

12.1.1 Examples: Photonic Integration

As we mentioned in Chapter 11, the use of an external switch to modulate a laser can result in lower laser chirp than can modulating the laser directly. By monolithically integrating a Mach-Zehnder interferometer modulator with a distributed Bragg reflector laser, as shown in Figure 12.1, any alignment and coupling problems that would arise if we were to do this with a bulk-optical connection are eliminated. A device such as this, fabricated in the InGaAs/InGaAsP/InP system and emitting at 1.55 μm, has been operated with a bandwidth of 3 GHz [9]. The DBR laser lends itself naturally to integration because at least one of the cleaved facets can be replaced by a grating; an alternative approach is to dry etch the laser facets [10].

A laser or optical amplifier [11] may also be monolithically integrated with a photodiode, using either a specialized directional coupler, as shown in Figure 12.2, or by dry etching a laser facet and monitoring the light from that side of the laser [12]. The coupler in the figure is designed so that only a fraction of the power is diverted to the detector, so that this may be used to monitor laser or amplifier output power. Through the use of appropriate feedback circuitry, the photodiode signal may then be used to stabilize the power emitted from the primary branch of the coupler. Monolithic integration obviates the need for external coupling of a photodiode for this purpose.

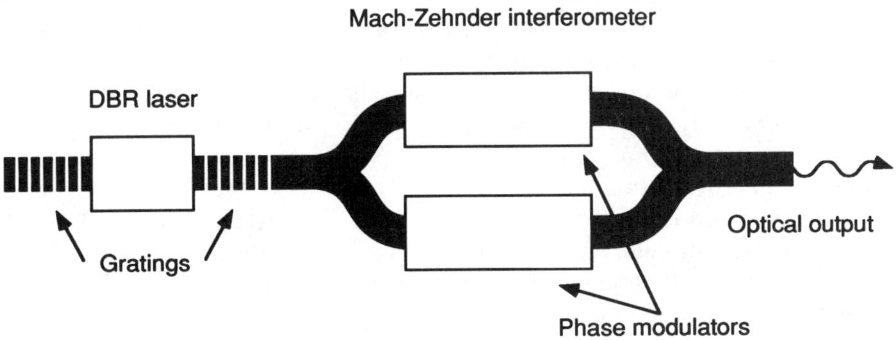

Figure 12.1 Schematic sketch of a DBR laser monolithically integrated with a Mach-Zehnder interferometer modulator. The interferometer is used for high-speed optical signal modulation.

[2]For a discussion of performance requirements, see, for example, Carney, J. K., and L. D. Hutcheson, "GaAs-Based Integrated Optoelectronic Circuits: Design, Development and Applications," in *Integrated Optical Circuits and Components*, Hutcheson, L. D. (ed.), New York: Marcel Dekker, 1987, Chapter 7.6.

Figure 12.2 A monitor photodiode monolithically integrated with a DBR laser. The directional coupler is designed so that only a small fraction of the power is diverted to the detector, which ensures that the output power of the laser can be monitored and, given a feedback circuit, stabilized.

By increasing the complexity somewhat, more benefits of monolithic integration come to light. Optical communications occasionally employ wavelength-division multiplexing (WDM), in which several signals of distinct wavelength may be modulated individually and all are sent over a single optical fiber. An integrated scheme such as the one shown in Figure 12.3(a) may be used as a transmitter: each laser is at a slightly different wavelength from the others and has its own modulator. All of the output signals are fed through an integrated amplifier and coupled off-chip into a single output fiber. A WDM transmitter with 16 channels, spaced in wavelength by 0.6 nm around $\lambda = 1.55$ μm, has been demonstrated [13], monolithic integration allowing the fabrication of all the functions shown in the figure on a 4×6 mm chip.

At the opposite end of a WDM fiber link, an integrated optical chip can be used to decode the various wavelengths input from a fiber and may employ a scheme as shown in Figure 12.3(b). A curved grating mirror at the end of a slab waveguide acts as a two-dimensional monochromator and directs each input wavelength to an appropriate detector, thereby demultiplexing the input signal [14,15]. Similar schemes using integrated aspheric lenses [16] or variable absorption of detectors on a waveguide employing the quantum-confined Stark effect [17] have been demonstrated, as has a complete transceiver [18] for both WDM transmission and reception.

For data transmission applications, monolithic integration allows simultaneous fabrication of lasers of closely spaced wavelength, considerably simplifies the alignment of lasers to modulators and mirrors to detectors, and reduces the external-coupling requirement to that of the single optical fiber. The use of monolithically integrated photonic circuits of this type for such high-performance systems as required for telecommunications will likely increase significantly in the near future. Low cost, but high volume, consumer optoelectronics applications, such as "fiber to the home" or individual customer terminals [19], will also likely stimulate rapid development of this technology, and the potentially large market will force a reduction of the costs of such fully integrated systems.

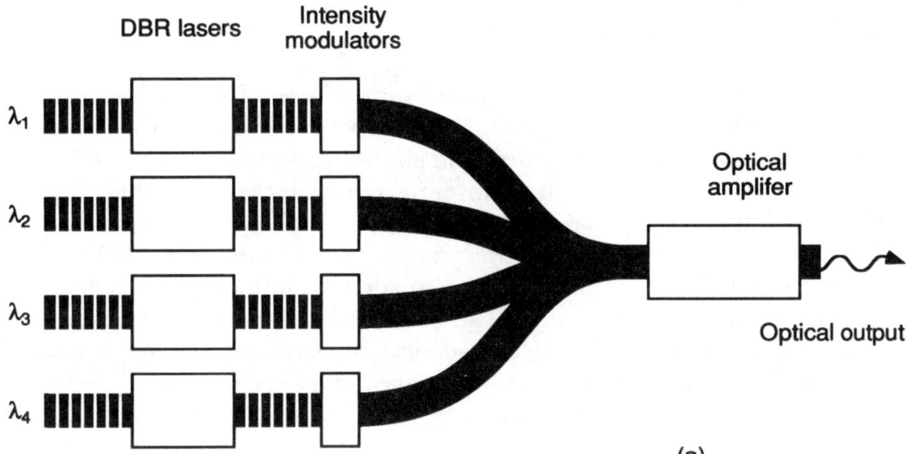

(a)

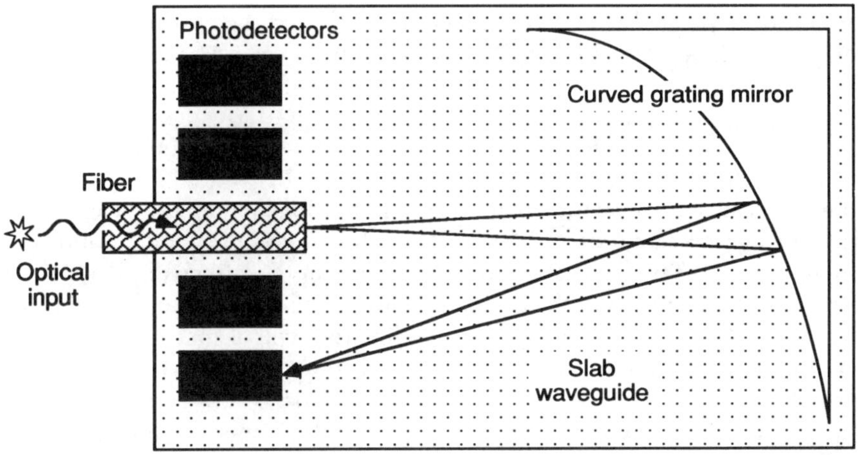

(b)

Figure 12.3 Monolithically integrated wavelength-division multiplexing (WDM) transmitters and receivers: (a) a WDM transmitter, with an array of DBR lasers, each at a distinct wavelength and with individual modulator; and (b) a WDM receiver using a grating mirror to direct the different wavelengths from the input fiber onto the appropriate detector.

12.1.2 Examples: Electronics Integration

An extension of optical device integration is the addition of electronics, usually as drivers for laser-based transmitters or amplifiers for photoreceivers [20]. To manufacture a monolithic structure with optical and electronic devices, added complexity in layer growth is usually necessary. In a transmitter circuit, for example, the heterostructure layers for the laser have been grown over those required for the fabrication of a MESFET [21], yielding an entire transmitter that operates at a modulation frequency of 3 GHz. Likewise, an MSM photodiode monolithically integrated with a two-stage HEMT-based amplifier [22] has permitted a 14-GHz receiver bandwidth. Silicon may also be employed for detector applications, so that the ability to monolithically integrate a MOSFET-based transimpedance amplifier with p-i-n photodiodes [23] has allowed the fabrication of complete optical receiver chips using standard Si processes.

The advantage of monolithic electronics and photonics integration in the above three examples is predominantly the reduction of electrical wiring and packaging parasitics, elements that tend to reduce high-speed performance. Proper microwave layout can be accomplished at the chip design stage, so that the optical and electrical devices may be optimally coupled and the package becomes a less critical factor.

12.1.3 Examples: Hybridization

By the time we reach the level of integration required for a complete receiver, with either the optical processing required for WDM or electronics for signal amplification, hybridization often begins to become attractive. For optical fiber applications, we will sooner or later be faced with the problem of attaching the fiber to the transmitter or receiver chip, where packaging issues will define the type of material used for the fabrication of an entire system. In Figure 12.4, we show a schematic example of a hybrid optoelectronic transceiver [24], using a III-V-based laser transmitter and receiver mounted on a Si substrate; the latter is used for fiber alignment, through the use of V-grooves, electrical power and signal routing, thermal contact, and optical interconnect between the active chips and the fibers.

We see that for a complex system of this type, the use of hybridization solves a number of problems simultaneously. Optimal technologies may be employed for the transmitter, receiver, and substrate; the fiber alignment is simplified; and advanced Si electronics, not shown in the figure, may be employed for electrical signal processing. In addition, the good thermal and physical properties of silicon may prove to be an advantage with regard to robustness and reliability. The use of a hybridization technology also allows the use of III-V optics for interchip communication in Si-based electronic systems [25], as shown in Figure 12.5. Optical data transfer, either through the use of Si waveguide structures as shown in the figure, or through free-

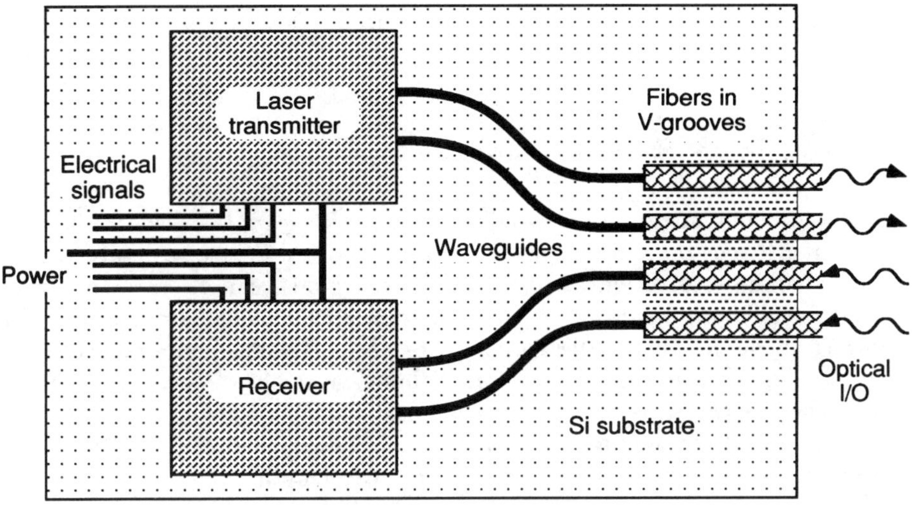

Figure 12.4 A hybrid Si/III-V transceiver module. The laser transmitter and receiver are fabricated in III-V technology and mounted on a Si substrate. The silicon acts as a physical device mount, alignment structure for the fibers, and thermal heat sink, and it allows electrical signal distribution. Si-based waveguides are used to optically connect the III-V chips with the fibers.

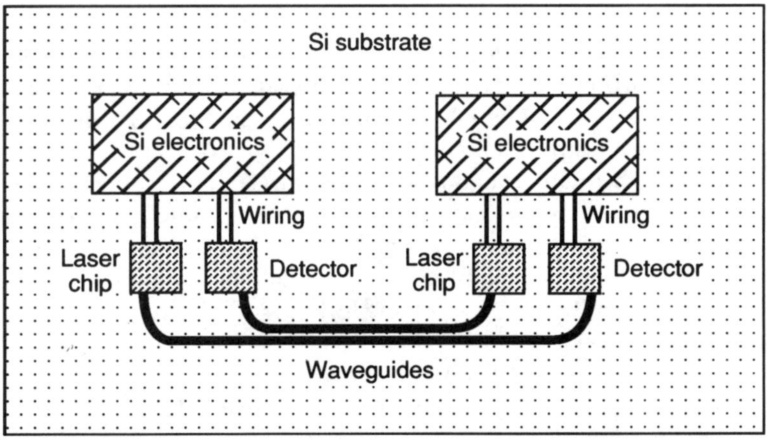

Figure 12.5 Small III-V-based transmitter and receiver chips may be used to route data between silicon electronic circuits in an otherwise completely Si-based system.

space transmission [26], permits added flexibility and enhanced performance for data transfer in a more complex electronic system.

12.2 WAVEGUIDE-BASED INTEGRATION

Most monolithically integrated photonic circuits are waveguide based. While the use of waveguides opens numerous options for useful optical device structures, some of which we shall meet at the end of this section, waveguides can also present a fundamental problem: at the operating wavelength, some portions of a particular waveguide need to be transparent, while others must be absorbing. The detector attached to a long waveguide coupler, as in Figure 12.2, for example, has to absorb at the wavelength of interest, whereas the waveguide of the coupler itself must be transparent, lest no light reaches the detector. Likewise, the waveguides attached to the DBR of Figure 12.1 must be transparent at the laser emission wavelength.

Thus, the successful fabrication of a waveguide-based monolithically integrated optical circuit requires either a means to couple light between two different waveguides, each with a different absorption edge, or a technique to generate different effective bandgaps within a single waveguide. In the following sections, we will look at how one achieves transparency and also address the question of how devices along a single waveguide may be electrically isolated.

12.2.1 Evanescent and Butt Coupling

If our MBE- or MOCVD-grown semiconductor device layer structure consists of transparent and absorbing waveguide regions, separated vertically, we can use several techniques to selectively couple the light from one to the other. Evanescent coupling relies on the evanescent optical field that extends outside the waveguide core and into the cladding, in a manner analogous to the directional coupler of Figure 8.16. As seen in the detector structures of Figure 12.6, we employ here a coupling in the vertical rather than in the lateral direction; detectors integrated with waveguides are particularly suitable for this type of coupling, as one basically needs only to bring an absorber into proximity of the waveguide mode.

The InGaAs absorber of Figure 12.6(a) is selected to absorb at a longer wavelength, say 1.55 μm, than the InGaAsP waveguide, which can be chosen to have an absorption edge at 1 μm [27]. Through etch removal of the absorber over the remainder of the waveguide, a single detector region can be defined, where a bias then leads to detector current. In addition, properly selecting the refractive index of the intermediate matching layer, by using InGaAsP rather than InP, for example, allows efficient coupling of the optical field into the absorber. Up to 90% absorption over a length of only 42 μm has been seen with a structure of this type [28].

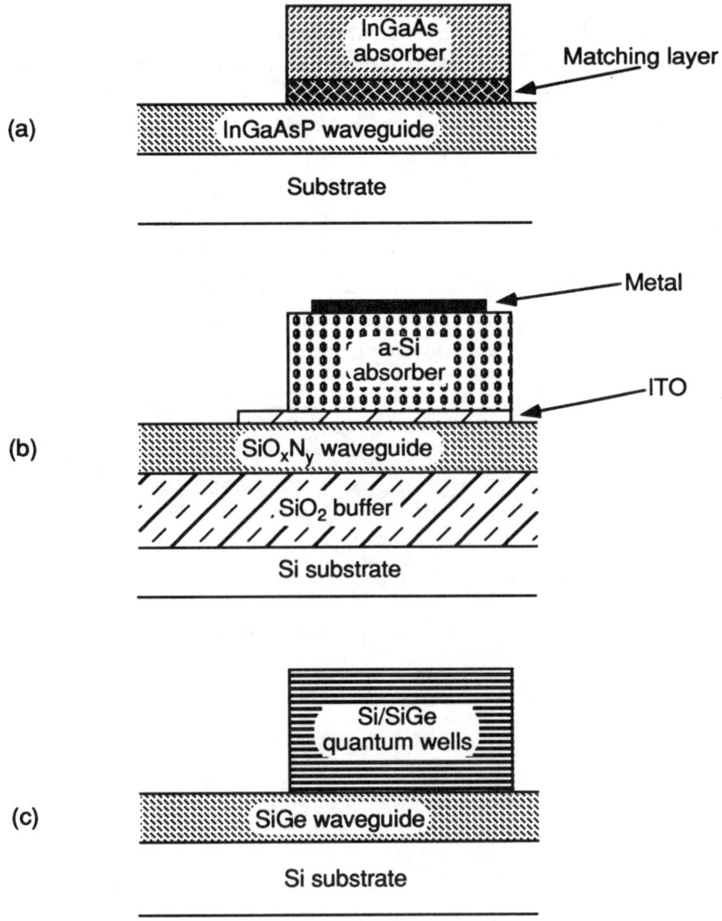

Figure 12.6 Monolithically integrated waveguide/detector structures relying on evanescent coupling: (a) an InGaAs absorber on a InGaAsP waveguide, (b) amorphous-silicon detector on an SiO$_x$N$_y$ waveguide, and (c) a Si/SiGe multi-quantum-well stack on a SiGe waveguide.

The same scheme can be used for an amorphous-silicon detector patterned onto an oxynitride waveguide, as seen in Figure 12.6(b). The use of transparent indium tin oxide (ITO) between waveguide and absorber permits the application of an electrical bias, and this configuration has been seen to absorb with a coefficient, $\alpha \sim 5 \cdot 10^{-3}$ dB/μm [29], implying a length of 2 mm required for 90% absorption. Using a silica waveguide with a tapered cladding thickness [30], the length of detector overlap can be defined by the region of reduced cladding, leading to higher absorption coefficients in the range of 0.15 dB/μm. We can also fabricate an evanescently coupled integrated waveguide detector using SiGe: in Figure 12.6(c), the $Si_{0.98}Ge_{0.02}$ waveguide core is capped by a Si/SiGe multi-quantum-well stack, where the quantum-well width can be varied to tune the absorption edge [31].

As we see, the efficiency of evanescent coupling varies widely and, while suitable for detectors and waveguides, it is less useful for other optical devices. We can, however, more directly couple between two different waveguides both grown or deposited on a single substrate; many of these butt-coupling techniques require multiple MOCVD or MBE growth steps, usually following an etch, or deposition of an alternative waveguide material. In Figure 12.7(a), for example, we see a pair of InGaAsP waveguides grown on an InP substrate onto which a step has been etched [32]; the upper waveguide has an absorption edge at 1.5 μm and the lower, at 1.3

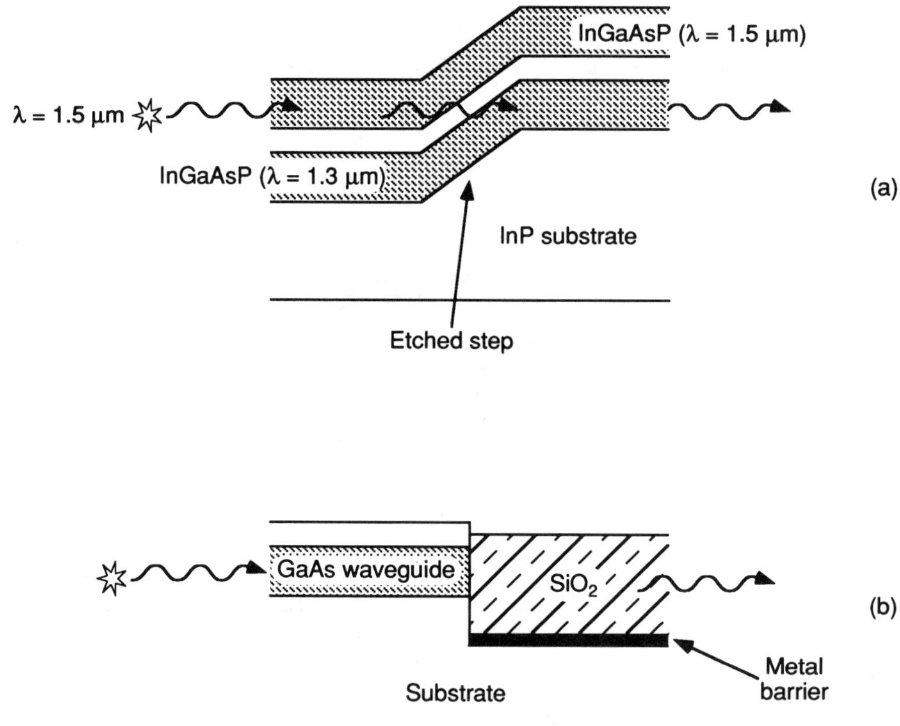

Figure 12.7 Butt coupling between waveguides of varying composition: (a) InGaAsP waveguides grown over an etched step in the InP substrate. The upper waveguide, which might be the active region of a laser, has an absorption edge at 1.5 μm, the lower, at 1.3 μm; light can couple from the upper into the lower. (b) The GaAs waveguide has been completely removed and butt coupled to a deposited SiO_2 dielectric waveguide; the metal layer is for optical confinement because the refractive index of SiO_2 is lower than that of GaAs.

μm. Thus, the upper may be used as the active region of a laser, the light from which is butt coupled into the transparent lower waveguide. Coupling efficiency of about 60% has been demonstrated. A second, downward step further to the right would allow coupling the light back into the 1.3-μm guide, at which point a detector could be fabricated.

Butt coupling is also useful if the III-V material is etched and replaced by a nonabsorbing dielectric, as shown in Figure 12.7(b); the SiO_2 waveguide is deposited over a Au/Cr metal layer that hinders transfer of the light into the lower GaAs layer, which has a higher refractive index [33]. Coupling losses are in the range of 1 dB. The identical scheme can use a dielectric waveguide sandwich of Ta_2O_5/SiO_2 [34] or polymers such as polystyrene on PMMA [35]; the choice of material for these structures is application dependent, and can give considerable flexibility.

12.2.2 Selective Growth

In the III-V system, the MOCVD growth of waveguide layers may be locally affected by dielectric masks or patterns on the substrate surface. Using SiO_2 as a masking layer during growth, for example, it has been found that the growth rate is a function of the mask stripe width, so that waveguides and quantum wells of slightly differing thicknesses may be grown in the same growth step [36]; the effect is illustrated schematically in Figure 12.8. The width of the SiO_2 stripe outside of the growth region is the relevant factor (i.e., not the width of the gap in which the growth takes place) because the migration of growth species across the mask to the growth area is that which enhances the growth rate. Waveguide regions that experienced a slightly lower growth rate have slightly thinner quantum wells, and hence a slightly shorter absorption wavelength. Thus, they are transparent for laser radiation generated in the waveguide regions that had a slightly higher growth rate. A single waveguide is

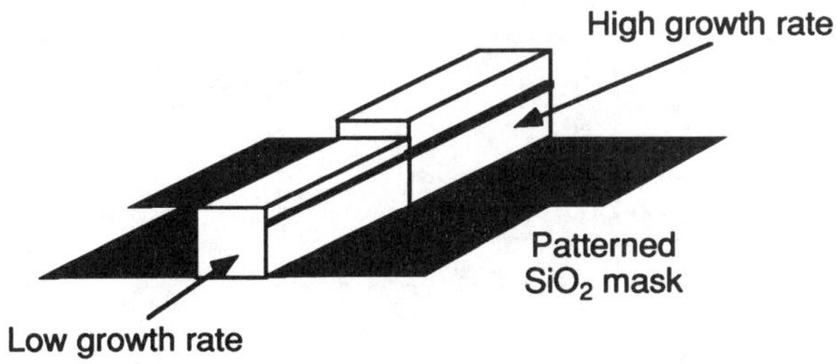

Figure 12.8 Selective area growth; the regions with the wider SiO_2 mask outside the growth area grow at a slightly higher rate than the regions with a narrower mask. No growth occurs on the SiO_2 layers. The region with a lower growth rate will have thinner quantum wells and thus have an absorption edge at a higher energy.

defined during growth to have absorbing and transparent regions for a particular wavelength, the coupling between the regions being a form of seamless butt coupling.

Using this approach, a DFB laser has been integrated with a modulator using InGaAs quantum wells in InGaAsP [37]. Using mask widths of 4 and 10 μm, the emission wavelengths were found to be 1.54 μm and 1.48 μm in the two regions, indicating that a shift of 60 nm could be achieved during growth. This game can be driven further, for example in the fabrication of various wavelength laser sources for WDM applications [38]; by using several different mask widths during InGaAs/GaAs growth, a laser wavelength spacing of about 12 nm—each device attached to a transparent waveguide—has been demonstrated.

Growth rate variations are also found on patterned or nonplanar semiconductor surfaces. For the MBE growth of InGaAs on GaAs, for example, the growth rate varies due to In migration across an etched surface, moving from facets on which the growth rate is low to those on which growth is preferred [39]. Optical wavelength variations of 70 nm have been achieved with this approach, which permitted monolithic integration of a laser, waveguide, and quantum-well modulator in a single growth step. Periodic surface corrugations have also been shown to increase the MBE growth rate of GaAs/AlGaAs, allowing energy shifts of 44 meV, which is also employed for integration of a laser array with transparent waveguides [40].

12.2.3 Quantum-Well Intermixing

An attractive alternative to selective growth for the generation of transparent and absorbing regions in a single waveguide is quantum-well intermixing, which permits a selective shift of the waveguide absorption edge by controlled interdiffusion of the quantum wells after growth [41]. The technique, also referred to as compositional disordering, is schematically illustrated in the energy band diagrams of Figure 12.9. The as-grown GaAs quantum well of Figure 12.9(a) has a sharp interface with the

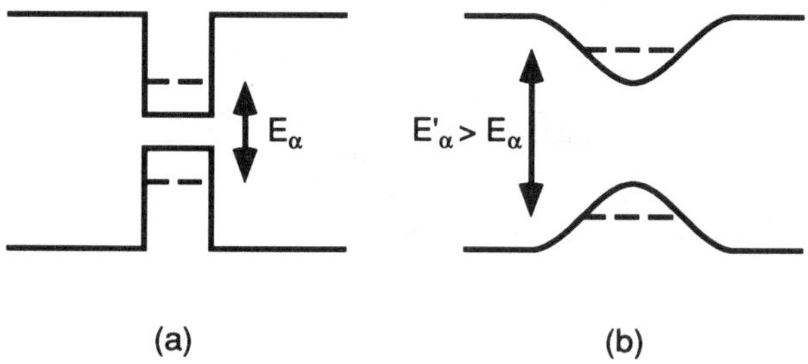

Figure 12.9 A schematic representation of quantum-well intermixing: (a) the as-grown well has sharp interfaces and an effective bandgap of E_α, and (b) after high-temperature anneal, the intermixed well has a "smeared out" profile and a higher bandgap energy, $E'_\alpha > E_\alpha$.

AlGaAs barrier material. During a high-temperature anneal step, the Al and Ga of the well and barrier interdiffuse, forming the "smeared out" profile of Figure 12.9(b), and effectively increasing the bandgap; if properly controlled, the quantum well does not disappear during intermixing, but merely changes its shape, as shown.

A remaining problem is how to pattern the process so that some waveguide regions are intermixed and others are not. Interdiffusion can be promoted by the incorporation of certain impurities into the quantum-well region, an approach known as impurity-induced disordering. Following patterned implantation or surface diffusion, yielding the required impurity concentration on the order of 10^{18} cm^{-3}, a high temperature anneal at anywhere from 600° to 950° is performed to achieve intermixing. The presence of Si or Zn [42] in GaAs/AlGaAs has been shown to significantly enhance the interdiffusion coefficient, so that anneals at temperatures as low as 600° result in disordered quantum-well structures. The same is achieved with P implanted into InGaAsP [43], which has also been used to define buried waveguides, as the refractive index of a disordered superlattice changes in addition to the absorption edge. Because Si or Zn are dopants in III-Vs, the use of neutral impurities has the advantage that the pn junction is not shorted after impurity incorporation. B and F, not active dopants in these materials, have also been used for intermixing [44], as have H^+ (proton) implants [45]. Because appreciable intermixing only occurs in the regions where the impurities are diffused or implanted, and intermixing increases the bandgap, we can define transparent regions of a waveguide as those in which impurities are present. Thus, the monolithic integration of, for example, a transparent waveguide with an active laser is straightforward [46].

Quantum-well interdiffusion may also be promoted by impurity-free processes, namely through the presence of vacancies in the substrate, leading to the nomenclature dielectric cap annealing. The use of various surface encapsulants, such as SiO_2, Si_3N_4, or the group IIA fluorides, during high-temperature anneal, incites or inhibits vacancy formation, so that dielectric patterning before anneal allows us to define intermixed and nonintermixed regions [47]. A silicon dioxide surface layer generates vacancies during anneal, which diffuse into the quantum-well region and promote intermixing, thereby raising the effective bandgap. Silicon nitride tends to inhibit vacancy formation at high temperatures, so wells under this layer are less intermixed; using these materials, the formation of a laser integrated with a modulator whose bandgap was shifted into the region of maximal chirp [48] has been demonstrated. It has also been shown [49] that the group IIA fluorides, particularly SrF_2, are more effective in inhibiting intermixing than Si_3N_4; using SiO_2 and SrF_2 covered regions, and rapid thermal anneals of 960° for 90 s, differential absorption edge shifts of up to 60 nm have been demonstrated [50].

Bandgap shifting by post-growth quantum-well intermixing has the advantage that the absorption edge can be tuned selectively over the wafer surface, requiring only a single waveguide structure. Important in its application is the maintenance of suitable device properties after intermixing. Excitonic absorption and electroabsorption are maintained after rapid thermal annealing [51], and the quantum-confined Stark effect for modulation is still active [52]. Lasers fabricated in rapid thermal annealed regions of the wafer tend to have almost a factor two lower threshold current density but also slightly lower quantum efficiency [53]. Arrays of wavelength-tuned lasers, with emission wavelength shifts of over 10 nm, have been fabricated by impurity-free disordering techniques [54]. Intermixing has also proven useful in the generation of transparent waveguides for DBR laser reflectors [55] and for intracavity electroabsorption modulators monolithically integrated in a Fabry-Perot laser [56].

12.2.4 Electrical Isolation

The generation of transparent and absorbing regions along a waveguide solves the problem of optical coupling between devices. The electrical coupling along the waveguide, unfortunately, is usually quite good at the outset. Because lasers, modulators, and detectors usually operate under very different bias conditions, we require a means to isolate them electrically from one another while maintaining optimal optical coupling. Approaches such as the dry etching of thin slits into the waveguide [57] or depositing an insulating dielectric on an etched waveguide butt joint [58] have been employed, although the optical losses and reflections at these interfaces can be problematic.

A useful means to achieve electrical isolation is through implantation-induced damage of the waveguide. Implantation of one of several species, usually H^+ or He^+, but also F^+ and O^+, into a region of the waveguide significantly increases the electrical resistance due to damage of the crystal lattice [59]. This damage also increases the optical losses, so the extent of implantation must be chosen to give the necessary electrical isolation with minimal optical degradation. Implantation of H^+ at a dose of 4×10^{15} cm^{-2} into a 2-μm-wide stripe along an AlGaAs/GaAs waveguide, for example, can yield isolation resistances on the order of several 100 MΩ with negligible additional optical loss. A 1-μm-thick waveguide requires three implants, at energies of 40, 80, and 120 keV to uniformly isolate along the waveguide depth. Post-implant anneals of 500° to 600° are necessary for maximum effectiveness in increasing electrical resistance [60].

12.3 MONOLITHIC DEVICES AND SYSTEMS

Having seen how waveguide-based monolithic integration can take place, we turn to a brief discussion of a few additional device structures useful for monolithically integrated systems and provide a curt overview of the types of monolithic device integration that have been accomplished.

12.3.1 Other Integratable Device Structures

Waveguide Tapers

The use of integrated optical circuits for fiber communications applications will always eventually require coupling to a fiber. While we will examine the use of hybrid structures to accomplish this in the next section, improved coupling efficiency can be attained through the use of monolithically integrated structures for matching the mode size between waveguide and fiber. Most III-V-based single-mode waveguides yield a mode with a full-width at half-maximum (FWHM) size of 1 to 2 μm, and usually it is of elliptical shape. Single-mode fibers for $\lambda = 1.55$ μm typically have a circular, 8-μm-diameter core. Mode matching between these two extrema can be accomplished through the use of a waveguide taper, either in the lateral or vertical direction. A controlled decrease in the cross-sectional area of the waveguide implies that optical confinement decreases and the size of the mode grows. A taper can be formed by lateral etching, forming a trapezoidal waveguide as seen from above [61], or vertically through dry-etch removal of increasing portions of the waveguide core [62], as shown in Figure 12.10. Waveguide tapers of this type are useful for decreasing fiber coupling losses.

Optical Isolators

The stability of a laser is dependent, among other factors, upon a freedom from optical feedback; feedback due to reflections from interfaces or devices further down the waveguide can lead to wavelength instabilities of a monolithically integrated laser. An optical isolator, which essentially permits light to propagate in only one direction, employs a nonreciprocal polarization rotator, meaning a material that rotates the polarization of a propagating optical field in the *same direction*, regardless of whether the wave propagates in the **+z** or **−z** direction. The Faraday effect, central

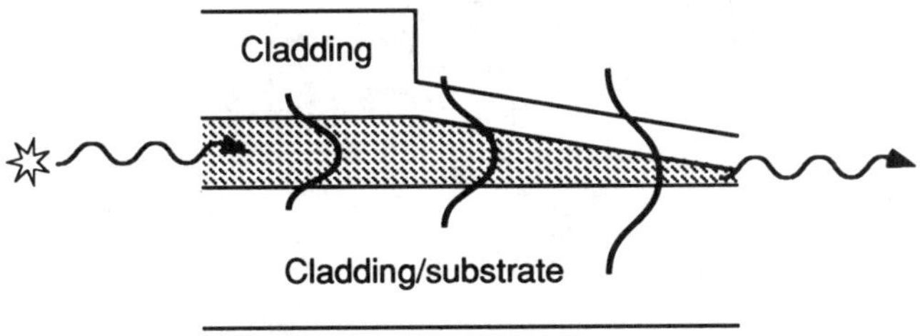

Figure 12.10 A waveguide taper for expansion of the mode size; this structure is formed by controlled dry-etch removal of the waveguide core and regrowth of the cladding.

to the operation of isolators, uses an applied magnetic field to accomplish polarization rotation. Materials such as garnet (YIG) are usually employed.

A practical optical isolator uses the Faraday rotator and two polarizers on either end, as shown schematically in Figure 12.11. The input polarizer, nearest the laser, is set vertically (0°) and the YIG rotator rotates the optical field by 45°; the output polarizer is set to 45°. Reflected light returns through the output polarizer, is rotated by a further 45°, and thus emerges from the YIG polarized at 90°. Because the input polarizer is orthogonal (0°), no light is transmitted through it in the reverse direction and the laser is effectively isolated. For integrated optical applications, optical isolation is complicated by the need for the Faraday material, the magnetic field, and two polarizers. It has been demonstrated [63] that (La,Ga):YIG can be grown by LPE means and patterned by Ar ion beam etching into integrated ridge waveguide structures; isolation values of –21 dB have been achieved. Although a hybrid system is still required to implement a complete isolator, the ability to integrate the Faraday rotator is significant for further developments.

Lenses

Slab waveguides form the basis for numerous monolithically integrated applications, such as the WDM structures of Figure 12.3. Through the use of dry-etch techniques, which generate vertical profiles—typically ion milling or chemically assisted ion beam etching—two-dimensional lens structures may be etched into a slab waveguide [64]. A shallow etch to slightly change the effective waveguide index [65] or an etch with dielectric refill [66] can be employed. Using geodesic, aspheric, Luneberg, or Fresnel lens structures, the etched and refilled patterns can be used for focusing an optical beam in a slab waveguide onto an output detector or coupling fiber, or for the collimation of an input laser beam. The large slab regions including the lens can

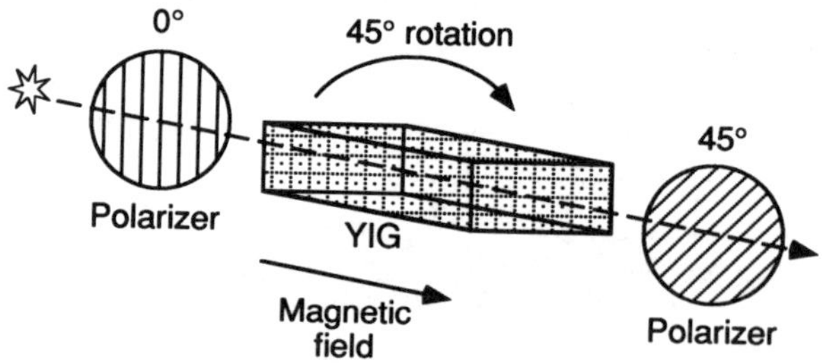

Figure 12.11 An optical isolator using a nonreciprocal YIG rotator and two polarizers. The YIG, in the presence of a magnetic field, rotates the optical field in the same direction regardless of propagation direction. Thus, a reflected wave incident from the right, polarized at 45°, will be rotated to 90° and thus not transmitted by the leftmost polarizer.

be made transparent by a quantum-well intermixing technique [67]. Etched two-dimensional lenses have been used, for example, with SAW transducers to fabricate an integrated optical spectrum analyzer [68].

Gratings

Diffractive elements with useful properties as lenses, couplers, or reflectors, gratings are suited to numerous integrated optical applications [69]. Many two-dimensional lensing functions can be performed by gratings, as can wavelength-selective reflection and beam deflection. With an eye to the impending discussion concerning hybridization, gratings can also be used to deflect the optical fields normal to the wafer surface and thus serve as output couplers. Grating couplers may be used to focus [70] or electrically steer [71] an output beam, functions that are useful for coupling from a monolithically integrated optical chip to other structures in a three-dimensional optical system. A monolithically integrated version of a confocal scanning microscope, employing a grating focusing beam splitter and a focusing grating coupler for vertical emission [72], as an example, has shown the utility of integrated grating structures.

Micromachining

Using a combination of dry and wet selective etch techniques plus etch-stop and sacrificial layers, sophisticated mechanical structures may be formed in the III-V and Si material systems, the latter being considerably more advanced in this regard. Silicon micromachining has permitted the fabrication of complex micro-electromechanical systems, and the inclusion of optics allows added flexibility in the design of microsystems with electrical, physical, and optical components.[3] By constructing a Si-based waveguide over a diaphragm etched from the wafer backside, and configuring this as one arm of an optical Mach-Zehnder interferometer, a monolithically integrated pressure sensor may be fabricated. The sensor's performance is further enhanced by the integration of photodetectors and CMOS readout electronics [73]. The photoelastic properties of the silicon oxynitride membrane result in index shifts, which are translated by the interferometer into intensity changes [74]. In GaAs/AlGaAs, etched cantilevers whose position may be determined optically through integrated lasers and photodiodes have been fabricated [75] and function as monolithically integrated optical vibration sensors. Micromachining techniques, particularly when involving both III-V and Si materials, will likely have increasing relevance in the future, and the full capabilities of these technologies with regard to integrated optical applications have only begun to be appreciated.

[3]For an overview of machined optical semiconductor structures, see Deimel, P. P., "Micromachining Processes and Structures in Micro-Optics and Optoelectronics," *Journal of Micromechanics and Microengineering*, Vol. 1, 1991, pp. 199–222.

12.3.2 Monolithic System Implementations

Using some of the device structures and integration techniques we have discussed, a wide variety of monolithically integrated implementations of useful optical systems has been demonstrated. OEICs have focused on the use of monolithically integrated electronics for driving modulators with MESFETs [76], lasers with MESFETs, and bipolar transistors [77]; PICs have integrated detectors with other optical structures for the fabrication of complete receivers [78]. The development of OEICs for high-speed telecommunications and data transfer is a very active research area, one bound to experience appreciable growth.

In the area of optical sensors, the monolithic integration of lasers, modulators, detectors, and interferometers proves promising for physical, biological, or chemical sensor applications [79]. Lasers and detectors of different emission and absorption wavelengths have also been integrated on a single chip [80], as have vertical cavity lasers and photothyristors for free-space optical logic applications [81]. In Si, detectors and waveguides have been combined with standard CMOS electronics [82], demonstrating the compatibility of existing transistor technology with optical applications.

12.4 HYBRID OPTICAL SYSTEMS

A hybrid system takes individual semiconductor optical components and mounts them together on a single platform in order to perform an optical function. Hybridization allows the designer to chose the optimal material for each subunit of an integrated optical system, permitting the selection of the most advanced technology for any particular application. The price is the need to develop further technologies for the alignment, adhesion, and packaging of a hybrid system. As we have mentioned previously, the choice of a hybrid approach over a monolithic one depends on the intended application, available technology, and performance requirements.

Most hybridization work has centered around the combination of III-Vs with silicon. The III-V materials excel as light emitters, for active optics such as modulators and for long-wavelength detectors. Si is attractive due to its advanced electronics and micromachining capabilities, passive optics, detectors for $\lambda < 1.1$ μm, good thermal properties, and relatively low price. As a result, many hybrid developments use Si as an optical bench, permitting accurate alignment of both III-V components and fibers, plus increasing functionality through the use of standard silicon electronics and passive optics. This use of silicon as a physical and electronic support has resulted in the terminology Si optical bench, Si motherboard, and Si waferboard, all of which amount to much the same thing.

Silicon has long been used as an optical package, primarily for fiber alignment in anisotropically etched V-grooves [83]. These, combined with micromachined surface features for mechanical alignment, integrated electronics, passive Si waveguide optics, and machined actuator structures, can yield highly functional systems; a generic assembly of these functions is sketched in Figure 12.12. The Si optical bench idea is one enjoying considerable popularity at present and, due to the potentially wide scope of relevance, will most likely see considerable technological development in the near term.

12.4.1 Alignment

The V-grooves for fiber placement have seen extensive use and are easily obtained in Si: anisotropic chemical etching from the (100) surface will stop on the (111) planes, which are always at 54.74° from the (100) wafer surface, as we saw in Figure 5.9. Thus, the depth and width of the grooves may also be accurately determined photolithographically, implying that the height position of a fiber may be easily defined.

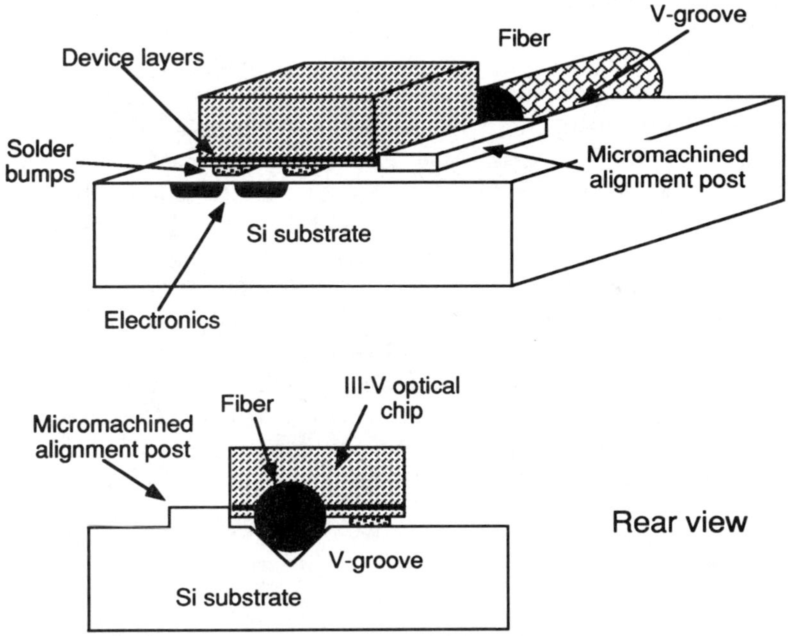

Figure 12.12 A schematic sketch of a hypothetical hybrid III-V/Si system. The III-V optical chip is mounted upside down with the device layers on the Si surface. Alignment is assisted by the use of a micromachined alignment post; electrical contact is made by the solder bumps between Si and the III-V chip. A fiber, sitting in a recessed V-groove etched into the Si, is coupled to the optical input for optical input/output, as seen more clearly in the rear view at the bottom of the figure.

A complete Si/III-V hybrid system typically requires optical alignment to and coupling between the III-V chip, fibers, and the silicon substrate. Due to the advanced machining capabilities in Si, and to some extent in III-Vs, it is possible to fabricate physical alignment structures in both material systems. A selection of etched slots, pedestals, and stops can facilitate the accurate placement of one chip relative to the other [84] so that the alignment tolerances are defined photolithographically. Alternatively, lithographic registration features using fiducial marks on the substrate may be employed [85] or, through the use of electrostatically actuated micromirrors, the coupling between laser and fiber may be optimized by active means [86]. We often distinguish between active alignment, during which the laser is operated and alignment is optimized for maximum output, as in the latter technique, and passive alignment, when physical movement is restricted without any capability for optimization, as in the first two examples.

Aligning a III-V laser to Si optics, such as a waveguide, can be more demanding because the waveguide dimensions tend to be smaller than those of the fiber core; most of the successful approaches are active. Using simple piezoelectric manipulation of an AlGaAs laser and thermocompression bonding onto a substrate with an Al_2O_3 waveguide, coupling efficiencies of 5% have been demonstrated [87]. Employing physical alignment structures, taking the form of either a recess etch before cleaving [88] or an etched alignment ridge [89], considerably simplifies alignment, although coupling losses of −15 dB are still registered for the former. Physical alignment structures on the Si carrier have been successfully used to couple a III-V laser into a silicon waveguide with excess coupling losses of −7.5 dB, allowing operation with hybridly mounted driver electronics at 2.5 GHz [90].

Machined alignment structures on a Si surface can also be used to couple an active III-V device vertically into a Si waveguide, through the use of Au-plated angle-etched surface mirrors [91]; losses of 0.9 dB are demonstrated. In addition, relatively complex systems may be assembled, such as a complete optical isolator including a YIG nonreciprocal waveguide, two polarizers, and a half-wave plate. These components are aligned by the use of a machined Si substrate and function as a complete hybridized optical isolator, with 2- to 3-dB excess loss and an isolation of 30 dB [92].

12.4.2 Bonding

Once an active optical III-V chip is aligned to structures on a Si surface, electrical, thermal, and physical contact must be made, the former permitting signals and power to be transferred and the latter insuring that the hybrid chip remains in place. Various forms of laser spot welding, soldering, and gluing with UV-curable epoxies have been employed, where adhesives have been of particular use for fixing fibers [93]. A useful technique for mounting hybridized substrates is flip-chip bonding, an approach with a long history [94] that combines physical and electrical contact with self-alignment.

Small bumps of a Pb/Sn solder are deposited over Ti/Pt/Au or Cr/Cu/Au contact pads on one of the III-V or Si substrates shown in Figure 12.13(a). The photolithographically defined contact pad patterns are designed so that they overlap when the chips are sandwiched together. One of the chips is "flipped" so that the two are brought into face-to-face contact. The solder then reflows at a temperature of about 350°, and surface tension brings the contact pads into accurate lateral alignment, as seen in Figure 12.13(b). This self-alignment implies that the chips can be brought into roughly the correct location before solder reflow then pulls the chips into the position defined by the contact pads. Using this approach, optical coupling of fibers with a waveguide together on a silicon substrate has been demonstrated with 1-μm lateral tolerances and coupling losses of 1 dB, without the need for accurate initial positioning [95].

Further added flexibility in hybrid chip bonding is achieved if the substrate of the III-V-based structures is removed so that only the active device layers remain on a thin semiconductor sheet of several micron thickness [96]. In an approach known as epitaxial lift-off [97], an AlAs release layer, which etches rapidly in HF, is used

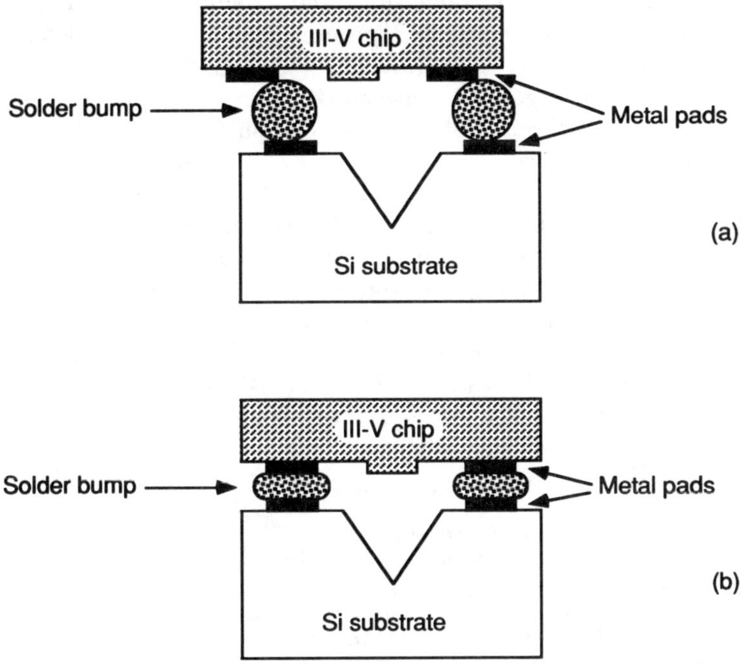

Figure 12.13 Flip-chip bonding between a III-V waveguide chip and a Si V-groove. Initial positioning (a) is not critical when the two are brought into contact because after solder reflow (b), the surface tension pulls the chips into alignment.

to underetch the surface layers; these remaining thin-film device layers may be attached to a Si platform and fixed by Van der Waals forces [98]. Alternatively, a silicon substrate may be bonded to a GaAs-based chip by a Au-based eutectic, and the GaAs substrate will again be removed by selective etch means [99]. Epitaxial lift-off has been used, for example, in the fabrication of a hybrid Si/InGaAs/InP resonant cavity photodiode and LED, where the III-V portion was bonded to the Si/SiO_2 quarter wave stack mirror, the latter attractive due to its superior thermal conductivity [100]. The light weight of the thin-film III-V device layer also allows hybridization with mechanical structures [101], which points the way to optomechanical systems based on III-Vs, Si, and polymer materials.

12.4.3 Heteroepitaxy

All of the problems associated with alignment and bonding in a hybrid system would be alleviated if it were easy to epitaxially grow, say, GaAs on a Si substrate. Such an undertaking is hindered by the 4% lattice mismatch between Si and GaAs, which leads to grown films with considerable stress and a high density of dislocations. The high density of dislocations results in a prominence of dark line defects in lasers, which degrade device lifetime. The lattice mismatch, and the significant difference in thermal coefficient of expansion between GaAs and Si, implies that growth of a minimal volume of a III-V material might provide suitable device layers.

Despite these difficulties, some advances have been made in the heteroepitaxial growth of device-quality III-V material on Si substrates. Lasers have been fabricated in MOCVD-grown material in Si V-grooves [102], the structure giving rise to very small lasing volumes. Using selective-growth techniques, GaAs/AlGaAs-based lasers and GaAs-based MESFET driving electronics have been grown in two growth steps on Si substrates [103]. Passive optical structures, such as switches based on the electro-optic effect [104], are somewhat less problematic, due to the reduced relevance of dislocation density; in this case, suitable GaAs/AlGaAs layers of several micron thickness can be grown by MOCVD on a Si substrate.

REFERENCES

[1] Stoppard, T., *Rosencrantz and Guildenstern are Dead*, London: Faber and Faber, 1967.

[2] Miller, S. E., "Integrated Optics: An Introduction," *Bell System Technical Journal*, Vol. 48, 1969, pp. 2,059–2,069.

[3] Kogelnik, H., "An Introduction to Integrated Optics," *IEEE Transactions on Microwave Theory and Techniques*, Vol. MTT-23, 1975, pp. 2–16.

[4] Evtuhov, V., and A. Yariv, "GaAs and GaAlAs Devices for Integrated Optics," *IEEE Transactions on Microwave Theory and Techniques*, Vol. MTT-23, 1975, pp. 44–57.

[5] Koch, T.L., and U. Koren, "Semiconductor Photonic Integrated Circuits," *IEEE Journal of Quantum Electronics*, Vol. 27, 1991, pp. 641–653.

[6] Wakao, K., "Optoelectronic and Photonic Integrated Circuits," in *Waveguide Optoelectronics*,

Marsh, J. H., and R. M. De La Rue, (eds.), Dordrecht, Holland: Kluwer Academic, 1992, Chapter 10.

[7] Margalit, S., and A. Yariv, "Integrated Electronic and Photonic Devices," in *Semiconductors and Semimetals*, Tsang, W. T. (ed.), Vol. 22E, Orlando: Academic Press, 1985, Chapter 2.

[8] Nichols. D. T., W. S. Hobson, D. Vakhshoori, J. D. Wynn, G. J. Zydzik, R. A. Morgan, N. K. Dutta, D. L. Sivco, A. Y. Cho, and R. E. Leibenguth, "Optical Transmission Systems Employing Vertical Cavity Surface Emitting Lasers and Monolithically Integrated Photoreceivers," *Applied Physics Letters*, Vol. 65, 1994, pp. 3,054–3,056.

[9] Zucker, J. E., K. L. Jones, M. A. Newkirk, R. P. Gnall, B. I. Miller, M. G. Young, U. Koren, C. A. Burrus, and B. Tell, "Quantum Well Interferometric Modulator Monolithically Integrated with 1.55 μm Tunable Distributed Bragg Reflector Laser," Electronics Letters, Vol. 28, 1992, pp. 1,888–1,889.

[10] Vettiger, P., M. K. Benedict, G.-L. Bona, P. Buchmann, C. Cahoon, K. Dätwyler, H.-P. Dietrich, A. Moser, H. K. Seitz, O. Voegeli, D. J. Webb, and P. Wolf, "Full-Wafer Technology—A New Approach to Large-Scale Laser Fabrication and Integration," *IEEE Journal of Quantum Electronics*, Vol. 27, 1991, pp. 1,319–1,331.

[11] Liou, K.-Y., U. Koren, E. C. Burrows, M. Oron, B. I. Miller, M. Young, G. Raybon, and C. A. Burrus, "Operation of Integrated InGaAsP-InP Optical Amplifier-Monitoring Detector with Feedback Control Circuit," *IEEE Photonics Technology Letters*, Vol. 2, 1990, pp. 878–880.

[12] Saito, H., and Y. Noguchi, "Coupling Efficiencies between Laser Diodes and Photo Diodes of InGaAsP/InP LD-PD Devices with Inclined PD Facets Considering PD Absorption," *Japanese Journal of Applied Physics*, Vol. 32, 1993, pp. 812–818.

[13] Young, M. G., U. Koren, B. I. Miller, M. A. Newkirk, M. Chien, M. Zirngibl, C. Dragone, B. Tell, H. M. Presby, and G. Raybon, "A 16×1 Wavelength Division Multiplexer with Integrated Distributed Bragg Reflector Lasers and Electroabsorption Modulators," *IEEE Photonics Technology Letters*, Vol. 5, 1993, pp. 908–910.

[14] Soole, J. B. D., A. Scherer, H. P. LeBlanc, N. C. Andreadakis, R. Bhat, and M. A. Koza, "Monolithic InP/InGaAsP/InP Grating Spectrometer for the 1.48-1.55 μm Wavelength Range," *Applied Physics Letters*, Vol. 58, 1991, pp. 1,949–1,951.

[15] Cremer, C., N. Emeis, M. Schier, G. Heise, G. Ebbinghaus, and L. Stoll, "Grating Spectrograph Integrated with Photodiode Array in InGaAsP/InGaAs/InP," *IEEE Photonics Technology Letters*, Vol. 4, 1992, pp. 108–110.

[16] Verdiell, J.-M., T. L. Koch, M. G. Young, B. I. Miller, U. Koren, and B. Tell, "WDM Receiver with Integrated Optical Preamplifier, Aspheric Lens and Grating Filter," *Electronics Letters*, Vol. 29, 1993, pp. 992–993.

[17] Moss, D., F. Ye, D. Landheer, P. E. Jessop, J. G. Simmons, H. G. Champion, I. Templeton, and F. Chatenoud, "Ridge Waveguide Quantum-Well Wavelength Division Demultiplexing Detector with Four Channels," *IEEE Photonics Technology Letters*, Vol. 4, 1992, pp. 756–759.

[18] Williams, P. J., P. M. Charles, I. Griffiths, N. Carr, D. J. Reid, N. Forbes, and E. Thom, "WDM Transceiver OEICs for Local Access Networks," *Electronics Letters*, Vol. 30, 1994, pp. 1,529–1,530.

[19] Matz, R., J. G. Bauer, P. Clemens, G. Heise, H. F. Mahlein, W. Metzger, H. Michel, and G. Schulte-Roth, "Development of a Photonic Integrated Transceiver Chip for WDM Transmission," *IEEE Photonics Technology Letters*, Vol. 6, 1994, pp. 1,327–1,329.

[20] Koren, U., "Optoelectronic Integrated Circuits", in *Optoelectronic Technology and Lightwave Communications Systems*, (Lin, C. ed.), New York: Van Nostrand Reinhold, 1989, pp. 677–694.

[21] Nichols, D. T., J. Lopata, W. S. Hobson, N. K. Dutta, P. R. Berger, Sivco, D. L. and Cho, A. Y., "Monolithic GaAs/AlGaAs Optical Transmitter Circuit Using a Single Growth Step," *Electronics Letters*, Vol. 30, 1994, pp. 490–491.

[22] Hurm, V., M. Ludwig, J. Rosenzweig, W. Benz, M. Berroth, R. Bosch, W. Bronner, A. Hülsmann, K. Köhler, B. Raynor, and J. Schneider, "14 GHz Bandwidth MSM Photodiode AlGaAs/GaAs

HEMT Monolithic Integrated Optoelectronic Receiver," *Electronics Letters*, Vol. 29, 1993, pp. 9–10.

[23] He, Y. S., L. D. Garrett, K.-H. Lee, and J. C. Campbell, "Monolithically Integrated Silicon nMOS *PIN* Photoreceiver," *Electronics Letters*, Vol. 30, 1994, pp. 1,887–1,888.

[24] Jackson, K. P., E. B. Flint, M. F. Cina, D. Lacey, Y. Kwark, J. M. Trewhella, T. Caulfield, P. Buchmann, C. Harder, and P. Vettiger, "A High-Density, Four-Channel, OEIC Transceiver Module Utilizing Planar-Processed Optical Waveguides and Flip-Chip, Solder-Bump Technology," *IEEE Journal of Lightwave Technology*, Vol. 12, 1994, pp. 1,185–1,191.

[25] Yamada, Y., M. Yamada, H. Terui, and M. Kobayashi, "Optical Interconnections Using a Silica-Based Waveguide on a Silicon Substrate," *Optical Engineering*, Vol. 28, 1989, pp. 1,281–1,287.

[26] Nagata, T., T. Tanaka, K. Miyake, H. Kurotaki, S. Yokoyama, and M. Koyanagi, "Micron-Sized Optical Waveguide for Optoelectronic Integrated Circuit," *Japanese Journal of Applied Physics*, Vol. 33, 1994, pp. 822–826.

[27] Emeis, N., M. Schier, L. Hoffmann, H. Heinecke, and B. Baur, "High Speed Waveguide-Integrated Photodiodes Grown by Metal Organic Molecular Beam Epitaxy," *Electronics Letters*, Vol. 28, 1992, pp. 344–345.

[28] Deri, R. J., W. Doldissen, R. J. Hawkins, R. Bhat, J. B. D. Soole, L. M. Schiavone, M. Seto, N. Andreadakis, Y. Silberberg, and M. A. Koza, "Efficient Vertical Coupling of Photodiodes to InGaAsP Rib Waveguides," *Applied Physics Letters*, Vol. 58, 1991, pp. 2,749–2,751.

[29] Kapser, K., and P. P. Deimel, "Lateral Coupling between a Silicon-Oxynitride Waveguide and an Amorphous Si Photodiode," *Journal of Applied Physics*, Vol. 71, 1992, pp. 3,614–3,616.

[30] Mak, G., D. M. Bruce, and P. E. Jessop, "Waveguide-Detector Couplers for Integrated Optics and Monolithic Switching Arrays," *Applied Optics*, Vol. 28, 1989, pp. 4,629–4,636.

[31] Splett, A., T. Zinke, K. Petermann, E. Kasper, H. Kibbel, H.-J. Herzog, and H. Presting, "Integration of Waveguides and Photodetectors in SiGe for 1.3 μm Operation," *IEEE Photonics Technology Letters*, Vol. 6, 1994, pp. 59–61.

[32] Remiens, D., B. Rose, M. Carre, and V. Hornung, "GaInAsP/InP Integrated Ridge Laser with a Butt-Jointed Transparent Optical Waveguide Fabricated by Single-Step Metalorganic Vapor-Phase Epitaxy," *Journal of Applied Physics*, Vol. 68, 1990, pp. 2,450–2,453.

[33] Wu, M. C., Y. J. Chen, and P.-T. Ho, "GaAs/AlGaAs Quantum Well Laser With Monolithically Integrated Extended Glass Waveguide Cavity," *Electronics Letters*, Vol. 27, 1991, pp. 1,954–1,956.

[34] Alferov, Z. I., S. A. Gurevich, S. Y. Karpov, E. L. Portnoi, and F. N. Timofeev, "Monolithically-Integrated Hybrid Heterostructure Diode Laser with Dielectric-Film Waveguide DBR," *IEEE Journal of Quantum Electronics*, Vol. QE-23, 1987, pp. 869–881.

[35] Bouadma, N., J. Liang, R. Levenson, S. Grosmaire, P. Boulet, and S. Sainson, "Integration of a Laser Diode with a Polymer-Based Waveguide for Photonic Integrated Circuits," *IEEE Photonics Technology Letters*, Vol. 6, 1994, pp. 1,188–1,190.

[36] Galeuchet, Y. D., and P. Röntgen, "Selective Area MOVPE of GaInAs/InP Heterostructures on Masked and Non-Planar (100) and {111} Substrates," *Journal of Crystal Growth*, Vol. 107, 1991, pp. 147–150.

[37] Kato, T., T. Sasaki, K. Komatsu, and I. Mito, "DFB-LD/Modulator Integrated Light Source by Bandgap Energy Controlled Selective MOVPE," *Electronics Letters*, Vol. 28, 1992, pp. 153–154.

[38] Lammert, R. M., T. M. Cockerill, D. V. Forbes, and J. J. Coleman, "Dual-Channel Strained-Layer InGaAs-GaAs-AlGaAs WDM Source with Integrated Coupler by Selective-Area MOCVD," *IEEE Photonics Technology Letters*, Vol. 6, 1994, pp. 1,167–1,169.

[39] Brovelli, L. R., R. Germann, J. P. Reithmaier, H. Jäckel, H. P. Meier, and H. Melchior, "Transform-Limited Picosecond Optical Pulses from a Mode-Locked InGaAs/AlGaAs QW Laser with Integrated Passive Waveguide Cavity and QW Modulator," *IEEE Photonics Technology Letters*, Vol. 5, 1993, pp. 896–899.

[40] Chang-Hasnain, C. J., E. Kapon, J. P. Harbison, and L. T. Florey, "Integrated External Cavity

Quantum Well Laser Array Using Single Epitaxial Growth on a Patterned Substrate," *Applied Physics Letters*, Vol. 56, 1990, pp. 429–431.

[41] Marsh, J. H., "Quantum Well Intermixing," *Semiconductor Science and Technology*, Vol. 8, 1993, pp. 1,136–1,155.

[42] Laidig, W. D., N. Holonyak, M. D. Camras, K. Hess, J. J. Coleman, P. D. Dapkus, J. Bardeen, "Disorder of an AlAs-GaAs Superlattice by Impurity Diffusion," *Applied Physics Letters*, Vol. 38, 1981, pp. 776–778.

[43] Zucker, J. E., K. L. Jones, B. Tell, K. Brown-Goebeler, C. H. Joyner, B. I. Miller, and M. G. Young, "InGaAsP/InP Quantum Well Buried Heterostructure Waveguides Produced by Ion Implantation," *Electronics Letters*, Vol. 28, 1992, pp. 853–855.

[44] Marsh, J. H., S. I. Hansen, A. C. Bryce, and R. M. De La Rue, "Applications of Neutral Impurity Disordering in Fabricating Low-Loss Optical Waveguides and Integrated Optical Devices," *Optical and Quantum Electronics*, Vol. 23, 1991, pp. S941–S957.

[45] Redinbo, G. F., and H. G. Craighead, "Proton Implantation Intermixing of GaAs/AlGaAs Quantum Wells," *Journal of Applied Physics*, Vol. 74, 1993, pp. 3,099–3,102.

[46] Thornton, R. L., J. E. Epler, and T. L. Paoli, "Monolithic Integration of a Transparent Dielectric Waveguide into an Active Laser Cavity by Impurity-Induced Disordering," *Applied Physics Letters*, Vol. 51, 1987, pp. 1,983–1,985.

[47] Guido, L. J., N. Holonyak, K. C. Hsieh, R. W. Kaliski, W. E. Plano, R. D. Burnham, R. L. Thornton, J. E. Epler, and T. L. Paoli, "Effects of Dielectric Encapsulation and As Overpressure on Al-Ga interdiffusion in $Al_xGa_{1-x}As$-GaAs Quantum-Well Heterostructures," *Journal of Applied Physics*, Vol. 61, 1987, pp. 1,372–1,379.

[48] Ribot, H., K. W. Lee, R. J. Simes, R. H. Yan, and L. A. Coldren, "Disordering of GaAs/AlGaAs Multiple Quantum Well Structures by Thermal Annealing for Monolithic Integration of Laser and Phase Modulator," *Applied Physics Letters*, Vol. 55, 1989, pp. 672–674.

[49] Gontijo, I., T. Krauss, J. H. Marsh, and R. M. De La Rue, "Postgrowth Control of GaAs/AlGaAs Quantum Well Shapes by Impurity-Free Vacancy Diffusion," *IEEE Journal of Quantum Electronics*, Vol. 30, 1994, pp. 1,189–1,195.

[50] Zappe, H. P., D. Hofstetter, and H. E. G. Arnot, "Quantum Well Engineering for Semiconductor Integrated Optical Sensors," *Nanofabrication Technologies and Device Integration, SPIE Proceedings*, Vol. 2213, 1994, pp. 205–216.

[51] Ralston, J. D., W.J. Schaff, D. P. Bour, and L. F, Eastman, "Room-Temperature Exciton Electroabsorption in Partially Intermixed GaAs/AlGaAs Quantum Well Waveguides," *Applied Physics Letters*, Vol. 54, 1989, pp. 534–536.

[52] Seidel, W., and P. Voisin, "Calculations of Interdiffusion and the Quantum-Confined Stark Effect in GaAs-GaAlAs Quantum Wells," *Semiconductor Science and Technology*, Vol. 8, 1993, pp. 1,885–1,888.

[53] Zhang, G., J. Näppi, A. Ovtchinnikov, H. Asonen, and M. Pessa, "Effects of Rapid Thermal Annealing on Lasing Properties of InGaAs/GaAs/GaInP Quantum Well Lasers," *Journal of Applied Physics*, Vol. 72, 1992, pp. 3,788–3,791.

[54] O'Brien, S., J. R. Shealy, F. A. Chambers, and G. Devane, "Tunable (Al)GaAs Lasers Using Impurity-Free Partial Interdiffusion," *Journal of Applied Physics*, Vol. 71, 1992, pp. 1,067–1,069.

[55] Hirata, T., M. Maeda, M. Suehiro, and H. Hosomatsu, "GaAs/AlGaAs GRIN-SCH-SQW DBR Laser Diodes with Passive Waveguides Integrated by Compositional Disordering of the Quantum Well Using Ion Implantation," *Japanese Journal of Applied Physics*, Vol. 29, 1990, pp. L961–L963.

[56] O'Brien, S., J. R. Shealy, and G. W. Wicks, "Monolithic Integration of an (Al)GaAs Laser and an Intracavity Electroabsorption Modulator Using Selective Partial Interdiffusion," *Applied Physics Letters*, Vol. 58, 1991, pp. 1,363–1,365.

[57] Inoue, H., H. Sano, and S. Nishimura, "Field-Induced Refractive Index Change Dependence on

Incident Light Wavelength and Polarization to an InGaAs/InAlAs Multi-Quantum Well," *Integrated Photonics Research*, New Orleans, LA, April 13–16, 1992, pp. 338–339.
[58] Kato, K., S. Hata, A. Kozen, S. Oku, S. Matsumoto, and J. Yoshida, "22-GHz Photodiode Monolithically Integrated with Optical Waveguide on Semi-Insulating InP Using Novel Butt Joint Structure," *Electronics Letters*, Vol. 28, 1992, pp. 1,140–1,142.
[59] Zucker, J. E., K. L. Jones, B. I. Miller, M. G. Young, U. Koren, B. Tell, and K. Brown-Goebler, "Interferometric Quantum Well Modulators with Gain," *IEEE Journal of Lightwave Technology*, Vol. 10, 1992, pp. 924–932.
[60] Pearton, S. J., F. Ren, S. N. G. Chu, C. R. Abernathy, W. S. Hobson, and R. G. Elliman, "Defects and Ion Redistribution in Implant-Isolated GaAs-Based Device Structures," *Journal of Applied Physics*, Vol. 74, 1993, pp. 6,580–6,586.
[61] Zengerle, R., H. Brückner, H. Olzhausen, and A. Kohl, "Low-Loss Fibre-Chip Coupling by Buried Laterally Tapered InP/InGaAsP Waveguide Structure," *Electronics Letters*, Vol. 28, 1992, pp. 631–632.
[62] Koch, T. L., U. Koren, G. Eisenstein, M. G. Young, M. Oron, C. R. Giles, and B. I. Miller, "Tapered Waveguide InGaAs/InGaAsP Multiple-Quantum-Well Lasers," *IEEE Photonics Technology Letters*, Vol. 2, 1990, pp. 88–90.
[63] Sugimoto, N., Y. Katoh, and A. Tate, "Magneto-Optic Buried Channel Waveguides for a 45, Nonreciprocal Waveguide Rotator," *Applied Physics Letters*, Vol. 63, 1993, pp. 2,744–2,746.
[64] Nishihara, H., M. Haruna, and T. Suhara, *Optical Integrated Circuits*, New York: McGraw Hill, 1989, Chapter 9.6.
[65] Vu, T. Q., and C. S. Tsai, "Ion-Milled Waveguide Lenses and Lens Arrays in GaAs," *IEEE Journal of Lightwave Technology*, Vol. 7, 1989, pp. 1,559–1,566.
[66] Minot, M., and C. C. Lee, "A New Guided-Wave Lens Structure," *IEEE Journal of Lightwave Technology*, Vol. 8, 1990, pp. 1,856–1,865.
[67] Hirata, T., M. Suehiro, M. Hihara, M. Dobashi, and H. Hosomatsu, "Demonstration of a Waveguide Lens Monolithically Integrated with a Laser Diode by Compositional Disordering of a Quantum Well", *IEEE Photonics Technology Letters*, Vol. 5, 1993, pp. 698–700.
[68] Vu, T. Q., C. S. Tsai, and Y. C. Kao, "Integration of a Curved Hybrid Waveguide Lens and Photodetector Array in GaAs Waveguide," *Applied Optics*, Vol. 31, 1992, pp. 5,246–5,254.
[69] Suhara, T., and H. Nishihara, "Integrated Optics Components and Devices Using Periodic Structures," *IEEE Journal of Quantum Electronics*, Vol. QE-22, 1986, pp. 845–867.
[70] Suhara, T., K. Okada, T. Saso, and H. Nishihara, "Focusing Grating Coupler in AlGaAs Optical Waveguide," *IEEE Photonics Technology Letters*, Vol. 8, 1992, pp. 903–905.
[71] Vasey, F., F. K. Reinhart, R. Houdré, and J. M. Stauffer, "Spatial Optical Beam Steering with an AlGaAs Integrated Phased Array," *Applied Optics*, Vol. 32, 1993, pp. 3,220–3,232.
[72] Sheard, S., T. Suhara, and H. Nishihara, "Integrated-Optic Implementation of a Confocal Scanning Optical Microscope," *IEEE Journal of Lightwave Technology*, Vol. 11, 1993, pp. 1,400–1,403.
[73] Adams, S., U. Hilleringmann, and K. Goser, "Plasma Etching Techniques for CMOS Compatible Micromachining based on Integrated Optics," in *Micro System Technologies 1992: Proceedings of the 3rd International Conference on Micro, Electro, Opto, Mechanic Systems and Components*, Reichl, H. (ed.), Berlin, October 21–23, 1992, pp. 217–226.
[74] Fischer, K., J. Müller, R. Hoffmann, F. Wasse, and D. Salle, "Elastooptical Properties of SiON Layers in an Integrated Optical Interferometer Used as a Pressure Sensor," *IEEE Journal of Lightwave Technology*, Vol. 12, 1994, pp. 163–169.
[75] Uenishi, Y., H. Tanaka, and H. Ukita, "Monolithic Integration of Microbeam Resonators and Laser Diodes Using AlGaAs/GaAs Micromachining," in *Micro System Technologies 1992: Proceedings of the 3rd International Conference on Micro, Electro, Opto, Mechanic Systems and Components*, Reichl, H. (ed.), Berlin, October 21–23, 1992, pp. 313–322.
[76] Ade, R. W., D. E. Bossi, R. P. Basilica, and J. M. Berak, "Monolithic Integration of GaAs-Waveguide

Optical Intensity Modulators with MESFET Drive Electronics," *Electronics Letters*, Vol. 28, 1992, pp. 702–704.

[77] Evaldsson, P. A., T. A. Vang, G. W. Taylor, S. K. Sargood, P. A. Kiely, and P. Cooke, "Monolithic Integration of Lasers with FET and Bipolar Transistors in Inversion Channel Technology," *Electronics Letters*, Vol. 29, 1993, pp. 60–62.

[78] Deri, R. J., "Monolithic Integration of Optical Waveguide Circuitry with III-V Photodetectors for Advanced Lightwave Receivers," *IEEE Journal of Lightwave Technology*, Vol. 11, 1993, pp. 1,296–1,313.

[79] Zappe, H. P., H. E. G. Arnot, and R. E. Kunz, "Technology and Devices for Hybrid and Monolithic Integration of Integrated Optical Sensors," *Sensors and Actuators* A, Vol. 41-43, 1994, pp. 141–144.

[80] Tsang, J. S., C. P. Lee, K. L. Tsai, and H. R. Chen, "Vertical Integration of a GaAs/AlGaAs Quantum-Well Laser and a Long-Wavelength Quantum-Well Infra-Red Photodetector," *Electronics Letters*, Vol. 30, 1994, pp. 450–451

[81] Zhou, P., J. Cheng, C. F. Schaus, S. Z. Sun, C. Hains, K. Zheng, A. Torres, D. R. Myers, and G. A. Vawter, "High-Performance Latchable Optical Switch and Logic Gates Based on the Integration of Surface-Emitting Lasers and Photothyristors," *Applied Physics Letters*, Vol. 59, 1991, pp. 2,504–2,506.

[82] Hilleringmann, U., and K. Goser, "Results of Monolithic Integration of Optical Waveguides, Photodiodes and CMOS Circuits on Si," *Microelectronic Engineering*, Vol. 19, 1992, pp. 211–214.

[83] Crow, J. D., L. D. Comerford, J. S. Harper, M. J. Brady, and R. A. Laff, "Gallium-Arsenide Laser-Array-on-Silicon Package," *Applied Optics*, Vol. 17, 1978, pp. 479–485.

[84] Armiento, C. A., M. Tabasky, C. Jagannath, T. W. Fitzgerald, C. L. Shieh, V. Barry, M. Rothman, A. Negri, P. O. Haugsjaa, and H. F. Lockwood, "Passive Coupling of InGaAsP/InP Laser Array and Singlemode Fibres Using Silicon Waferboard," *Electronics Letters*, Vol. 27, 1991, pp. 1,109–1,111.

[85] Cohen, M. S., M. F. Cina, E. Bassous, M. M. Oprysko, and J. L. Speidell, "Passive Laser-Fiber Alignment by Index Method," *IEEE Photonics Technology Letters*, Vol. 3, 1991, pp. 985–987.

[86] Solgaard, O., M. Daneman, N. C. Tien, A. Friedberger, R. S. Muller, and K. Y. Lau, "Optoelectronic Packaging Using Silicon Surface-Micromachined Alignment Mirrors," *IEEE Photonics Technology Letters*, Vol. 7, 1995, p. 41–44.

[87] Joppe, J. L., A. J. T. De Krijger, and O. F. J. Noordman, "Hybrid Integration of Laser Diode and Monomode High Contrast Slab Waveguide on Silicon," *Electronics Letters*, Vol. 27, 1991, pp. 162–163.

[88] Friedrich, E. E. L., M.G. Öberg, B. Broberg, S. Nilsson, and S. Valette, "Hybrid Integration of Semiconductor Lasers with Si-Based Single-Mode Ridge Waveguides," *IEEE Journal of Lightwave Technology*, Vol. 10, 1992, pp. 336–339.

[89] Jones, C. A., K. Cooper, M. W. Nield, J. D. Rush, R. G. Waller, J. V. Collins, and P. J. Fiddyment, "Hybrid Integration of a Laser Diode with a Planar Silica Waveguide," *Electronics Letters*, Vol. 30, 1994, pp. 215–216.

[90] Mino, S., K. Yoshino, Y. Yamada, M. Yasu, and K. Moriwaki, "Optoelectronic Hybrid Integrated Laser Diode Module Using Planar Lightwave Circuit Platform," *Electronics Letters*, Vol. 30, 1994, pp. 1,888–1,890.

[91] Crookes, C. G., I. R. Croston, and C. R, Pescod, "Selfaligned Integrated Silica on Silicon Waveguide-Photodiode Interface," *Electronics Letters*, Vol. 30, 1994, pp. 1,002–1,003.

[92] Terui, H., A. Tate, Y. Yamada, N. Sugimoto, Y. Katoh, A. Sugita, Y. Inoue, S. Ando, and T. Sawada, "Hybrid Integrated Isolator Array Formed on Silica-Based Optical Waveguide," *Proceedings of the 20th European Conference on Optical Communications (ECOC)*, Florence, Italy, 1994, pp. 213–216.

[93] Yamada, Y., F. Hanawa, T. Kitoh, and T. Maruno, "Low-Loss and Stable Fiber-to-Waveguide Connection Using UV Curable Adhesive," *IEEE Photonics Technology Letters*, Vol. 4, 1992, pp. 906–908.

[94] Miller, L. F., "Controlled Collapse Reflow Chip Joining," *IBM Journal of Research and Development*, Vol. 13, 1969, pp. 239–250.

[95] Wale, M. J., and C. Edge, "Self-Aligned Flip-Chip Assembly of Photonic Devices with Electrical and Optical Connections," *IEEE Transactions on Components, Hybrids and Manufacturing Technology*, Vol. 13, 1990, pp. 780–786.

[96] Chan, W. K., A. Yi-Yan, and T. J. Gmitter, "Grafted Semiconductor Optoelectronics," *IEEE Journal of Quantum Electronics;* Vol. 27, 1991, pp. 717–725.

[97] Yablonowitch, E., T. Gmitter, J. P. Harbison, and R. Bhat, "Extreme Selectivity in the Lift-Off of Epitaxial GaAs Films," *Applied Physics Letters*, Vol. 51, 1987, pp. 2,222–2,224.

[98] Yablonowitch, E., D. M. Hwang, T. J. Gmitter, L. T. Florez, and J. P. Harbison, "Van der Waals Bonding of GaAs Epitaxial Liftoff Films onto Arbitrary Substrates," *Applied Physics Letters*, Vol. 56, 1990, pp. 2,419–2,421.

[99] Venkatasubramanian, R., and M. L. Timmons, "Optoelectronic Properties of Eutectic-Metal-Bonded (EMB) GaAs-AlGaAs Structures on Si Substrates", *Solid-State Electronics*, Vol. 37, 1994, pp. 1,809–1,815.

[100] Salvador, A., F. Huang, B. Sverdlov, A. E. Botchkarev, and H. Morkoç, "InP/InGaAs Resonant Cavity Enhanced Photodetector and Light Emitting Diode with External Mirrors on Si," *Electronics Letters*, Vol. 30, 1994, pp. 1,527–1,529.

[101] Wilkinson, S. T., Y. W. Kim, N. M. Jokerst, and M. G. Allen, "Integration of Thin Film Optoelectronic Devices onto Micromachined Movable Platforms," *IEEE Photonics Technology Letters*, Vol. 6, 1994, pp. 1,115–1,118.

[102] Hasegawa, Y., T. Egawa, T. Jimbo, and M. Umeno, "Fabrication of Low-Threshold AlGaAs/GaAs Patterned Quantum Well Laser Grown on Si Substrate," *Japanese Journal of Applied Physics*, Vol. 32, 1993, pp. L997–L999.

[103] Egawa, T., T. Jimbo, and M. Umeno, "Monolithic Integration of AlGaAs/GaAs MQW Laser Diode and GaAs MESFET Grown on Si Using Selective Regrowth," *IEEE Photonics Technology Letters*, Vol. 4, 1992, pp. 612–614.

[104] Yuasa, T., Y. Nagashima, T. Murase, T. Jimbo, and M. Umeno, "GaAs/AlGaAs Single Quantum Well Optical Switch Fabricated on Si Substrate," *Japanese Journal of Applied Physics*, Vol. 32, 1993, pp. L1,055–L,1057.

Appendix A

List of Variables

Notes:

- If the units traditionally used are not mks, the traditional units are given.
- a.u. = arbitrary units.
- If no units are given, the variable is dimensionless.
- If a single variable has several meanings (such as L for various lengths), the definition should be clear from the context.

A	Area	m^2
α	Absorption coefficient or loss (waveguides)	cm^{-1} or dB/cm
a	Lattice spacing	nm
α_0	Background (residual) loss	cm^{-1}
A_{21}	Einstein A coefficient	s^{-1}
a_{curve}	Radiation loss from waveguide curve	cm^{-1}
α_{fc}	Free carrier absorption or loss	cm^{-1}
α_{sc}	Scattering absorption or loss	cm^{-1}
α_T	Sum of all losses in a waveguide, detector	cm^{-1}
α'	Extinction coefficient for evanescent wave	m^{-1}
$\mathscr{B}$	Bandwidth	Hz
B	Magnetic flux density	Wb/m^2
β	In a waveguide: **z**-directed propagation constant, $\equiv k_z$	m^{-1}
β	Real propagation constant in a material	m^{-1}
B_{12}, B_{21}	Einstein B coefficients	m^3/Js^2
C	Capacitance	F
c	Speed of light in vacuum	m/s

χ	Susceptibility	
χ_{atom}	Susceptibility due to lattice atoms	
χ_e	Susceptibility due to electrons	
CR	Contrast ratio	
$\chi_{\mathrm{R,I}}$	Real, imaginary part of the susceptibility	
D	Electric flux density	$\mathrm{C/m^2}$
$\Delta\alpha$	Induced change in absorption	$\mathrm{cm^{-1}}$
$D_c(E)$	Density of states, conduction band	$\mathrm{J^{-1}cm^{-3}}$
D_{eqw}	Two-dimensional density of states	$\mathrm{J^{-1}cm^{-3}}$
Δf	Bandwidth	Hz
$\Delta\lambda_0$	Mode spacing in wavelength	m
D_n	Electron diffusion constant	$\mathrm{cm^2/s}$
$\Delta \mathrm{n}$	Induced refractive index change in a modulator (Chapter 11)	
$\Delta\nu$	Linewidth	Hz
$\Delta \mathrm{n}$	Refractive index step between waveguide core and cladding	
$\Delta\nu_{\mathrm{FP}}$	Free spectral range or mode spacing in frequency	$\mathrm{s^{-1}}$
D_{p}	Hole diffusion constant	$\mathrm{cm^2/s}$
$D_v(E)$	Density of states, valence band	$\mathrm{J^{-1}cm^{-3}}$
$\mathcal{E}$	Electric field	V/m
E	Energy	J or eV
ϵ	Dielectric constant	
ϵ'	Complex dielectric constant	
ϵ_0	Permitivity of free space	F/m
E_A	Acceptor energy level	eV
E_c	Conduction band energy	J
E_{c0}	Minimum conduction band energy	eV
E_D	Donor energy level	eV
E_F	Fermi energy	eV
E_{fc}	Conduction band quasi-Fermi level	J
E_{fv}	Valence band quasi-Fermi level	J
E_g	Energy gap	eV
ϵ_m	Permitivity	F/m
E_ν	Photon energy	J
$\mathcal{E}_{\mathrm{peak}}$	Peak electric field in the depletion region	V/m
E_{ph}	Phonon energy	eV
$\epsilon_{R,I}$	Real, imaginary part of dielectric constant	

E_v	Valence band energy	J
E_{v0}	Minimum valence band energy	eV
$\mathcal{F}$	Finesse	
F	Force	N
Φ	Optical field	Js/m^3
f	Frequency	Hz
ϕ	Phase or phase shift of a wave	rad
$f(E)$	Fermi-Dirac distribution function	
F_{fk}	Franz-Keldysh effect coefficient	m^2/V^2
Φ_ν	Optical field energy density	J/m^3
Φ_p	Optical photon flux	$m^{-2}s^{-1}$
Γ	Confinement factor	
γ	Damping term	s^{-1}
γ	Gain	cm^{-1}
$g(\nu)$	Lineshape function	Hz^{-1}
γ_D	Differential gain	cm^{-1}
γ_{peak}	Peak gain	cm^{-1}
Γ_{qw}	Confinement factor in quantum wells	
H	Magnetic field	A/m
η	Efficiency (in general)	
h	Half the waveguide core width, $t = 2h$	m
h	Planck's constant	J s
η_d	Differential efficiency	
η_i	Internal quantum efficiency	
η_{wp}	Wall-plug efficiency	
I	Current	A
i	Index	
$\mathcal{I}(z)$	Intensity	a.u.
I_0	PN junction reverse saturation current (Chapter 2)	A
$\mathcal{I}_0$	Surface or input intensity	a.u.
I_d	Dark current	A
I_{kt}	Thermal current	A
$\mathcal{I}_\nu$	Optical field energy intensity	J/m^2s
I_{pd}	Photodiode current	A
I_{sig}	Time average cptical signal current	A
J	Current density	A/m^2
j	$\sqrt{(-1)}$	
J_n	Electron current density	A/cm^2
J_p	Hole current density	A/cm^2
J_{pd}	Photodiode current density	A/cm^2
J_{th}	Threshold current density	A/cm^2

J_{tr}	Transparency current density	A/cm^2
K	Spring constant	N/m
κ	Coupling coefficient	m^{-1}
k	Wavevector, propagation constant	m^{-1}
k_0	Propagation constant in free space	m^{-1}
k_B	Boltzmann constant	J/K
$k_{c,g,s}$	Propagation constant in waveguide cap, core, and substrate, respectively	m^{-1}
$k_{x,y,z}$	**x**-, **y**-, **z**-directed propagation constant	m^{-1}
$k_{x\text{eff}}$	**x**-directed propagation constant in a two-dimensional waveguide	m^{-1}
k_{xL}, k_{xR}	**x**-directed propagation (decay) constant outside a two-dimensional waveguide	m^{-1}
Λ	Grating period	m
L	Physical length or spacing	m
l	Index	
λ	Wavelength	m
λ_0	Vacuum wavelength	m
L_c	Coherence length	m
L_{cp}	Physical overlap length for waveguide coupler	m
λ_g	Wavelength corresponding to the bandgap	m
λ_n	Wavelength in a material, $\equiv \lambda$	m
L_π	Coupling length for waveguide coupler	m
L_π	Coupling length in directional couplers	m
M	Multiplication factor for avalanche	
m	Mass	kg
m	Waveguide mode index	
m^*	Effective mass	kg
m_0	Electron rest mass	kg
μ_0	Permeability of free space	H/m
m_c^*	Conduction band effective mass	kg
m_e	Effective electron mass factor	
m_{hh}	Effective heavy hole mass factor	
m_{lh}	Effective light hole mass factor	
μ_m	Permeability	H/m
μ_n	Electron mobility	$cm^2/V\ s$
μ_p	Hole mobility	$cm^2/V\ s$
m_v^*	Valence-band effective mass	kg

N	Waveguide effective index	
n	Electron density	cm^{-3}
ν	Frequency	Hz
n	Index of refraction	
n′	Complex refractive index	
ν_0	Resonant frequency	Hz
n_0	Unmodulated refractive index	
N_A	Acceptor concentration	cm^{-3}
n_a	Refractive index outside waveguide (air)	
N_A^-	Ionized acceptor concentration	cm^{-3}
n_{atom}	Density of atoms in a lattice	cm^{-3}
N_c	Effective density of states, conduction band	cm^{-3}
$n_{c,g,s}$	Refractive index of waveguide cap, core, and substrate, respectively	
N_D	Donor concentration	cm^{-3}
N_D^+	Ionized donor concentration	cm^{-3}
n_e	Electron density	cm^{-3}
n_{eff}	Effective index in one dimension	
NEP	Noise-equivalent power	W
n_i	Intrinsic carrier concentration	cm^{-3}
n_i	Refractive index of material for incident wave at a boundary	
N_{ph}	Phonon density	cm^{-3}
$n_{R,I}$	Real, imaginary part of the refractive index	
N_s	More lightly doped side of a pn junction	cm^{-3}
n_t	Refractive index of material for transmitted wave at a boundary	
n_{tr}	Transparency carrier concentration	cm^{-3}
N_v	Effective density of states, valence band	cm^{-3}
N_{wg}	Two-dimensional waveguide effective index	
p	Hole density	cm^{-3}
p	Momentum	kg m/s
P_{in}	Optical input power	W
P_{out}	Optical output power	W
Q	Charge	C
q	Electron charge	C

θ_B	Bragg reflection angle	rad
θ_c	Critical angle	rad
θ_i	Angle of incidence	rad
θ_m	Waveguide reflection angle for mode m	
θ_{pe}	External polarization (Brewster) angle	rad
θ_{pi}	Internal polarization (Brewster) angle	rad
θ_r	Angle of reflection	rad
θ_t	Angle of transmission	rad
R	Reflectance (Chapter 6)	
$\mathscr{R}$	Responsivity	A/W
R	Waveguide curve radius	m
ρ	Charge density	C/m^3
r	Position	m
r_{41}	Linear electro-optic coefficient	m/V
r_{41qw}	Quantum confined Stark effect linear coefficient	m/V
r_{TE}, r_{TM}	TE and TM reflection coefficient	
S	Oscillator strength (Chapter 9)	m^2/s
S	Poynting vector	W/m^2
σ	Conductivity	Ω^{-1}
σ	Transition cross section (Chapter 9)	m^2
S'	Power flow vector	W
S/N	Signal-to-noise ratio	
s_{33}	Quadratic electro-optic coefficient	m^2/V^2
s_{33qw}	Quantum confined Stark effect quadratic coefficient	m^2/V^2
S_{GH}	Goos-Hänchen shift	m
T	Temperature	K
T	Transition distance for waveguide curve (Chapter 8)	m
T	Transmittance (in Chapter 6)	
t	Quantum-well or waveguide-core thickness	m
t	Time (only in the form $j\omega t$)	s
T_0	Laser temperature coefficient	K
T_e	Electron temperature	K
τ_n	Electron lifetime	s
τ_p	Hole lifetime	s
τ_{sp}	Spontaneous lifetime	s
t_{TE}, t_{TM}	TE and TM transmission coefficient	
V	Voltage	V

v	Velocity	m/s
V_a	Externally applied voltage on a pn junction	V
V_{bd}	PN-junction breakdown voltage	V
V_{bi}	PN junction built-in potential	V
V_n	PN junction voltage drop, n-side	V
V_p	PN junction voltage drop, p-side	V
V_π	Voltage for a phase shift of π in a Mach-Zehnder interferometer	
W	Depletion region width/absorption length	m
$\mathscr{W}$	Potential energy	J
ω	Angular frequency	rad/s
w	Half the waveguide ridge or rib width	m
ω_0	Resonance frequency	rad/s
$\mathscr{W}_e$	Energy density in the electric field	J/m^3
$\mathscr{W}_h$	Energy density in the magnetic field	J/m^3
W_i	Electron transition rate per atom	s^{-1}
W_n	Depletion region width: n-side	m
W_p	Depletion region width: p-side	m
W_{sp}	Spontaneous radiation transition rate	s^{-1}
$\mathscr{W}_T$	Total energy density	J/m^3
x	Mole fraction in III-V alloy	
x,y,z	Cartesian directions	m
y	Mole fraction in III-V alloy	
Z	Impedance	Ω
Z_0	Impedance of free space	Ω

Appendix B

List of Useful Physical Constants

c	Speed of light in vacuum	$2.998 \cdot 10^{8}$	m/s
ϵ_0	Permitivity of free space	$8.854 \cdot 10^{-12}$	F/m
h	Planck's constant	$6.626 \cdot 10^{-34}$	J s
$\hbar$	Planck's constant / 2π	$1.055 \cdot 10^{-34}$	J s
k_B	Boltzmann constant	$1.381 \cdot 10^{-23}$	J/K
m_0	Electron rest mass	$9.110 \cdot 10^{-31}$	kg
μ_0	Permeability of free space	$4\pi \cdot 10^{-7}$	H/m
m_p	Proton rest mass	$1.673 \cdot 10^{-27}$	kg
q	Electron charge	$1.602 \cdot 10^{-19}$	C

About the Author

Hans P. Zappe was born in France and fulfilled a lifelong interest in the culinary arts by studying electrical engineering at the Massachusetts Institute of Technology (B.S. and M.S. in 1983) and the University of California, Berkeley (Ph.D. in 1989), with an emphasis on silicon microfabrication and device technology. He has worked at the IBM General Technology Division (Essex Junction, VT) and the Fraunhofer Institute for Applied Solid State Physics (Freiburg, Germany), having been introduced to the wondrous world of III-Vs at the latter. Presently, he is on the scientific staff of the Paul Scherrer Institute (Zurich, Switzerland) and when not developing monolithically integrated semiconductor optical systems for sensors and metrology with a group of enthusiastic scientists and students, he is pedaling through Europe in search of the perfect *brioche.*

Index

The Artech House Optoelectronics Library

Brian Culshaw, Alan Rogers, and Henry Taylor, *Series Editors*

Acousto-Optic Signal Processing: Fundamentals and Applications, Pankaj Das

Amorphous and Microcrystalline Semiconductor Devices, Optoelectronic Devices, Jerzy Kanicki, editor

Bistabilities and Nonlinearities in Laser Diodes, Hitoshi Kawaguchi

Coherent Lightwave Communication Systems, Shiro Ryu

Electro-Optical Systems Performance Modeling, Gary Waldman and John Wootton

Elliptical Fiber Waveguides, R. B. Dyott

The Fiber-Optic Gyroscope, Hervé Lefèvre

Field Theory of Acousto-Optic Signal Processing Devices, Craig Scott

Frequency Stabilization of Semiconductor Laser Diodes, Tetsuhiko Ikegami, Shoichi Sudo, Yoshihisa Sakai

Fundamentals of Multiaccess Optical Fiber Networks, Denis J. G. Mestdagh

Germanate Glasses: Structure, Spectroscopy, and Properties, Alfred Margaryan and Michael A. Piliavin

High-Power Optically Activated Solid-State Switches, Arye Rosen and Fred Zutavern, editors

Highly Coherent Semiconductor Lasers, Motoichi Ohtsu

Iddq Testing for CMOS VLSI, Rochit Rajsuman

Integrated Optics: Design and Modeling, Reinhard März

Introduction to Electro-Optical Imaging and Tracking Systems, Khalil Seyrafi and S. A. Hovanessian

Introduction to Glass Integrated Optics, S. Iraj Najafi

Introduction to Radiometry and Photometry, William Ross McCluney

Introduction to Semiconductor Integrated Optics, Hans P. Zappe

Laser Communications in Space, Stephen G. Lambert and William L. Casey

Optical Control of Microwave Devices, Rainee N. Simons

Optical Document Security, Rudolf L. van Renesse

Optical Fiber Amplifiers: Design and System Applications, Anders Bjarklev

Optical Fiber Sensors, Volume I: Principles and Components, John Dakin and Brian Culshaw, editors

Optical Fiber Sensors, Volume II: Systems and Applicatons, John Dakin and Brian Culshaw, editors

Optical Interconnection: Foundations and Applications, Christopher Tocci and H. John Caulfield

Optical Network Theory, Yitzhak Weissman

Optical Transmission for the Subscriber Loop, Norio Kashima

Optoelectronic Techniques for Microwave and Millimeter-Wave Engineering, William M. Robertson

Reliability and Degradation of LEDs and Semiconductor Lasers, Mitsuo Fukuda

Semiconductor Raman Laser, Ken Suto and Jun-ichi Nishizawa

Semiconductors for Solar Cells, Hans Joachim Möller

Single-Mode Optical Fiber Measurements: Characterization and Sensing, Giovanni Cancellieri

Ultrafast Diode Lasers: Fundamentals and Applications, Peter Vasil'ev